ELECTRICAL DRAFTING AND DESIGN

ELECTRICAL DRAFTING AND DESIGN

CHARLES W. SNOW, P. E.

Principal Electrical Engineer
United Engineers & Constructors Inc.
and
Evening School Instructor.
Wentworth Institute, Boston, Mass.

PRENTICE-HALL, INC., Englewood Cliffs, New Jersey

Library of Congress Cataloging in Publication Data

SNOW, CHARLES W date
Electrical drafting and design.

1. Electric drafting. 2. Electric engineering.
I. Title.
TK431.S57 621.3 75-20370
ISBN 0-13-247379-8

10 9 8

Printed in the United States of America

PRENTICE-HALL INTERNATIONAL, INC., *London*
PRENTICE-HALL OF AUSTRALIA, PTY. LTD., *Sydney*
PRENTICE-HALL OF CANADA, LTD., *Toronto*
PRENTICE-HALL OF INDIA PRIVATE LIMITED, *New Delhi*
PRENTICE-HALL OF JAPAN, INC., *Tokyo*
PRENTICE-HALL OF SOUTHEAST ASIA (PTE.) LTD., *Singapore*

To my wife, MARJORIE

CONTENTS

PREFACE

The demand for electrical energy continues to grow throughout the world as more and more uses are found for this most important servant of mankind. Power-generating plants must be continually designed and built. Distribution of power must be planned. Electrical systems for industry and commerce are on the drafting boards of engineering and design firms keeping thousands of men and women actively employed in challenging and rewarding work.

This text is directed at people who may have no formal background in the electrical field, yet it carries sufficient depth of instruction to be a valuable reference for those actively engaged in electrical careers. Mathematics throughout the book is limited to the solution of simple formulas requiring only an elementary knowledge of algebra.

Chapters are included on fundamentals of electricity and on drafting procedures. These fundamentals are developed chapter by chapter in explicit instruction in the design of electrical systems from generation to utilization of power.

Students of electrical engineering will find the theory courses of their classrooms translated into designs for electrical systems employing the generators, transformers, and similar apparatus that are part of their study curriculum.

Engineers, designers, and draftsmen active in design work will appreciate a text wherein a complete discussion is given to the practical production of electrical construction drawings in one text.

Those engaged in electrical construction from apprenticeship through responsibility for the installation and successful operation of the systems described will learn from the design side those considerations necessary to prepare drawings from which the system can be constructed. This exposure to the development of design drawings will prove invaluable in reading the drawings.

Chapters 1 and 2 start the reader at a low key into drafting and electrical fundamentals. Chapters 3 and 4 explain and give the basis for design through preparation of one-line diagrams. Elementary diagrams showing many control

1

DRAFTING PROCEDURES

1-1 Introduction

Drafting of electrical drawings is surprisingly simple! The drawings are generally composed of straight horizontal and vertical linework that require no special skills or experience. A fundamental knowledge in the use of a T square or its equivalent and drafting triangles is usually all that is needed and this can be easily acquired. No special techniques are generally required although a knowledge of orthographic (third angle) projection is handy for development of some views. Because most circles, curves, and other particular shapes can be made faster with a template, the need for drafting instruments is limited.

Knowledge of the *content* of the drawing is the more important factor in electrical design and drafting because the drafting capability can be quickly developed but the knowledge of what needs to be shown and how to show it takes time and experience. It is assumed that some fundamental drafting experience has been acquired or can be acquired; therefore this text directs itself primarily to drawing *content.* This chapter deals with drafting procedures related to portraying the content in practical drawings for electrical construction.

1-2 Drafting Materials

The same basic drawing materials are required to do electrical drawings that are required by other fields of drafting except, as previously stated, drawing instruments have limited use. For study and practice of the exercises contained in this text, the following materials will be needed:

1. Drawing board—not smaller than 22″ × 29″
2. T square—28″ or to suit board size
3. 45° triangle—10″
4. 30°/60° triangle—12″
5. Architects' scale, triangular type, 12″ with scales $\frac{1}{16}$″, $\frac{3}{32}$″, $\frac{1}{8}$″, $\frac{3}{16}$″, $\frac{1}{4}$″, $\frac{3}{8}$″, $\frac{1}{2}$″, $\frac{3}{4}$″, 1″, $1\frac{1}{2}$″ and 3″
6. Template with circles $\frac{1}{16}$″ through $1\frac{1}{2}$″, squares $\frac{3}{32}$″ through $\frac{5}{8}$″, triangles $\frac{3}{32}$″ through $\frac{3}{4}$″, and elipses $\frac{1}{8}$″ through $\frac{3}{4}$″ or similar

7. Lettering guide
8. Pencil pointer
9. Erasing shield
10. Eraser
11. Pencils or mechanical pencils and leads–No. 6H, 2H, H, or as required to suit drawing paper
12. Masking tape
13. Drafting cleaner
14. Dust brush
15. Clean cloth

1-3 Drawing Boards, T Squares, Parallel Straightedges, and Drafting Machines

Modern drafting rooms are generally equipped with standard full-sized drafting boards having spring-loaded mechanisms to provide for easy raising and lowering and tilting of the board surface. By stepping on a pedal, the board will elevate; pushing down on the board by hand will lower it and compress the spring. A hand lever permits tipping the board to a full flat or full vertical position. These easy controls accommodate all sizes of people and positions for drafting. Some people prefer standing while drafting and others draw while seated with the board set in a nearly full vertical position—like drawing on a wall. Some boards are made with electrical operators for raising and lowering. The base structure for the board is usually furnished with drawers and along the very bottom edge of the board there is usually a trough-type receptacle running the entire length for holding pencils, etc. Usually the boards are arranged so that a combination desk and reference table is available to the draftsman. A common arrangement is shown in Fig. 1-1.

The parallel straightedge or drafting machine has almost completely replaced

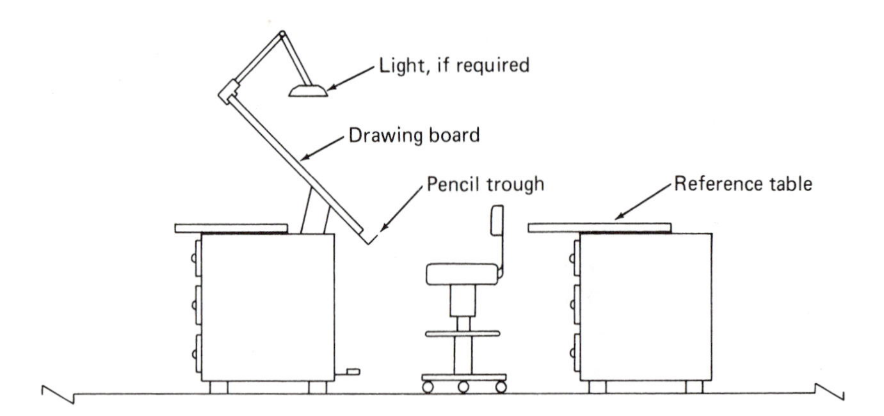

Figure 1-1 Typical Arrangement of Drafting Board

the familar T square. In many places the drafting machine has replaced both the parallel straightedge and the T square.

The drafting machine is mounted on the top of the board and made with either a single arm or combination of arms that holds the control head. Attached to the head are removable scales that also serve as the straightedges for drawing vertical and horizontal lines. The control head can be made to turn to any angle and this causes the attached scales to turn to the selected angle as well. These machines permit fast, accurate drafting with minimum fatigue to the draftsman. A typical modern type is shown in Fig. 1-2.

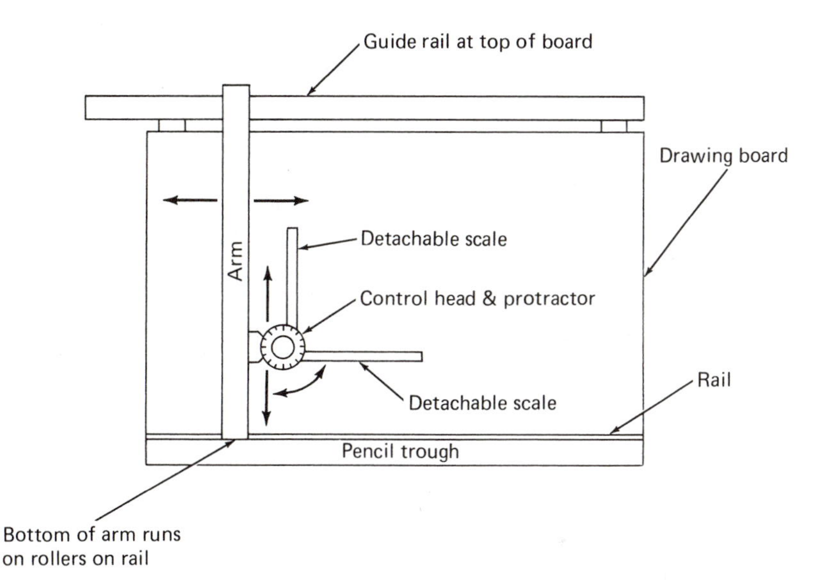

Figure 1-2 Modern Drafting Machine

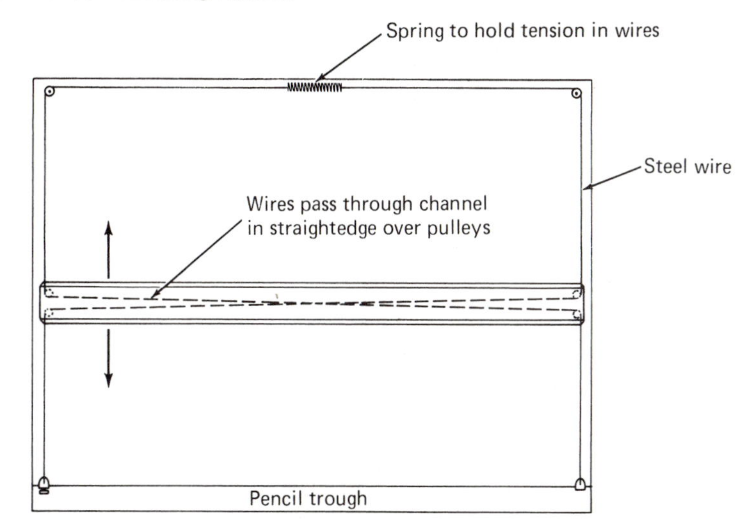

Figure 1-3 Parallel Straightedge on Board

The parallel straightedge is favored by some over the drafting machine where large drawings and long horizontal lines are common. This is often true of electrical drawings. The straightedge is secured to the drawing board by a system of slender steel wires that holds it in place and keeps it parallel to the bottom and top edge of the board. It maintains its parallel attitude automatically as it is moved up and down the board by the draftsman. Straight, true, horizontal lines can be drawn along its entire length. It usually is installed to cover the entire length of the board. A typical parallel straightedge is shown in Fig. 1-3.

If the parallel straightedge is used, vertical lines must be drawn by using triangles placed with their base against the top edge of the straightedge.

1-4 Drawing Board Covering for Electrical Drawings

Because many electrical drawings are in the form of diagrams, it is helpful to cover the drawing board surface with a smooth surface graph paper made especially for this purpose. When the blank drawing sheet is placed on the board over the graph paper covering, the grid lines are visible and therefore very useful as a guide to laying out diagrams. A common-sized graph paper is made in 1-in. grids divided in $\frac{1}{8}$-in. squares. Figure 1-4 shows this size.

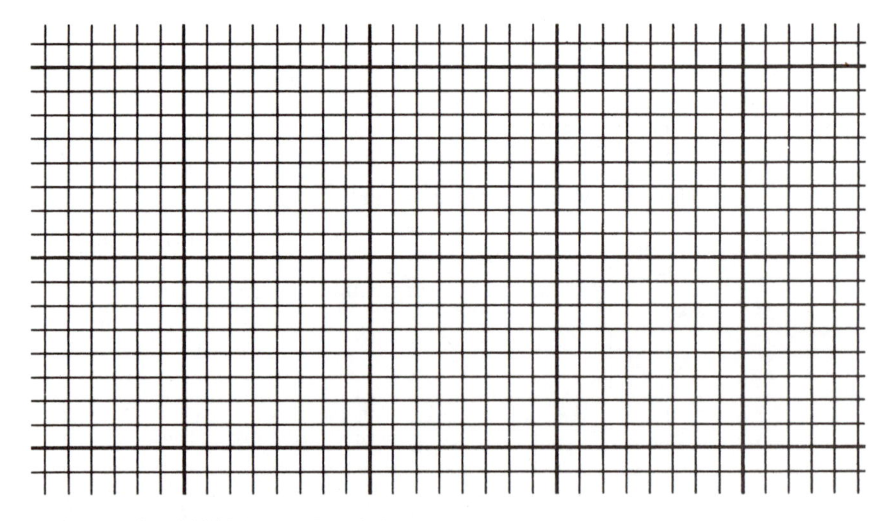

Figure 1-4 1″ Grid / $\frac{1}{8}$″ Square Board Cover

The grid paper of Fig. 1-4 is often light green in color and the surface finish is glossy so masking tape is easily removed without disturbing the surface. When put on the board, naturally care must be taken to be sure the lines of the grid line up exactly with the edge of the parallel straightedge or other drafting rule. The paper can be secured to the board with masking tape and replaced as required.

1-5 Use of Triangles

When using triangles for drawing vertical lines the triangle should be placed against the horizontal straightedge in such a way that the vertical line is drawn *up* the sheet whenever possible. In this way the light is usually better and the triangle can be held firmer. For the right-handed draftsman this means the edge being used for drawing the line should be nearest to the left side of the board. Care should be taken to make certain the pencil point is inclined into the base of the edge of the triangle. These directions are illustrated in Fig. 1-5.

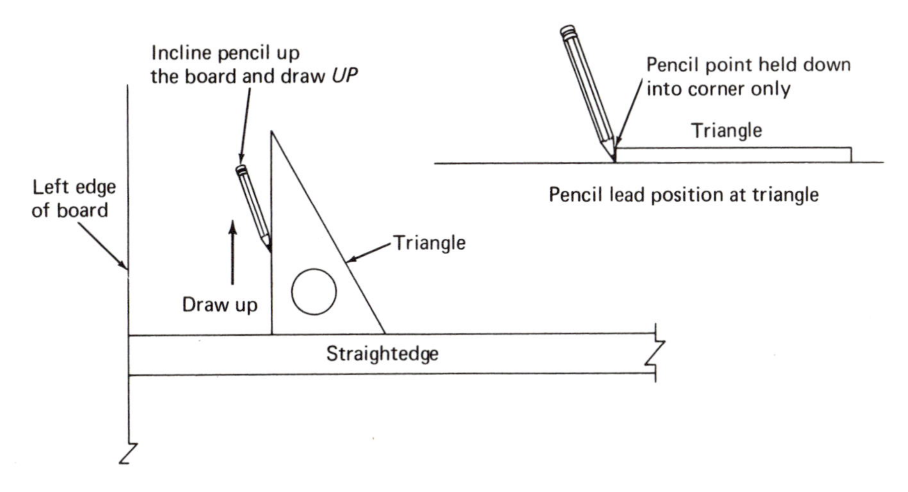

Figure 1-5 Correct Method for Drawing Vertical Lines

1-6 Drawing to Scale

All electrical layout drawings are drawn to scale. Accurate work with the scale is important for the designer and draftsman as a means of checking clearances, space requirements, etc., to the contractor because he usually has to scale the drawings to determine material requirements and also to the installing electrician in the field as a guide to locating equipment, etc. (A note is often included on drawings saying *not* to scale the drawings. This is done because scaled dimensions cannot be guaranteed as accurate and therefore the workman must be careful of interferences if he uses a scaled dimension.)

As a rule, electrical layouts are not drawn at scales smaller than $\frac{1}{8}'' = 1'\text{-}0''$. If smaller scales are used, the information to be shown probably covers a large area and a larger scale would require dividing the work up on several drawings. To avoid this the smaller scale can be used and where details need to be developed, smaller "part plans" can be shown at larger scales. Many electrical drawings are made at the scale $\frac{1}{4}'' = 1'\text{-}0''$. This usually permits the necessary amount of room for detail and is a fairly easy scale to use. As the scale becomes larger, the drafting is easier so the largest scale that can be used practically is to be desired.

The architects' scale, described in Sec. 1-2, is ideal for most electrical design work. (An engineers' scale with scales of 10, 20, 30, 40, 50, and 60 ft to the inch is needed for plot plans showing overhead and underground electrical systems.)

If a scale of $\frac{1}{8}'' = 1'\text{-}0''$ is used, the divisions of a foot marked on the end of the scale are 2 in. At $\frac{1}{4}'' = 1'\text{-}0''$, the divisions are 1 in.; at $\frac{1}{2}$-in. scale, $\frac{1}{2}$ in. can be read and the same is true for the $\frac{3}{4}$-in. scale. On the 1-in. and $1\frac{1}{2}$-in. scales, $\frac{1}{4}$ in. can be read, at the 3-in. scale, $\frac{1}{8}$ in. is marked off. Figure 1-6 shows parts of two of these scales enlarged to show the divisions.

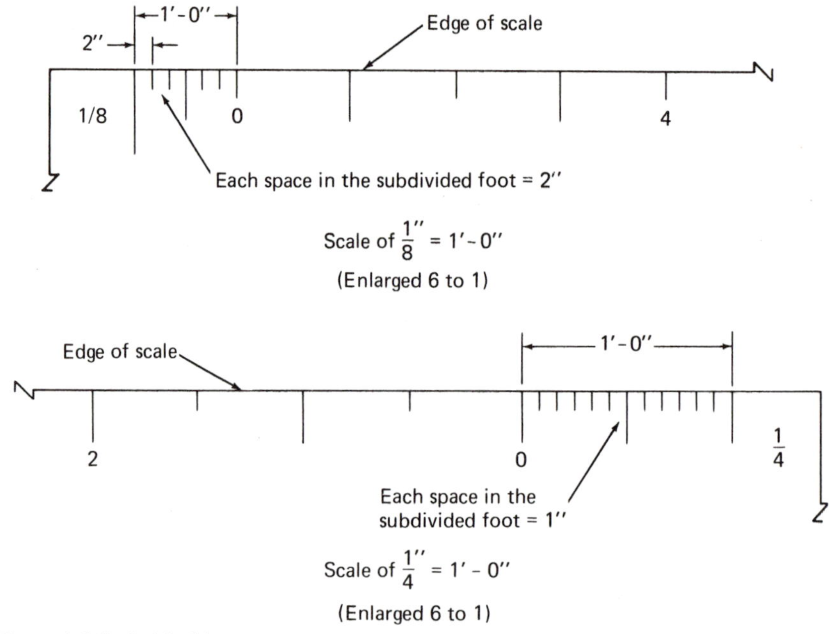

Figure 1-6 Scale Markings

When a dimension including both feet and inches must be laid out to scale on the drawing, it is usually easier to set the inch division mark at the end of the scale at the starting point of the dimension and then make a pencil mark *tick* (a little short line) at the foot dimension of the scale rather than trying to mark a tick accurately at the inch (or fraction of an inch) dimension. For example, say the scale being used is $\frac{1}{4}'' = 1'\text{-}0''$ and the dimension to be layed off is 8′ 4″. The scale should be laid onto the drawing with the 4-in. mark in the subdivided section in line with the point at which the dimension starts and then move down along the scale, in the direction of the measurement, and make a tick at the 8-ft line on the $\frac{1}{4}$-in. scale. This method is usually easier than trying carefully to make a tick at the 4-in. mark.

1-7 Choice and Use of Templates

There are several templates made expressly for use in drafting electrical drawings. Most of these templates include standard symbols used in wiring diagrams. The technique in the successful use of these is in being careful to follow the symbol cutout closely and to keep the pencil lead sharpened to the right size to draw a neat and clean-cut line. If the pencil is too dull, it may not fit into the cutout opening at all; if it is too sharp, it may slide around in the cutout. Also care must be taken to be sure the template is properly in line on the drawing to assure that the symbol will be correctly placed. If a template such as described in Sec. 1-2, is used, the symbols can be constructed in a few simple steps. For example, an incoming line symbol can be made from a triangle in the template; transformer windings are a series of half circles; safety switches and motor starters are made by using squares; disconnect devices are made with the triangle cutouts. Some of these simple symbols are shown in Fig. 1-7, using the template shown below.

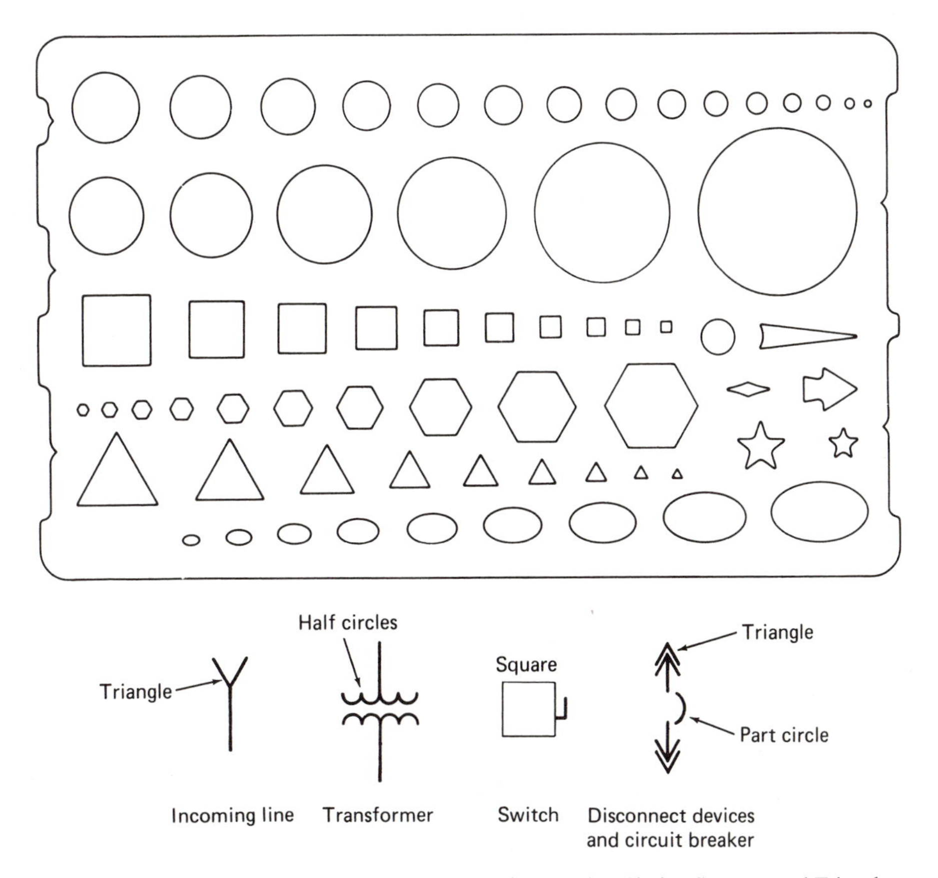

Figure 1-7 Examples of Symbols Made from Template Having Circles, Squares, and Triangles

1-8 Lettering Guides

Lettering guides are devices made from plastic and have several groups of countersunk holes arranged so that a hard, sharp-pointed pencil can be inserted through the holes and used to draw a series of lines properly spaced for the height of lettering required. The lines are drawn by moving the guide, using the pencil, across the drawing at the place where lettering is to be done. The guide runs against a straightedge of some sort when being used. It takes at least two moves of the pencil for one set of guidelines. These guides are sometimes made as part of standard small triangles. Another type has a rotating centerpiece that can be adjusted to suit the height of letter required. If the grid board covering described in Sec. 1-4 is used, the $\frac{1}{8}''$ spacing can be used instead of a lettering guide for $\frac{1}{8}''$ high letters. *Some form of guide should always be used* in order to keep lettering uniform in size.

1-9 Drawing Paper, Cloth, and Film

A wide variety of papers, cloths, and films are available for drafting use. The least expensive is an onionskin paper, more or less yellow in color and used primarily for sketching and studies. It lacks durability and a good transparency. It is easy to remove a piece from a roll to make a quick study sketch so it finds a fairly wide usage. (Architects use it extensively as study material.)

Vellum is a rag paper that has been treated to give it greater transparency. It found so much common use in drafting rooms that original drawings became known as *vellums*, no matter what the material might be.

Transparent cloth is a coated linen material. This has been used a great deal for original drawing and largely replaced vellum. The face side may vary in texture and some are more "toothy" than others, causing continual pencil sharpening. The reverse side is generally slick and smooth and will resist marking. This feature helps in file drawers where the drawings lay on top of each other because there is less tendency for pencil lines to smudge.

Films under several trade names are made for drafting. These materials are extremely durable. It is practically impossible to tear them and therefore they are favored for their resistance to wear. Line work can be done with either lead pencils or with special pencils made for drafting on film. Erasures on film may require a special eraser.

1-10 Pencils and Pencil Pointers

Manufacturers of pencils grade drawing pencils by number and letter code. Classifications are generally as follows:

Soft: 7B (softest) 6B, 5B, 4B, 3B, 2B. These pencils are too soft for drafting electrical drawings. Their use for such work results in smudged, rough lines that are hard to erase. They can be used in artwork and for shading where smudging can be controlled.

Medium: B (softest) HB, F, H, 2H, 3H. These grades are for general-purpose work in technical drawings. An H pencil is about as soft as needed for lettering and 2H and 3H are good for line work on most drawing papers and cloths. The softer grades are good for sketching, etc.

Hard: 4H (softest) 5H, 6H, 7H, 8H, 9H. These hard pencils may be needed for accurate line work on graphs, etc.; however, their use is restricted because the lines will be too light. A 6H or harder is a good pencil to use with a lettering guide because a light guideline is desired.

The pencils listed above are available as wood pencils or as pencil leads for use in a mechanical pencil holder.

Just as with any other tool that requires sharpening, a pencil does its best work when properly sharpened. A sandpaper block sharpener is still an excellent way to sharpen a wooden drafting pencil but it has been pretty much replaced by mechanical sharpeners. Because a drafting pencil should have a fairly long exposed lead, the manual mechanical sharpeners work better for the mechanical lead holders where it is possible to run the lead out to any length. Wooden pencils can be sharpened by electric sharpeners that have indicating lights to tell when the point is right.

1-11 Erasing Shields and Erasers

Among the sayings around drafting rooms is the one that says "if we run out of erasers, we're out of business" and another that says "don't get any more down in the morning than you can scrub off in the afternoon!" Changes are part of the design business and therefore expected. An erasing shield is a thin sheet of stainless steel with several shapes and sizes of slots and holes that allows the draftsman to expose only the part to be erased and shields all others.

For pencil lines, a firm eraser will generally do the erasing without injury to the drawing surface. Hard erasers can cut through either the paper or cloth if not used with care. Electric erasers are used a good deal but extreme care and the development of a technique is important to avoid rubbing a hole right through the paper.

1-12 Securing the Drawing to the Board and Keeping the Drawing Clean

When a drawing is to be started, it is important to line it up properly and secure it to the board in such a way as to cause the least interference with the drawing operation. First the board should be swept or wiped clean and then the sheet placed to the left-hand side of the board at a height to allow full use of the drafting straightedge or machine. It is usually the best practice to align the printed horizontal border line of the sheet with the straightedge rather than the edge of the sheet itself. Elements of the drawing will tend to relate for "squareness" and "level" to the printed border. (Of course the printed border and the edge of the sheet should be parallel but sometimes they are not.) If a grid sheet covering has been put on the board and aligned with the straightedge, then the border line on the sheet can be set to align with any of the grid lines (see Sec. 1-4).

The sheet should be secured at each of the four corners with a strip of masking tape placed diagonally across the corner. Further securing of the sheet is not generally necessary; however, additional strips at the top and bottom in the center can be placed if necessary. No tape should be put along either the right or left edge because they will interfere with movement of the straightedge.

As the drafting work develops, the triangles and other tools should be wiped clean with a cloth. A parallel rule can be picked up from the surface of the board for cleaning. The scales on a drafting machine can be wiped off by tipping the control head away from the board.

Drafting cleaners are simply cloth bags filled with eraser materials. These can be used somewhat like a sponge to help clean the drawing surface. Kneading the bag will cause small particles of erasing compound to drop on the drawing surface and some feel this helps to vacuum the excess pencil lead when swept away with a dust brush.

Some designers and draftsmen will protect areas of a drawing that have been completed by covering that section with a piece of paper. Only the area upon which they are working is uncovered and although this would seem to be a nuisance, the quality of a clean, neat drawing can be a work of art bringing pride to the person who cared enough to take some added precautions.

1-13 Lettering

Freehand lettering is definitely a talent with some and must become an acquired skill by the majority of others. If approached in a systematic way, the construction of the simplest letter forms can be learned first and practiced to perfection; then progress on to more difficult letters comes easier.

To start with, the letters I, H, T, L, E, and F are made with strictly horizontal and vertical lines. The strokes required to develop speed and neatness in making these letters are shown in Fig. 1-8.

Letters with diagonal lines are not difficult if the habit of making them with the right sequence of strokes is followed as shown in Fig. 1-9.

The widest letters are M and W and these are not any more difficult than those in Fig. 1-9. Letters requiring mastery of making an O are next and here only practice will help develop balanced figures (Fig. 1-10).

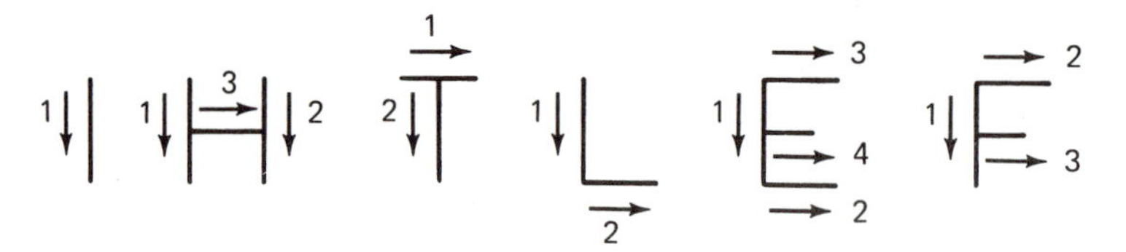

Figure 1-8 Strokes for Making Letters I, H, T, L, E, and F

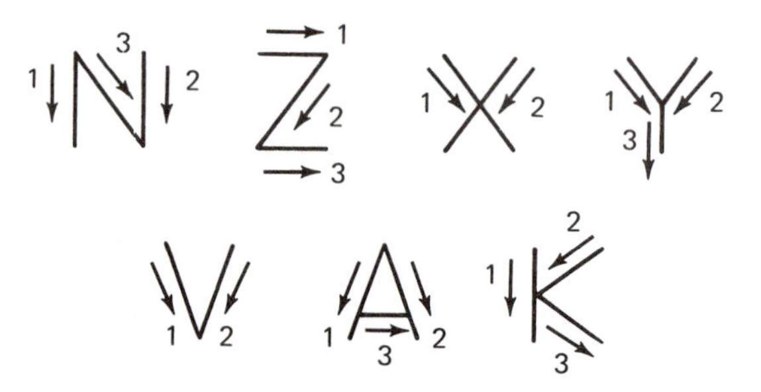

Figure 1-9 Strokes for Letters with Diagonal Lines

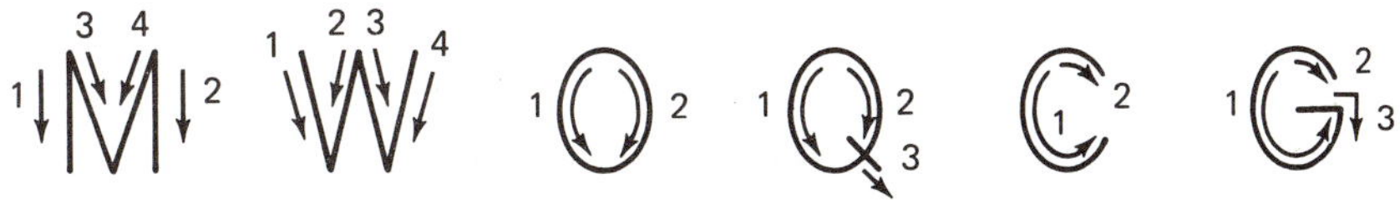

Figure 1-10 Strokes for the Widest Letters M and W and the "O" Family

The final and most difficult group is a combination of straight lines and curves; therefore good balance is the important requirement. These letters are shown in Fig. 1-11.

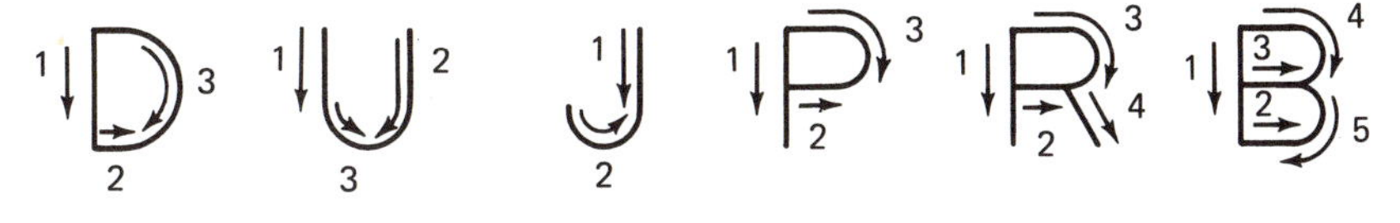

Figure 1-11

Undoubtedly an S is the hardest single letter to draw freehand and obtain good balance. This letter and most numbers are made almost completely with curved lines. The strokes are shown in Fig. 1-12.

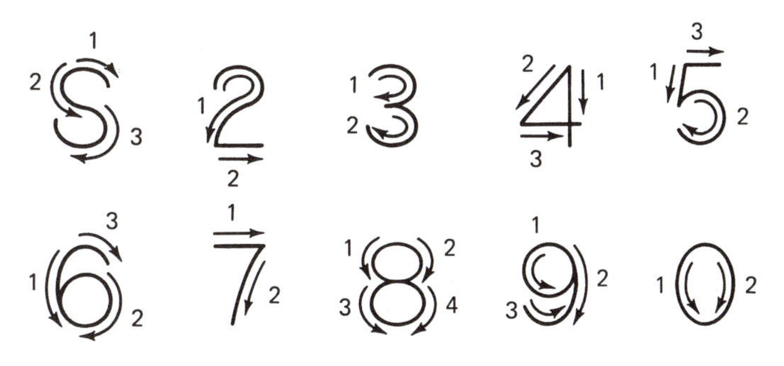

Figure 1-12

The examples of lettering shown are all vertical, capital letters. This style is the type most generally used on electrical drawings. The capital letters are desirable if any photographic reduction in size is made of the drawing because the letters are more open and can be read at the reduced size whereas lowercase letters tend to fill in when reduced and the letters a, e, etc., become blurred.

Templates and other mechanical lettering guides are available but of course are not so fast as hand lettering. These devices are almost always used for drawing titles and drawing numbers because in the large sizes the hand letters are difficult to make.

1-14 Line Characteristics

There are basically three weights of lines in all drawing: thick, medium, and thin. For electrical drawings these weights can be characterized to show specific elements of the drawing as follows:

One-line diagrams

Thick line	Thick line for main bus
Medium line	Medium line for feeders
Thin line	Thin line for control and relaying

Elementary diagrams

Medium line	Medium line throughout

Riser diagrams

Medium line	Medium line for electrical lines
Thin line	Thin line for equipment outlines

Interconnection and connection wiring diagrams

Medium line	Medium lines throughout for electrical lines (thick lines for busses can be used if desired)
Thin line	Thin lines for equipment outlines

Grounding drawings

Thick line	Thick line for main ground conductors or busses
Medium line	Medium line for grounding connection
Thin line	Thin line for building "out-of-function", equipment, etc.

Raceway and lighting layouts

Medium line	Medium line for all electrical work, called the "in-function"
Thin line	Thin line for building outline, etc., called the "out-of-function" or "shell"

Miscellaneous drawing

Medium line	Medium line for electrical work
Thin line	Thin line for nonelectrical work

These line characteristics, outlined above, will vary with the size and type of drawing. The fundamental idea in changing line weight is to make the electrical design work stand out clearly against the lighter background of the related surroundings. A well-executed drawing will make the electrical work "pop up from the sheet" when contrasts are maintained.

In addition to line weights, other standard line characteristics should be used with their proper weight as noted. Common miscellaneous line characteristics are as follows:

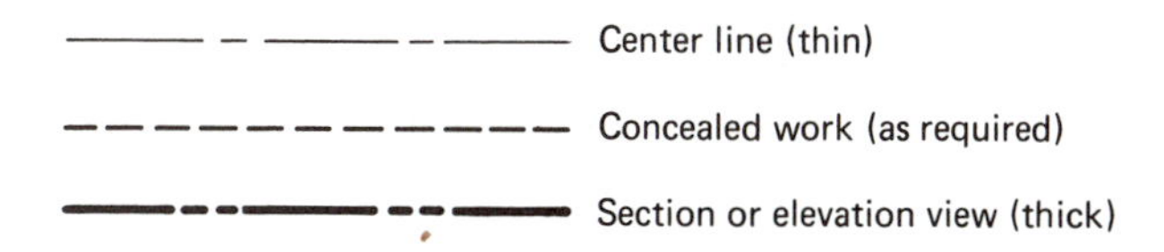

Line weight in itself should never become a symbol. For example, if thick lines are used for busses, they should also be identified with a note as to their electrical characteristic (voltage, phase, frequency, etc.).

1-15 Classification of Views

The principal views used in electrical drafting are

Plan: This is the view seen when looking from above down onto the area shown. For example, the layout of a floor is generally a *plan view*.

Elevation: This view is seen if the area shown is in the vertical position such as a wall area or the front of a building or structure. A reasonably good test of classifying a view as an elevation is to consider if the view could be photographed in the same way that a person standing might be photographed.

Section: This view requires that something must be cut in order to expose the view to be shown. As an example the *section view* through an underground electrical duct line would show the configuration of the conduits in the concrete encasement (see Fig. 9-2).

Detail: Details are generally used on electrical drawings to enlarge an area shown elsewhere at a smaller scale so that the detail can be seen. Detail views should *always* be oriented in the same direction as the view drawn at the smaller scale. Turning the view to some other quadrant can only cause confusion. Example of a detail is shown in Fig. 1-13.

The location on the drawing where elevations and section views are taken must be clearly identified. If possible, the views themselves should appear on the same sheet from which they are taken; when space is not available to do this, a common symbol for use in identifying the view and where it is drawn is shown in Fig. 1-14.

The drawing of the view should have a subtitle saying *Elevation* or *Section* and the circle part of the symbol of Fig. 1-14 should be repeated just below the view in order to identify it.

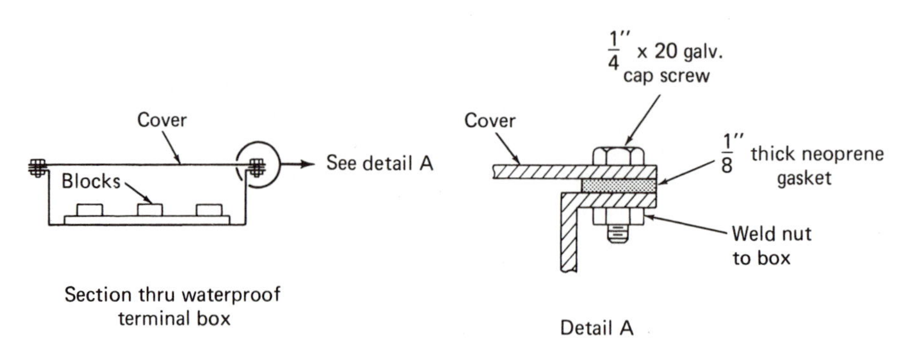

Figure 1-13 Typical Application of Detail View

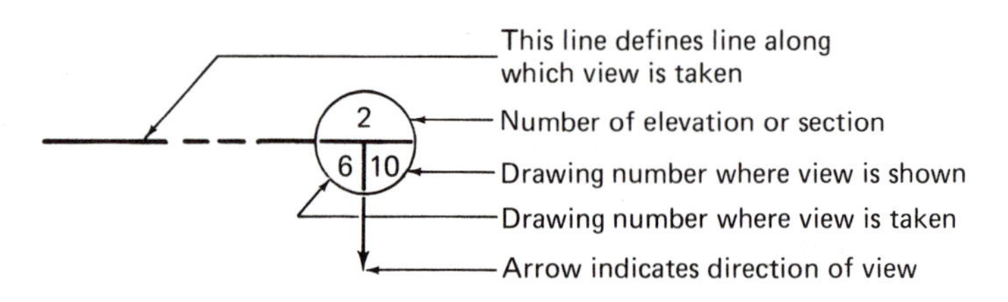

Figure 1-14 Typical Symbol for Identifying Section or Elevation View

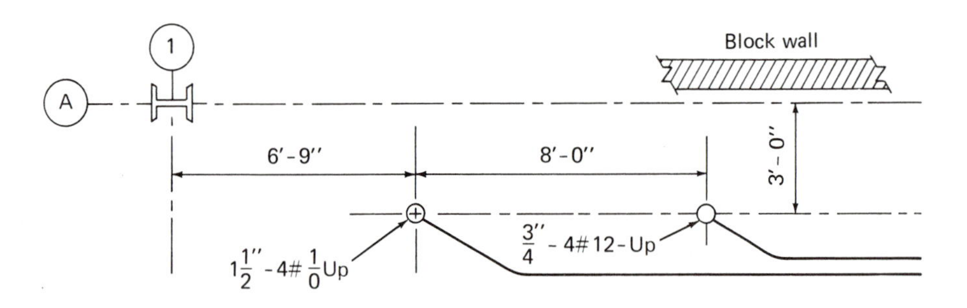

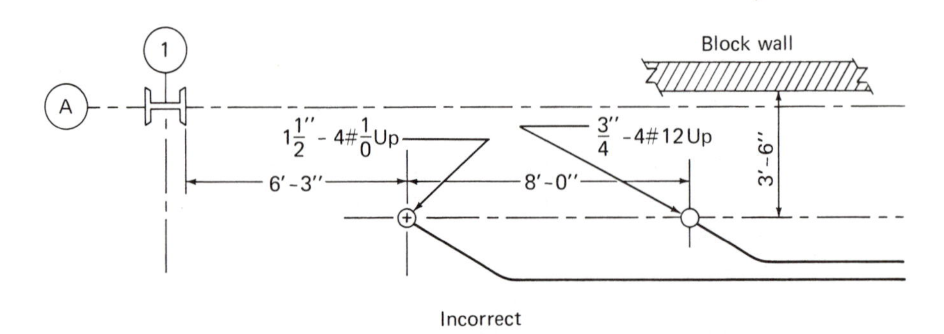

Figure 1-15 Correct and Incorrect Examples of Dimensioning and Labeling

1-16 Dimensioning and Labeling

Dimensions on electrical drawings should be referenced to column center lines, building lines, center line of major equipment, finished floor, etc. Dimensions to outside surfaces of columns, walls, and similar surfaces with variable positive position should be avoided. Dimensions up to and including 24 in. are often given in inches and larger dimensions are given in feet and inches. Care should be taken to avoid dimensions to small fractions of an inch. An electrician, in the field, can seldom set a device or piece of equipment to as close a tolerance as $\frac{1}{16}$". Sometimes dimensions just do not add up even if they relate to one another unless small fractions appear but these fractions should be avoided if possible. The easiest and fastest method for dimensioning is to place the dimension above the dimension line rather than to leave a break in the line for the dimension. Labels to identify lines, equipment, etc., should be located outside dimension lines and *never* across them. The correct and incorrect methods described above are shown in Fig. 1-15.

The use of long lines with arrowheads for labeling is preferred to crowding labels into congested spaces and very often this will also permit organization of several labels in a group, which is an aid in reading the drawing. As an example of this method, Fig. 1-16 shows several electrical conduits connecting into equipment mounted on a wall. By running long label lines outside the congestion the conduit runs are kept clear of any additional confusion.

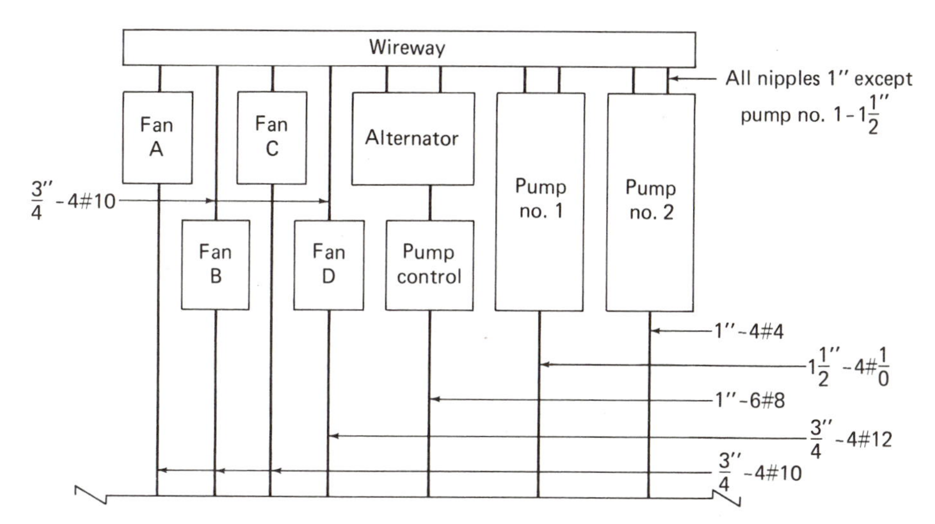

Figure 1-16 Labeling Is Done in Border Area of View to Avoid Congestion

1-17 The North Arrow and Arrowheads

All plan views should be oriented by either drawing a north arrow symbol onto the sheet or attaching a prepared decal north arrow symbol onto the back of the sheet. If at all possible layouts should be oriented on the sheet so north is toward the top of the drawing. Any part plans drawn on the same sheet should be oriented in the same north direction as the main plan. Figure 1-17 shows types of north arrows.

Arrowheads should be made sharp and never spread open. The correct and incorrect types are shown in Fig. 1-17.

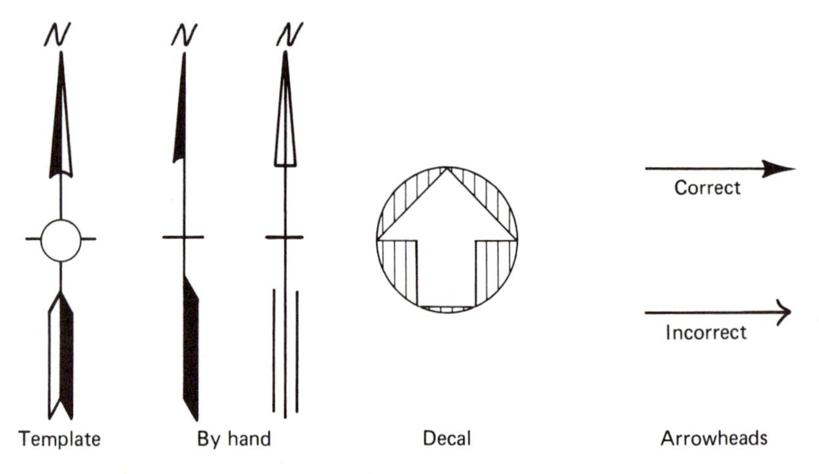

Figure 1-17 North Arrows and Arrowheads

1-18 Drawing Titles and Numbers

The titles and numbers of drawings are drafted into the title block on the drawing using templates or other mechanical devices for forming letters and numbers. It is usually necessary to try out the title spacing on a separate piece of paper before attempting to draft directly into the block. This trial run will permit centering the title properly because its overall length will be determined. Size of lettering will be dependent on the length of the title. Once a lettering size is chosen, all drawings in a set should use the same size even if some titles are short.

Drawing numbers are often assigned in blocks. For example, architectural drawings might be assigned numbers 1 through 99; structural, 100 to 199; mechanical, 200 to 299; and electrical, 300 to 399. The project prefacing number might be the same for all drawings and the block numbers used only as a suffix. For example, the one-line and relay diagram might be numbered 222-75-301. The 222 is the project; 75 might be the target year for construction; and 301 is the unique number for the electrical drawing.

1-19 Checking Drawings

Before a drawing is signed and issued for construction, it must be thoroughly checked. Checking procedures vary; however, a system that seems to work quite well is to draw a yellow line through anything that checks all right. A red line indicates an error requiring correction. The checker may choose to cover the area that is wrong with red and mark over with black pencil the correction to be made. If the correction is extensive, he may circle it in red and make a note explaining the error and correction to be made. Sometimes an attached sketch can be used. Some checkers will use a green pencil to note reference data used in their checking. Whatever method is used, the checking changes and instruction should be clear so corrections can be readily made.

Good checking requires a conscientious, methodical, step-by-step screening of all the information on the drawing. A costly interference in the field can be avoided by careful checking; therefore this responsibility should not be taken lightly. A list for checking is as follows:

1. Using latest reference material
2. Understanding design concepts
3. Drawing data agreeing with data from other related electrical drawings
4. Spelling
5. Proper abbreviations (spell words out if possible)
6. Dimensions
7. Interference of other designs (structural, mechanical, etc.)
8. Clearances sufficient for installation, removal, maintenance, etc., of equipment
9. Arrowheads directed to correct reference point
10. Conduit numbers and sizes agreeing with schedule if schedules are used
11. Size of wire and cables
12. Size of conduit or other raceway
13. Not more than three 90° bends in conduit runs (code does allow four)
14. Uniform lettering
15. Views properly identified and referenced
16. Scales indicated
17. North arrow shown if required
18. Notes read well, stand alone, easily understood
19. Correct title and number
20. Drawing is clean and legible

A checker should never take anything for granted! Improvements in design should be suggested but a checker should not try to redesign an accepted design.

1-20 Drawing Revisions

After a drawing is completed and issued for construction, changes may develop that require revisions to the drawing. Because the work shown on the drawing may be under construction, the contractor must be given clear information on the revisions. To do this the drawing is changed, the area where the change took place is circled, and a revision note is made in the revision block section of the drawing saying what was done. Circling is usually done on the back side of the drawing using either a relatively soft pencil or a thin line, black, pressure-sensitive graphic tape. When the contractor receives the revised drawing, he will compare it with the previous issue to determine the extent of the change. To identify the area on the drawing that has been revised clearly, the circled section has a flag-type symbol attached to identify the revision number. Figure 1-18 shows this method.

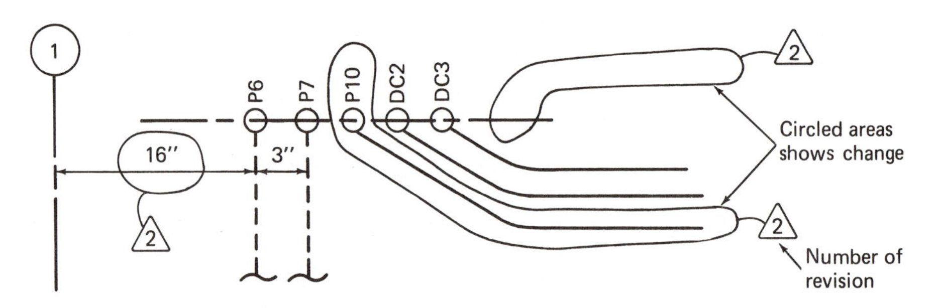

Figure 1-18 Method of Identifying Revisions

The circling is always removed when subsequent revisions are made. The pressure-sensitive tape strips off very easily, if that has been used. The triangle revision symbol may be removed or left as preferred.

The revisions shown in Fig. 1-18 must be "spelled out" in the revision block on the drawing. Figure 1-19 shows proper notations. Notice that in the dimen-

2	ADDED CONDUIT P10, DELETED CONDUIT DC4, CHANGED DIMENSION TO COL. 1	TRA	SAL	MBB	CWS	12/24/74
1	ADDED GENERAL NOTE NO. 10	PRT	JAS	MBB	CWS	5/24/73
NO.	DESCRIPTION OF REVISION	BY	CHKD.	APPD.	ENGR.	DATE

Figure 1-19 Revision Shown in Fig. 1-18 Described in Revision Block

sion change note it merely says the dimension was changed—not from what to what. Comparison of the revision 2 issue with the revision 1 issue will indicate the magnitude of the change.

Revision notes should be as brief as possible. Usually they can be described under the categories *added*, *deleted*, and *changed*.

Chapter Review Topics and Exercises

1-1 Why is the common T square not used as much in modern drafting rooms?

1-2 What would be the disadvantage of using a drafting machine where many long lines must be drawn?

1-3 How does grid paper help in doing electrical drawings?

1-4 If grid paper is used, why is it important to line the grid lines with the straightedge when attaching the grid paper to the board?

1-5 Why is it best to draw vertical lines up the board if possible?

1-6 How many subdivisions are needed at a scale of $\frac{1}{8}'' = 1'\text{-}0''$ to measure 8 in.? at $\frac{1}{2}'' = 1'\text{-}0''$ to measure $4\frac{1}{2}$ in.?

1-7 Why is it easier to lay off a scaled measurement by placing the subdivision part of the dimension at the starting point and locating the full foot dimension rather than the other way around?

1-8 How can a circle template be used to draw the symbol for a transformer?

1-9 How does the grid board covering help in lettering?

1-10 How is drafting cloth treated to prevent smudging?

1-11 Can a 2B grade pencil be used successfully on drafting of electrical drawings?

1-12 Describe the function of an erasing shield.

1-13 Letter five lines each of the I, H, T, L, E, and F letters with guide lines spaced $\frac{1}{4}$ in. apart using the strokes shown in the text.

1-14 Repeat Exercise 1-13 with N, Z, X, Y, V, A, and K.

1-15 Letter 10 lines of M, W, O, Q, C, and G; use $\frac{1}{4}$-in. high letters.

1-16 Letter 10 lines of D, U, J, P, R, and B; use $\frac{1}{4}$-in. high letters.

1-17 Letter 10 lines of S and all numbers; use $\frac{1}{4}$-in. high letters.

1-18 How are titles lettered?

1-19 Why is line weight important?

1-20 Describe an elevation view.

1-21 What is the purpose of a detail? How should it be shown?

1-22 What does a section line do? What does it look like?

1-23 What are some of the basic rules to be observed in dimensioning?

1-24 What is a label line? What is a good technique for using it to avoid congestion?

1-25 Is there any best way to make an arrowhead?

1-26 Which direction should the north arrow point on the drawing, if possible?

1-27 Do drawing numbers sometimes have code meanings?

1-28 What are several things (at least 10) to look for in checking an electrical drawing?

1-29 Describe how revisions are made and noted on the body of the drawing.

1-30 What categories of change should be noted in describing revisions?

1-31 Using a T square or other straightedge and triangles, draw the following figures:

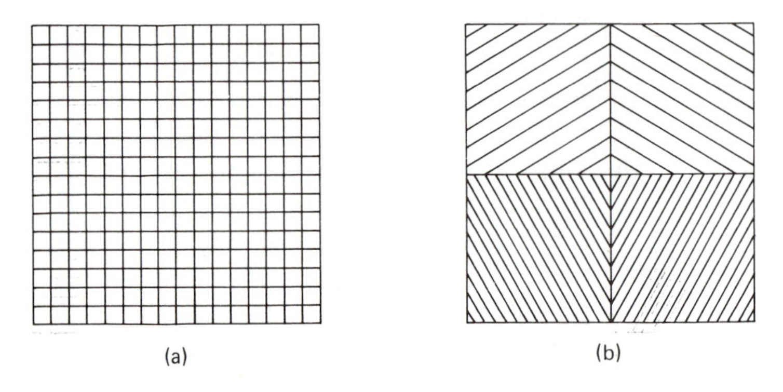

(a) (b)

Large squares are 4″ × 4″. For Fig. a, lay off $\frac{1}{4}$-in. divisions on bottom and left vertical line; using straightedge and triangle, complete grid. For Fig. b, lay off both a vertical and horizontal line through the center of the square. Make $\frac{1}{4}$-in. divisions up and down the vertical center line and use these as starting points for the 30° lines in the upper half and 60° lines in the lower half. (To complete the figure it will be necessary to extend the $\frac{1}{4}$-in. division marks along a temporary extension of the vertical line outside the square.)

Check accuracy of drafting to see if all squares are the same size in Fig. a and division of spaces are the same in Fig. b. Keep lines sharp and bright. Practice these figures or other similar ones to develop drafting skill.

2

FUNDAMENTALS OF ELECTRICITY

2-1 General

The material given in this chapter is basic to understanding the material presented in the subsequent chapters. It covers, in a simple way, the most important theoretical and practical elements of the subject of electricity. It is purposely held to an elementary level for those who are not familiar with electrical phenomena or terminology. Further study from the many good texts available is recommended.

2-2 The Electron Theory

Most scientists assume the existence of electricity as the actual entity (thing) of which all matter is composed. The electron theory assumes that all the different kinds of atoms are composed of *electrons* having a unit *negative* charge and *protons* having a unit *positive* charge. The electrons are relatively free to move and normally circulate about the protons, which form the nucleus (center) of the atom. Atoms of the various elements differ from each other only in the number and groupings of the electrons and protons. A simplified representation of the atomic structure of hydrogen with atomic number 1 and helium with atomic number 2 are shown in Fig. 2-1. Note that in the helium atom there is also present two neutrons that have neither a positive nor a negative charge.

Because all matter in an uncharged state has equal amounts of positive and

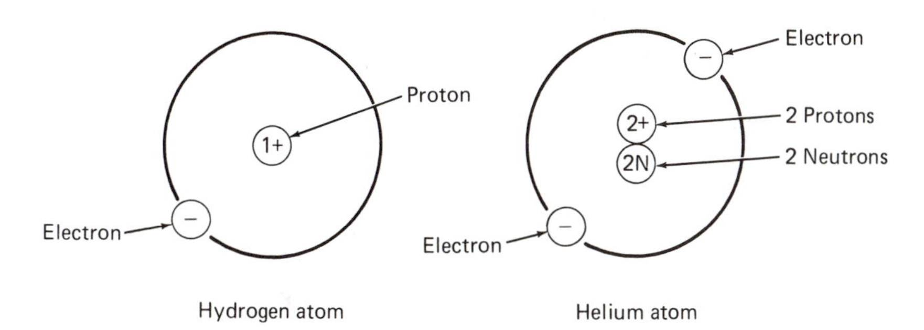

Figure 2-1 Simple Representation of Atomic Structures

negative electricity, it is necessary to provide some force to cause the electrons to line up or move in a definite direction. The ability to do work is dormant unless something causes the electrons to move. It is this motion of the electrons, which if confined to definite closed paths, makes it possible for electrical energy to be harnessed and put to work.

2-3 Electromotive Force

Because electrons are normally distributed evenly throughout a substance, a force or pressure called *electromotive force* (EMF) is required to detach them from the atoms and make them move in a definite direction. This force or pressure is called the *potential* or *voltage*. The unit used for measuring this force is called the *volt*. The higher the voltage, the greater the number of electrons that will be caused to flow. If an EMF is impressed across the ends of two wires that form a closed conducting path, the force will cause the electrons in the atoms to become detached and flow in a definite direction.

2-4 Sources of Electromotive Force

The two main sources of electromotive force are batteries and electric generators.

Batteries are constructed essentially of two dissimilar metallic plates immersed in a chemical solution called the *electrolyte*. This arrangement causes a transformation of chemical energy directly into electrical energy.

Electric generators produce an electromotive force through the relative motion between a magnetic field and an electric conductor. The conductor may be moving across a fixed magnetic field or conversely the magnetic field may be moved across the fixed conductor.

2-5 Electric Current

By applying a potential or voltage of sufficient strength, the electrons move; this results in a flow of electrons along a fixed path. This flow is called the *electric current*. The rate of flow is measured in *amperes*. The ampere is similar to the gallons-per-minute measurement in the flow of fluids. One ampere is equivalent to the flow of 6.25 million, million, million electrons per second.

2-6 Electric Conductors

Every substance allows the flow of electrons to take place; however, electrons flow very easily through certain materials such as copper, aluminum, iron, and other metals. Because the electron flow is easier when these metals are used, they have become the paths used to conduct electricity and are therefore called *electrical conductors*. Wires and cables are the most common form of conductors.

2-7 Electric Insulators

Where some materials are ideal for the flow of electrons, others are equally ideal in their use because they allow almost no electricity to pass through them. These materials are called *nonconductors, dielectrics,* or (more popularly) *insulators.* Such material as rubber, plastics, cotton, mica, porcelain, glass, fibers, dry paper, and air are used as insulators. When an insulator is continuous, such as the covering over a wire, it is called *insulation.*

2-8 Electric Resistance

Opposition to the flow of the electrons through any material is called *resistance.* The unit of measurement of resistance is the *ohm* (Ω). The best of conductors will have some resistance. Insulators, of course, have very high resistance. The unit of measurement of very low resistance is the *microhm,* which is equal to one-millionth of an ohm. The unit of measurement for very high resistance is the *megohm,* which is equal to one million ohms.

Resistance in a conductor varies directly as its length and inversely as its area. The greater the length, the greater the resistance. The greater the cross-sectional area, the smaller the resistance.

Resistance to the flow of electrons in electricity may be compared to the resistance of water flow through a pipe system. If the pipe is smooth inside and the run is straight, then the resistance to the flow of water is small and the water pressure at the discharge end of the pipe will be only slightly less than at the input end. If, however, the pipe is rough inside and has many bends and fittings for the water to flow through, then the resistance is high and the pressure drop from input to discharge is high. Similarly, a good conductor allows the electrons to flow with a small loss of voltage (pressure). The poor conductor acts just like the pipe system with higher resistance and a larger voltage drop is the result. The energy used up by the electrons being forced through the resistance is converted into heat. When wires and cables feel warm or even hot to the touch, it is an indication that they are too small in diameter to carry the current flow. This same principle is used beneficially when electrical heaters are designed. The wire used is chosen for its high resistance in order to generate heat purposely.

2-9 The Electric Circuit

The paths along which the electric current flows is called the *electric circuit.* In order to put the electricity to work, it has to be transmitted along these paths, which are generally formed of insulated conductors. Figure 2-2 shows a simple electric circuit. It consists of a generator, a motor, a switch, and a "path" of wire conductors that form the electric circuit. Understanding the electrical circuit is easier if compared to a similar circuit of piping for transmitting hydraulic energy using water. Figure 2-3 shows a water system consisting of a pump,

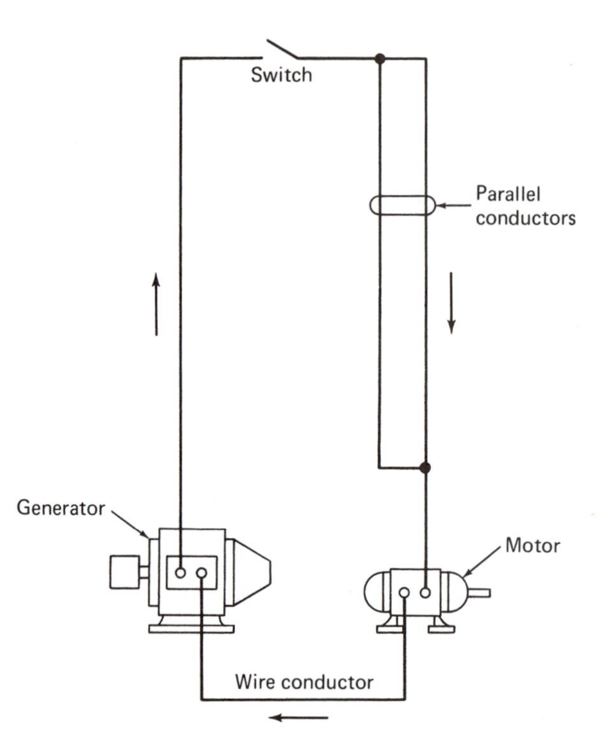

Figure 2-2 A System of Machines and Conductors for Transmitting Electric Power, Called an Electric Circuit

water turbine, pipes, and a valve that is similar in circuit arrangement to the electrical circuit. This is a commonly used analogy for explaining the electric circuit that is found in many texts describing the fundamentals of electricity.

The centrifugal pump is similar to the generator. Both must be driven by an outside source of power such as a motor or gasoline engine. The water turbine and the electric motor perform similar functions in that they are used to drive a piece of equipment that performs the work required. Of course, the switch and valve perform the same function in their relative circuits. The pipes are the water conductors and the wire conductors provide the path for the electric current flow.

In the water system, if the pipes are full of water and the valve is closed, then the pump will be able only to exert pressure in an effort to circulate the water and thus drive the water turbine. The same is true, of course, in the electric circuit. Closing the switch or opening the valve will allow the circuits to function. The water pressure is similar to voltage and water flow to current flow; resistance to water flow is produced by the piping combination and resistance to current flow depends on the wire characteristic and length.

The circuits shown are very simple. Complications are introduced when the two parallel paths are provided. Water or electricity will, of course, flow in both paths. Usually several motors, each controlled by its own switching

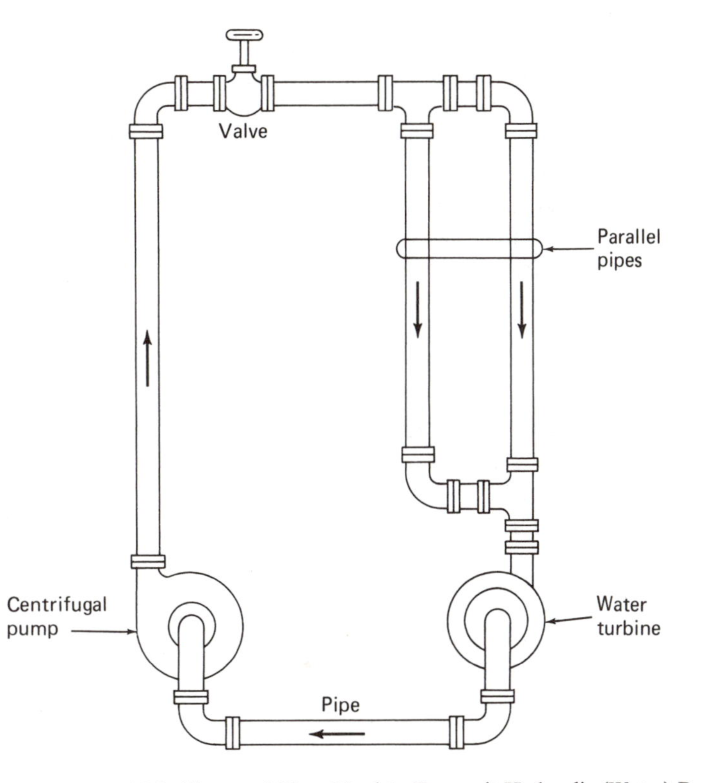

Figure 2-3 A System of Machines and Pipes Used to Transmit Hydraulic (Water) Power

device, are run by the same generator. Interconnecting circuits and networks of circuits complicate modern electrical systems in order to power the vast electrical demands.

2-10 Magnets and Magnetism

A *magnet* is any object that can attract iron or steel to it. This attraction ability is called *magnetism*. There are natural objects in the earth that have this ability. They are called *loadstones*. If handfuls of iron filings were scattered close to a loadstone, they would be attracted to it with the greatest concentration of filings occurring at the ends of the stone as shown in Fig. 2-4. The two ends that have the greatest attraction are called the *poles* of this natural magnet.

If the loadstone were suspended by a string attached around its center so

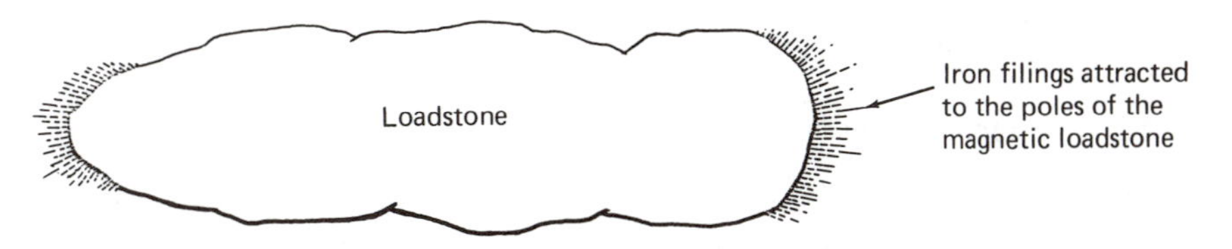

Figure 2-4 Natural Loadstone Magnet

that it were free to turn, it would immediately move so that an imaginary line drawn from pole to pole would line up with the earth's magnetic north and south poles. The end of the loadstone that pointed north could then be identified as the *north pole* and the end pointing south would be called the *south pole*. There is no practical use of the loadstone except that it provides evidence of magnetism in earthly elements and proves the existence, as a natural phenomenon, of the earth's magnetic field.

Artificial magnets have been made by man from pieces of iron or steel that did not previously have any magnetic ability. These magnets are generally of two types: permanent and temporary. Temporary magnets hold their magnetic qualities only as long as the magnetizing force is maintained. Permanent magnets, as their name implies, hold their magnetism for long periods of time. Interestingly, these magnets can lose their magnetism through exposure to too much heat or even from a severe jar.

2-11 The Magnetic Field

The easiest way to "see" a magnetic field is to perform a simple experiment with a bar magnet. The magnet is placed on a flat surface and covered with a sheet of white paper. Iron filings are sprinkled onto the paper and the resulting picture will be similar to that shown in Fig. 2-5.

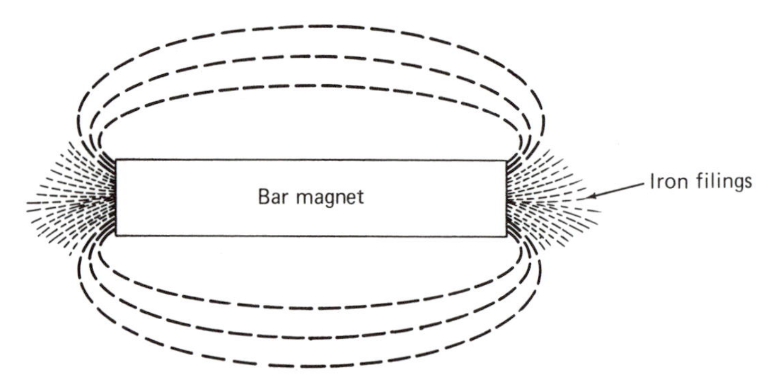

Figure 2-5 Iron Filings Surround Magnet to Show the Existence of a Magnetic Field

The way in which the iron filings arrange themselves around the magnet indicates the direction of the magnetic field. The strength of the field will vary at different points. Just like the loadstone, the concentration of filings will be at the ends or poles of the magnet. These lines (of course without the filings to "show" them, they are invisible) are called the *lines of magnetic flux* or simply *magnetic flux*.

2-12 Relationship between Electricity and Magnetism

Magnetism is especially important to the generation of electricity and to the use of electricity in motors and other devices.

First, magnetism can be produced by an electric current. Actually, every conductor carrying electricity has a magnetic field surrounding it. If the conductor is coiled in a helix (like a spring), the magnetic field will pass through the center of the helix and form loops of magnetic flux around the outside. If a core of iron or steel is placed in the center of the coil, the strength of the magnetic field is increased. This arrangement is shown in Fig. 2-6.

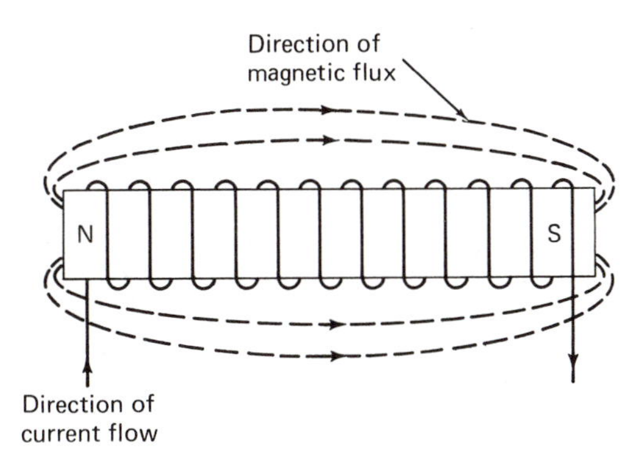

Figure 2-6 Construction of a Magnet Using an Electric Coil

The strength of the magnetic field of Fig. 2-6 depends on the current flowing in the wire and in the number of turns in the coil. Many turns and a strong current will create a strong field. The iron core in the center of the coil increases the strength of the field because iron has the peculiar property of being the easiest material in which to create a magnetic field.

If the electric magnet is made in the shape of a horseshoe as shown in Fig. 2-7, the magnetism will follow the iron core and pass across only the air gap marked G. The poles are identified as north and south (N and S) because this is the conventional way to identify the magnet. A mechanical force actually exists between the two poles that tends to draw them together.

The second important relationship between electricity and magnetism is the actual production of the electric voltage in a conductor moving across the magnetic field. The pressure that needs to be developed to force the electrons

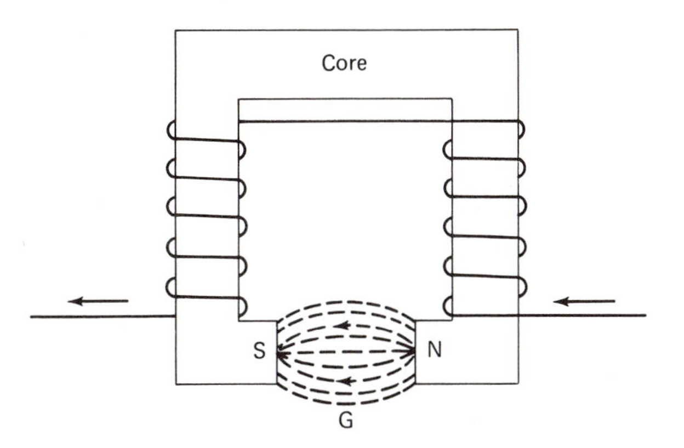

Figure 2-7 Magnetic Field across a Gap in a Horseshoe-Type Electric Magnet

to move from atom to atom along the fixed path (the conductor) is provided by moving the conductor across the lines of flux! This very simple procedure is all that is required to make the electrons flow as a *current* along the *conductor*, assuming that a closed *circuit* has been provided. This arrangement is illustrated in Fig. 2-8.

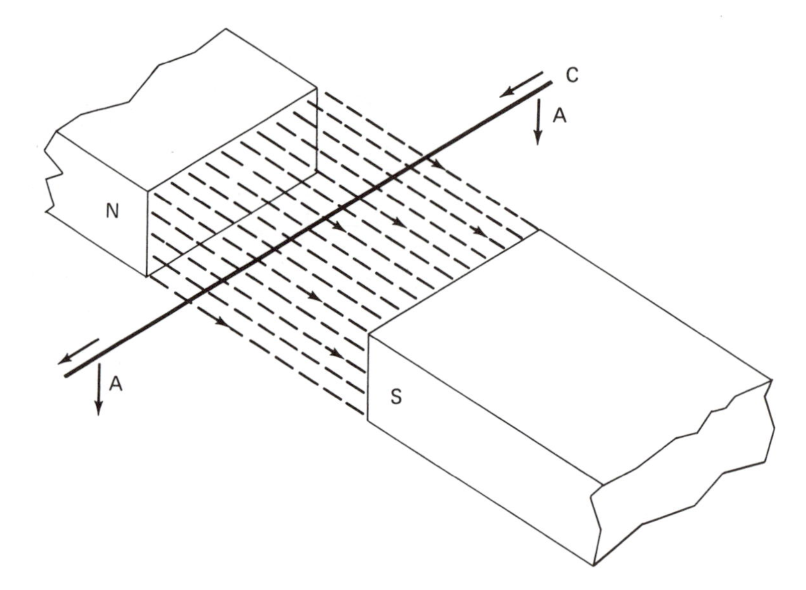

Figure 2-8 Movement of an Electric Conductor across a Magnetic Field Generates a Voltage in the Conductor

The conductor C in Fig. 2-8 is made to move, through some external force, across the magnetic field in the direction indicated by the arrows A–A. Voltage is produced in the direction of the arrows along the conductor if the flux is in the direction indicated, i.e., north to south. If the magnetic field is strong and the motion of conductor is fast, then a higher voltage will be produced than at a slower speed with a weaker field.

The same voltage would be produced if the conductor were held stationary and the magnetic field were made to move, by an external force, across the conductor.

2-13 Kinds of Current in Common Use

It is common to hear of *direct current* and *alternating current* with the latter being in more general use. Simply defined, the direct-current system has a voltage that remains at a constant value and the current flows in one direction. Thus the word *direct* is appropriate. In the alternating-current system, the voltage and current reverse their direction regularly from up to a maximum value in one direction to the same maximum value in the opposite direction. They *alternate* continuously. Figure 2-9 shows a comparative diagram of the

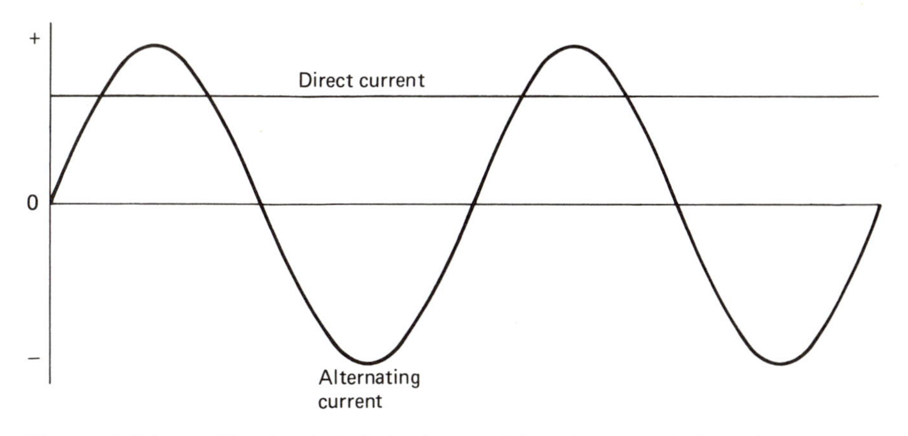

Figure 2-9 Diagram Showing the Relation between Direct Current and Alternating Current

two types of currents. The straight line represents the *direct current* and the wavy line shows the same value of the *alternating current* as compared to the direct current.

2-14 The Single-Phase Generator

Figure 2-10 shows a simplified diagram of a single-phase electrical generator. The horseshoe-type electric magnet is used to illustrate the magnetic field. The conductor which must pass across the magnetic lines of flux is formed into the

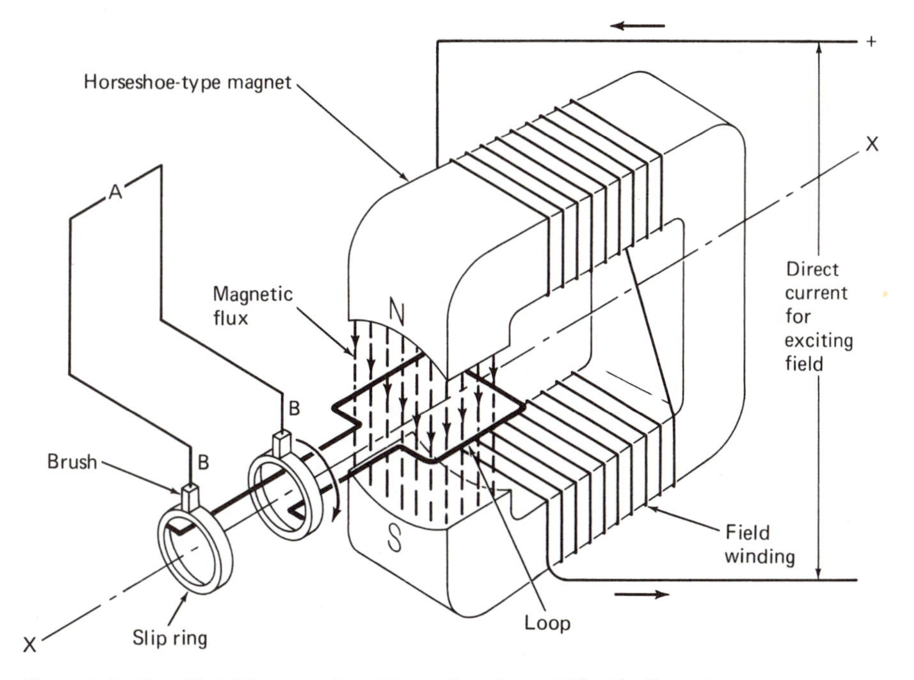

Figure 2-10 Simplified Diagram of an Alternating-Current Electric Generator

shape of a loop which can be imagined to rotate on the axis X–X. The ends of the conductor are connected to rings (called *slip rings*) mounted on the shaft and brushes (labeled B–B) ride on the rings so that the flow of electrons can flow from the rotating conductor into the circuit that is external to the machine. The brushes are usually made from blocks of carbon, which is a good conductor, and can be shaped to fit closely to the rings. They are held in place and pressed against the rings by spring devices mounted on the brush holders.

The magnetic flux is developed through the winding on the magnet core and is powered from an external direct-current source so that the flux in the gap remains steady and constant. The current in this magnet is called the *exciting current*.

The generator shown in Fig. 2-10 is a two-pole, single-phase, revolving-armature, alternating-current generator. The magnetic field, the coils of wire around the iron core, and the coil itself are called the *field* of the generator. The rotating loop is called the *armature*.

An external source of mechanical power such as a motor or gasoline engine or steam driven turbine is connected to the looped conductor in order to make it rotate. As the conductor "cuts" across the lines of flux, a voltage is developed

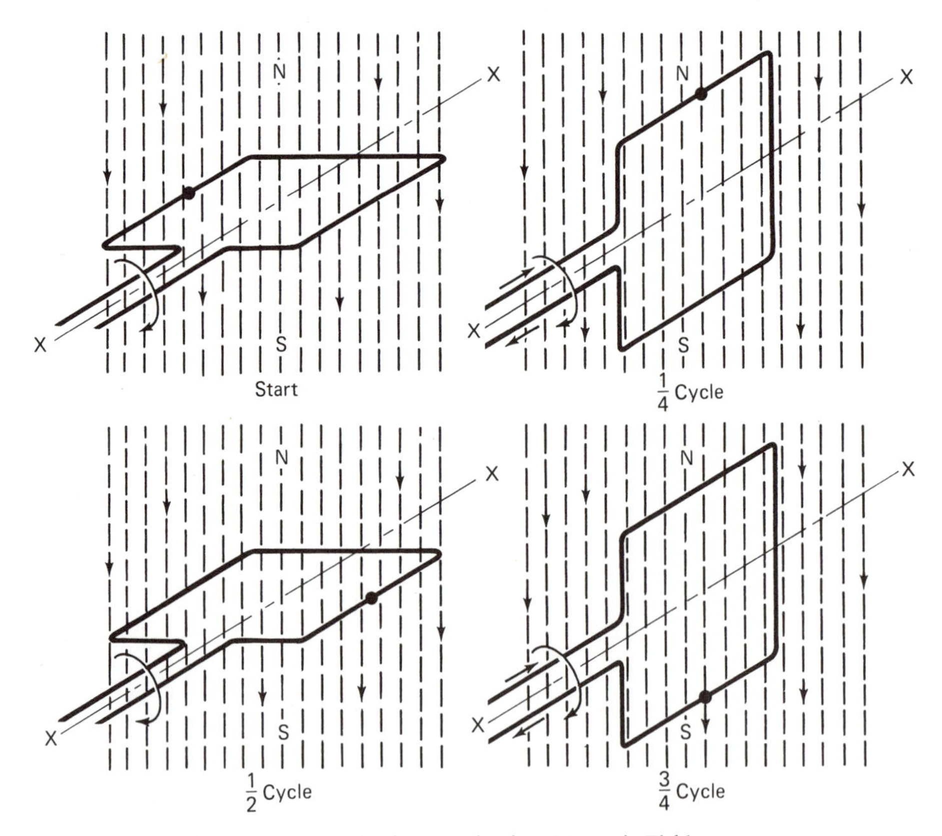

Figure 2-11 Diagram of Loop of Wire Rotating in a Magnetic Field

that causes the electrons to flow out to the slip rings, through the brushes, and into the circuit.

The series of diagrams in Fig. 2-11 show in more detail the position of the loop of the conductor in relation to the lines of magnetic flux in the gap. It is important to note the location of a particular spot on the conductor and follow it through the diagrams because the *value* of the voltage generated relates to *how many lines of flux the conductor is crossing at any given time.* For this reason the conductor is marked with a dot and following it through the following explanations will simplify comprehension of the text material. It is assumed that the conductor is rotating at a uniform speed in the direction of the curved arrows.

At the start, the dot on the conductor shows that the conductor is moving parallel to the lines of flux and therefore is not actually cutting across any of the lines; therefore the voltage is zero. As the conductor rotates toward the $\frac{1}{4}$-cycle point, it starts to cut some lines and a voltage begins to develop. At the $\frac{1}{4}$-cycle point, it is cutting across the maximum number of lines and therefore the maximum voltage is generated. As the dot swings toward $\frac{1}{2}$ cycle, it starts to cut fewer and fewer lines until at exactly $\frac{1}{2}$ cycle the voltage is zero again. The conductor (at the dot) swings down to $\frac{3}{4}$ cycle, cutting more lines all the time until at the $\frac{3}{4}$ point it is once again at a maximum value except this time in the opposite direction to that at $\frac{1}{4}$ cycle. It finally returns to zero at the completion of one revolution (or cycle) where it is at start once again for the next cycle. Figure 2-12 shows diagrammatically a "picture" of the voltage produced.

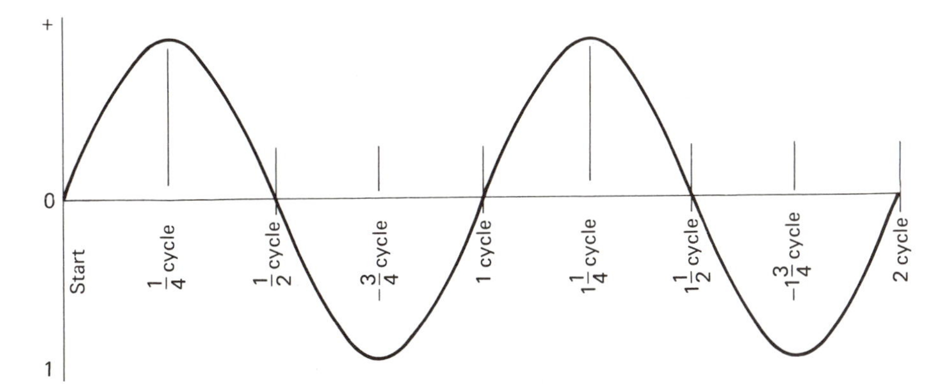

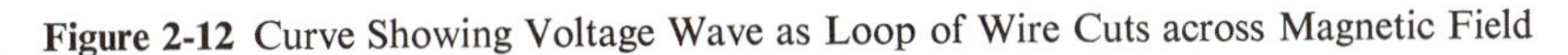
Figure 2-12 Curve Showing Voltage Wave as Loop of Wire Cuts across Magnetic Field

In following only the dot in the example given, it is important to bear in mind that the opposite side of the loop (from the dot side) also has a voltage produced in it as it passes across the lines of flux. The voltage in the two sections of the conductor add together and the voltage across the entire loop is the sum of the two.

The horizontal line in Fig. 2-12 represents time; the vertical line represents the value of voltage produced. To simplify the diagram, the peak values are represented only by the plus (+) and negative (−) signs and the time scale is marked off in parts of a cycle. (Cycles are a measure of time because it *takes time*, of course, for the conductor to rotate in each cycle.) The wavy line represents the rise and fall of the voltage produced and clearly shows that the voltage alternates in each cycle.

If the voltage completes 60 cycles in 1 sec, it is called 60-Hz voltage. (Formerly it was called 60-*cycle* voltage but the word *hertz* is now accepted as the correct International Standard term.) The current that this voltage will cause to flow is called 60-Hz current. This characteristic is called the *frequency* of the system.

2-15 Two-Phase Generator

In order to show the development of the three-phase generator, which is in common use throughout the electrical industry, it will help to understand the two-phase machine even though this machine is in little use.

The two-phase generator is merely a combination of two single-phase generators. Figure 2-13 shows the simplified arrangement of the loops of wire at 90° to each other; i.e., the circuit A loop is horizontal in the sketch and the circuit B loop is vertical along the axis X–X.

The armatures are mounted on the one shaft and therefore must revolve together and always be at right angles to each other. This arrangement means

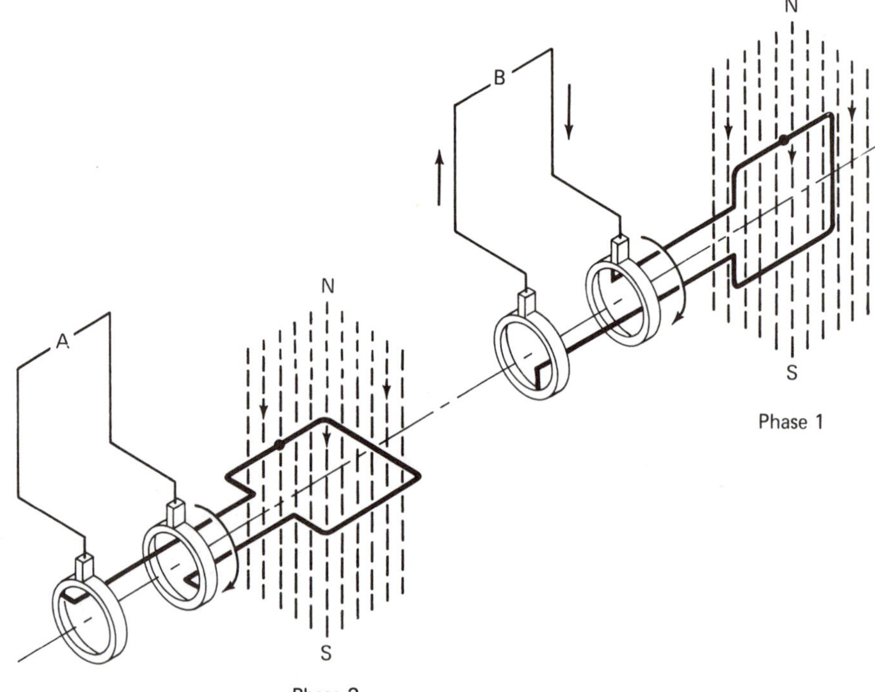

Figure 2-13 Arrangement of Armature Windings for Two-Phase Generator

that when the first loop is at zero voltage, the second loop will be at a maximum voltage. These zero–maximum relationships are always a $\frac{1}{4}$ cycle apart. The curve showing the voltages generated is illustrated in Fig. 2-14.

From this diagram it can be seen that phase 2 follows phase 1 and the voltage is always $\frac{1}{4}$ cycle behind due to the way in which the loops have been positioned on the shaft.

As a practical matter, the two windings are not actually displaced along the shaft but rather combined into the same shaft location as shown in Fig. 2-15.

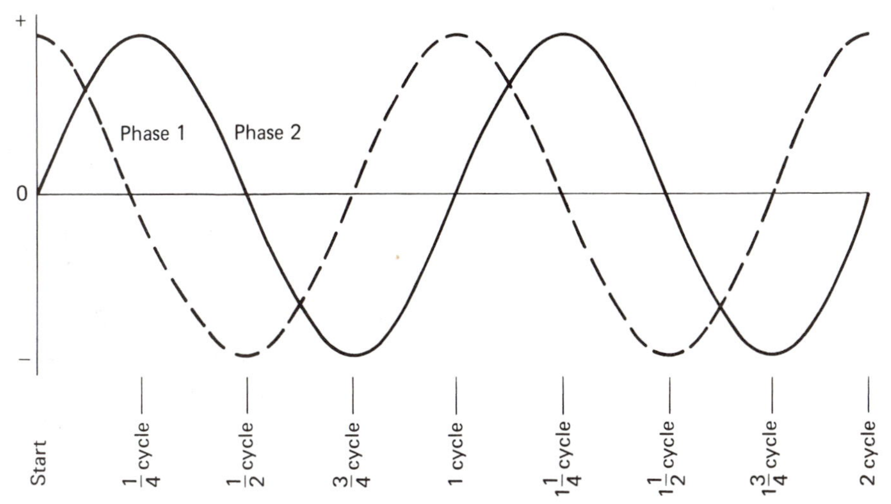

Figure 2-14 Curve Showing Voltages Produced in a Two-Phase Generator

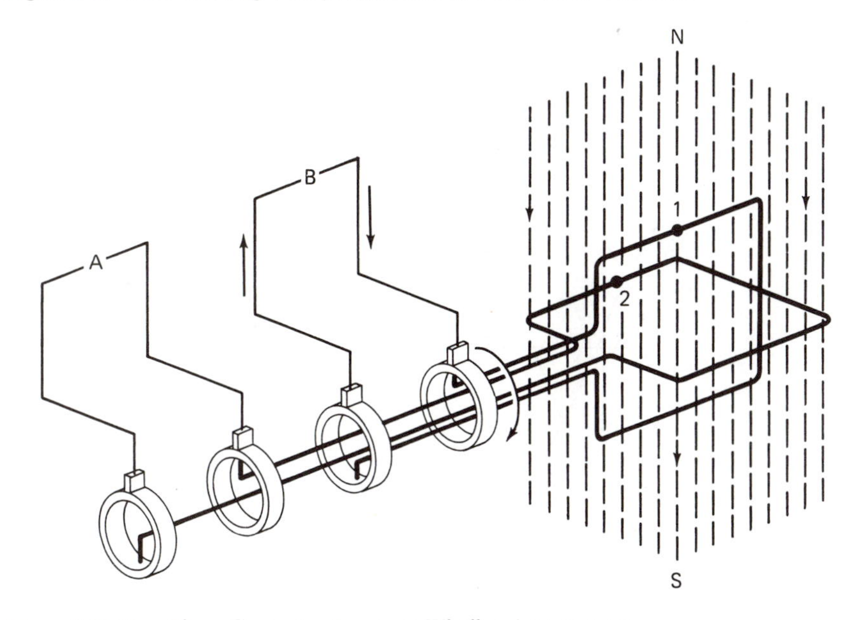

Figure 2-15 Two-Phase Generator Armature Winding Arrangement

2-16 Three-Phase Generator

As might be expected, through understanding of the single-phase and the two-phase generators, the three-phase generator is a combination of three loops mounted on a single shaft designed to rotate in the same magnetic field. This arrangement of coils is shown in simplified form in Fig. 2-16. Both ends of the coils are wired out through the slip rings.

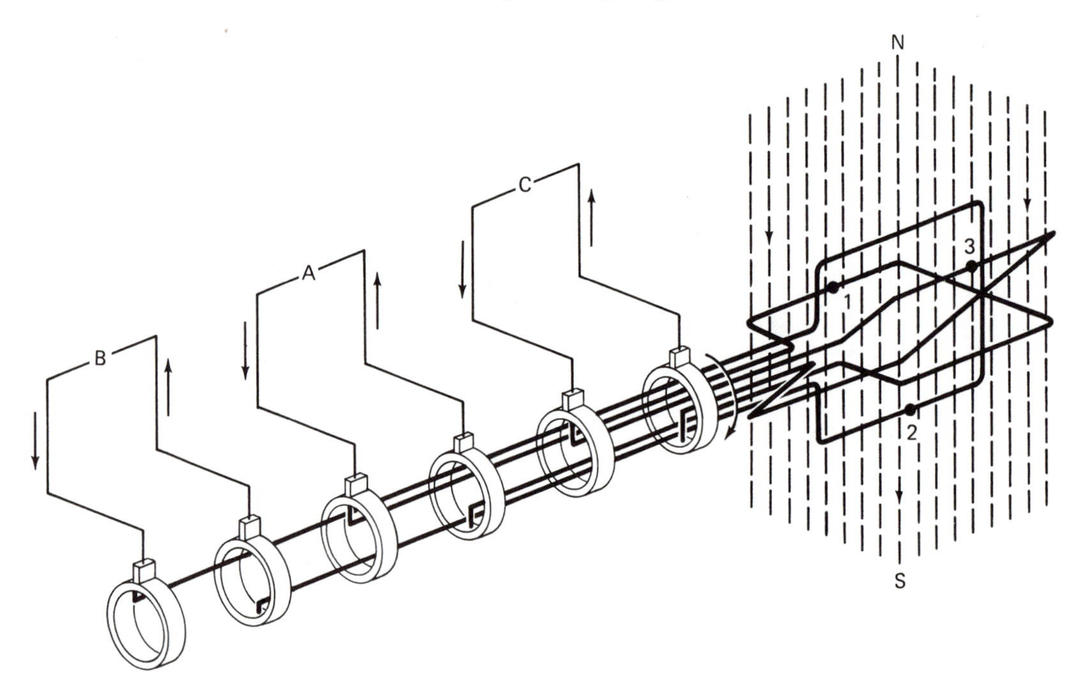

Figure 2-16 Coil Arrangement in the Armature of a Three-Phase Generator

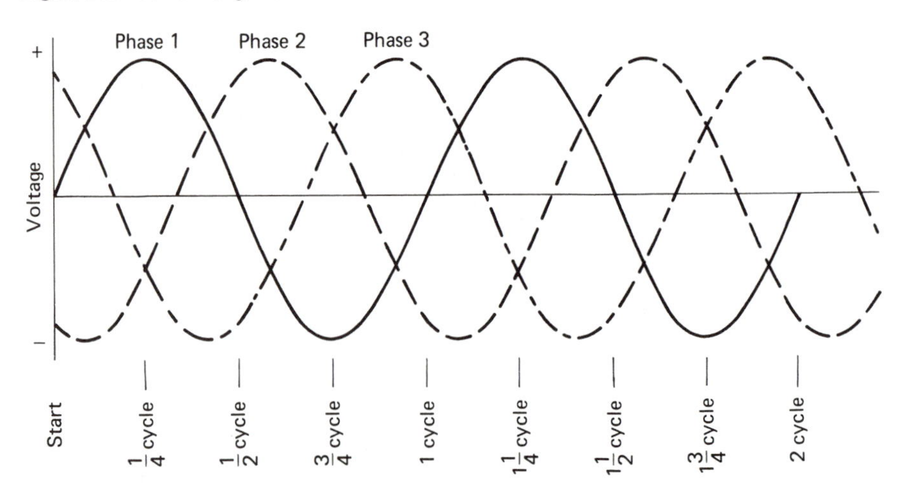

Figure 2-17 Curve Showing Voltage Produced in a Three-Phase Generator

Now the voltages produced follow each other just $\frac{1}{3}$ cycle apart due to the physical placement of the coils on the shaft and in relation to each other. When the voltage in phase 1 is approaching a positive maximum as indicated at the start point in Fig. 2-17, the voltage in phase 2 is at a negative maximum and the voltage in phase 3 is falling toward zero. The voltage waveforms for two cycles have been shown to indicate how the voltages continue at $\frac{1}{3}$ cycle apart from each other.

The windings are not always wired out through the slip rings. Quite often the desired circuit connections are made inside the generator so that only three wires are brought to the terminal box on the generator housing. This makes only three leads necessary for a three-phase winding with each lead serving two phases. Then each *pair* of leads acts like a single-phase circuit, substantially independent of the other phases. This internal connection of windings is illustrated in Fig. 2-18.

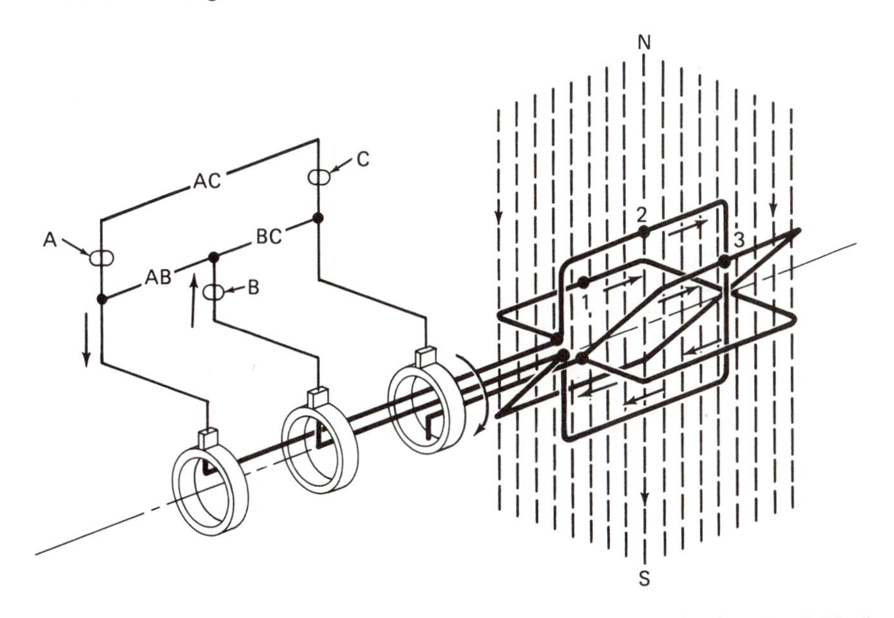

Figure 2-18 Three-Phase Generator with Internal Connections so that only Three Leads Need to Be Wired to Terminal Box on Outside of Machine Housing

The connections made in Fig. 2-18 are shown in diagrammatic form in Fig. 2-19. This shows the three coils arranged in a delta (Δ) connection. They could also be connected in a wye (Y) arrangement, which is also shown in Fig. 2-19. These connection types are common in three-phase systems.

Most power systems in use today produce and transmit three-phase power because it is best adapted to motors and provides the least expensive power distribution. More than three phases are possible, of course, but it is more expensive and there is no appreciable advantage to be gained. Some conversion equipment (alternating to direct current) does use six-phase arrangement of connections or machine windings.

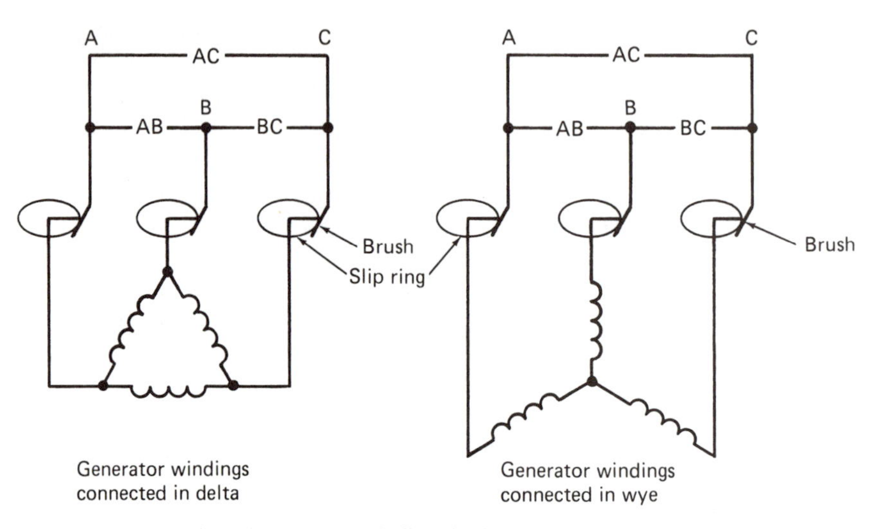

Figure 2-19 Connection of Armature Windings in Generator

2-17 The Transformer

One of the least complicated and therefore most reliable elements in the electrical circuit is the transformer. By its name, of course, it describes what it does. It is used only on alternating-current systems and changes the voltage either up or down as the system may require.

Transformers are made basically of three parts as follows:

1. Primary coil
2. Core of magnetic material
3. Secondary coil

Some transformers called *autotransformers* have only one winding, which is divided into two parts. This also changes the voltage but it does not insulate the two circuits. The two basic transformer types are shown in Fig. 2-20.

The means by which the transformer changes the voltage is essentially the same principle by which voltage is produced in a generator. The magnetic lines are cutting across the conductors and creating the voltage in the conductor. The principle is the same except that the voltage already exists in one coil and therefore has to be *induced* into the second coil. For this reason, the action of the transformer is called an *inductive* action. How does this take place? First, it must be remembered that every conductor carrying a current has a magnetic field surrounding it. In the alternating-current system, the voltage and current are building to maximum values and down to zero and then back to a maximum value in the opposite direction. Because this is happening with the voltage and current, then it can be readily seen that the magnetic field surrounding the conductor is rising and falling too. If conductors from a separate system are wound onto the same iron core as the system with the rising and falling mag-

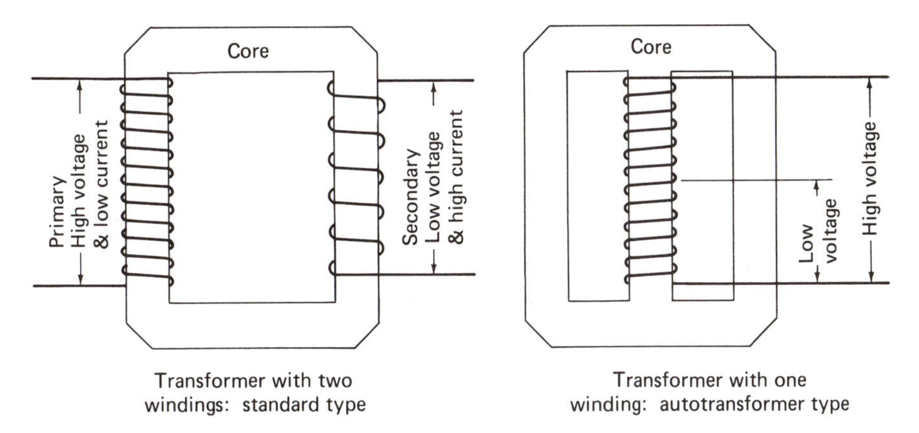

Figure 2-20 The Elementary Transformer

netic field, then those conductors are going to have the magnetic flux cut across them and a voltage will be *induced.*

The voltage is changed in exact proportion to the number of turns connected in series in each winding. For example, if the high-voltage winding has 1,000 turns and is connected to a 2,300-V circuit, a low-voltage winding of 100 turns will give a voltage of 230 V.

Transformers are usually wound for single-phase circuits and groups of three are connected together for three-phase transformation, although three-phase transformers can be built and are also in common use.

To keep the coils of the transformer cool, they are frequently immersed in oil or in a nonflammable coolant liquid in a tank. Distribution-type transformers can be seen in almost any neighborhood hanging from the poles. Of course, they vary in size in accordance with their rating but common sizes are about 24 to 30 in. in diameter and 3 to 4 ft high. Those on poles are generally oil-filled and they are used to transform the high distribution voltage down to 120 and 240 V for use in private residences.

2-18 Advantages of Alternating Current

The principal advantage of alternating current is its ability to be readily transformed from one voltage level to another. Alternating current can be generated at relatively low voltages and increased to high and very high voltages for modern power transmission through the use of the transformer. It can be easily understood that transformers cannot be used on direct current because the magnetic field surrounding the conductors remains steady and therefore cannot induce a voltage in any other conductors.

It is common on alternating-current systems to generate voltages of 18,000 or 24,000 V and step that voltage up to 115,000 or 230,000 V for long-distance transmission. (An old rule of thumb used to be to estimate 1,000-V per mile for transmission voltages levels. A line running 100 mi would probably be at 115,000-V, which is a standard voltage level. Actually, in practice, the 1,000-V per mi figure generally is too low.) The higher voltage is then stepped down to a lower voltage for local distribution and, of course, once or twice more to the voltage level for use in industry, commerce, homes, etc.

2-19 Generation of Direct Current

The direct-current generator produces a voltage in the same manner as the single-phase generator shown in Fig. 2-10; however, in the direct-current machine a switch called a *commutator* is used in place of the slip rings in the single-phase machine. A simplified direct-current generator is shown in Fig. 2-21.

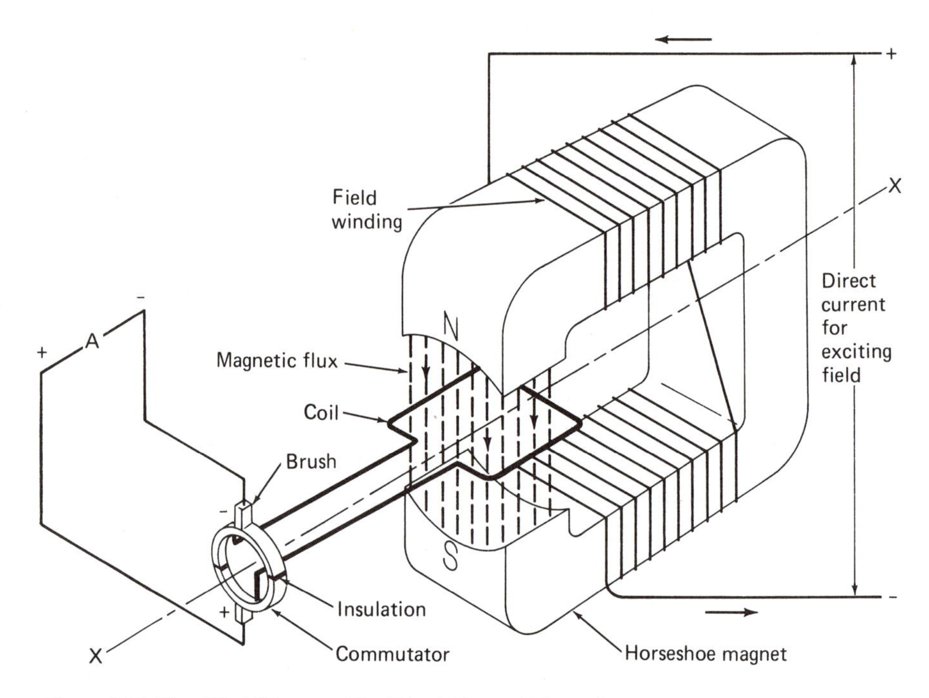

Figure 2-21 Simplified Diagram of a Direct-Current Generator

The commutator is used to reverse the connections to the revolving conductor at the precise instant that the current is reversing and "going below the line" to the opposite direction. By this means, the current is kept in one direction only. It can be seen in Fig. 2-21 that each end of the loop of wire is connected to a separate segment of the commutator. The brushes are placed against the commutator in such a location that the current always flows from the loop *into one brush* and from the other brush *into the loop*, regardless of which direction the current in the loop itself flows. In this way, the brushes carry current flowing in one direction only. The waveform produced is similar to that shown in Fig. 2-22.

The division between maximum voltage values is exaggerated in Fig. 2-22 to illustrate the one-direction condition. Actually, there are many loops and

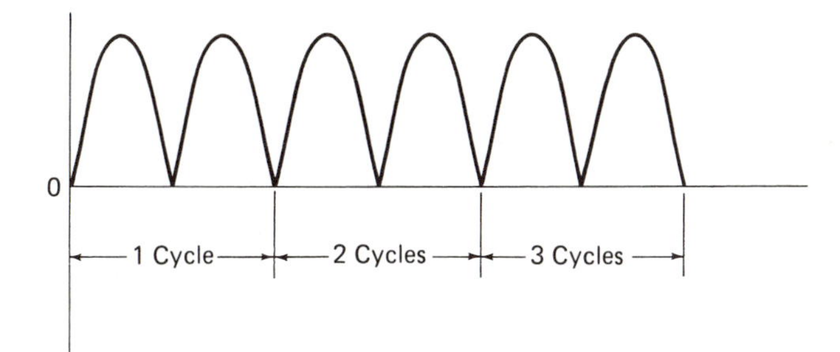

Figure 2-22 Curve of Direct-Current Waveform

many commutator sections that occur so often that they are extremely close to each other and therefore direct-current waves are generally shown as a straight horizontal line at a level near to the top of the peak values (see Fig. 2-9). Direct current is sometimes called *pulsating direct current* because of the fact that the voltage does rise and fall the same as in alternating-current systems; it does not reverse to the opposite direction because of the commutator.

2-20 Advantages of Direct Current

Direct-current systems are in service where adjustable speed motors are required such as in machine tools and especially in rail transportation. Direct-current motors are very desirable for this type of use.

Batteries require direct current for charging. Electroplating also requires direct current. A steady magnetic field is very useful in many places such as a magnetic chuck for a machine tool or a magnetic coupling for connecting drive motors to fans. Most elevator motors are direct-current drives and practically all trolley cars and rapid transit cars are direct-current drives. Even trains that pick up alternating current from overhead lines convert it on board the locomotive to direct current to power the motors on the train wheels because it is much better for this service.

Direct-current generators may be constructed as a complete unit itself, whereas the alternating-current machines require an external direct-current source for the excitation of the field windings. Because the direct-current generator generates direct current, it uses its own output power for excitation.

2-21 Multipolar Generators

Descriptions of generators in the foregoing text have incorporated only two-pole machines for reasons of simplification. Multipolar machines have more than a single pair of north and south poles. Most electric generators (and motors) have four or more poles. The number of poles is always even because for every north pole there must be a matching south pole. With the four-pole machine, the conductors pass by four poles (two pairs of poles), which will result in two complete cycles for every revolution. If the machine had six poles (three pairs), then in each revolution 3 cycles of voltage would be produced; i.e., 1 cycle for each pair of poles. If the generator has two pairs of poles (four poles in all) and is driven at 1,800 revolutions per minute (rpm), it will make two times 1,800 or 3,600 cycles per minute (cpm). Because frequency is always expressed in cycles per second (cps) instead of minutes, then dividing the 3,600 cpm by 60 sec per min, the correct frequency is 60 cps, which can be expressed as 60 Hz.

The general formula for frequency is

$$f = \frac{\text{rpm} \times P}{120}$$

where f = frequency
rpm = speed in revolutions per minute
P = number of poles

The most common frequency is 60 Hz; however, 25, 40, 50, and even others are used.

In the design of generators, the speed of the machine obviously has a direct effect on the number of poles required for a given frequency. Generators driven by water power and some steam engine drives are generally fairly slow moving and, therefore, the machine must have several poles, sometimes 30 or more. Modern, high-speed (3,600 rpm) generators, being driven by high-pressure steam turbines, have only two poles determined as follows:

$$P = \frac{f \times 120}{\text{rpm}}$$

$$P = \frac{60 \times 120}{3{,}600}$$

$$P = 2$$

Direct-current generators, which have no frequency limitation, generally have four poles. Small machines, however, may have only two poles and large ones often have six, eight, or even more poles.

2-22 Revolving Field Generators

Large alternating-current machines are built so that the magnetic field is wound onto the rotor and revolves. This arrangement is opposite to that shown in the preceding examples and diagrams. The reasons for this are simple and logical. Only two slip rings are required as opposed to three. The voltage for the direct-current excitation is much lower (usually less than 600 V dc) and with the generator windings on the fixed part of the machine, called the *stator*, they can be more effectively insulated for the higher voltages produced in them. Also, if these windings were constructed on the rotor, they would have to be resistant to the mechanical stresses of rotation, which with heavily insulated conductors would be a significant problem. Of course, the principle of voltage

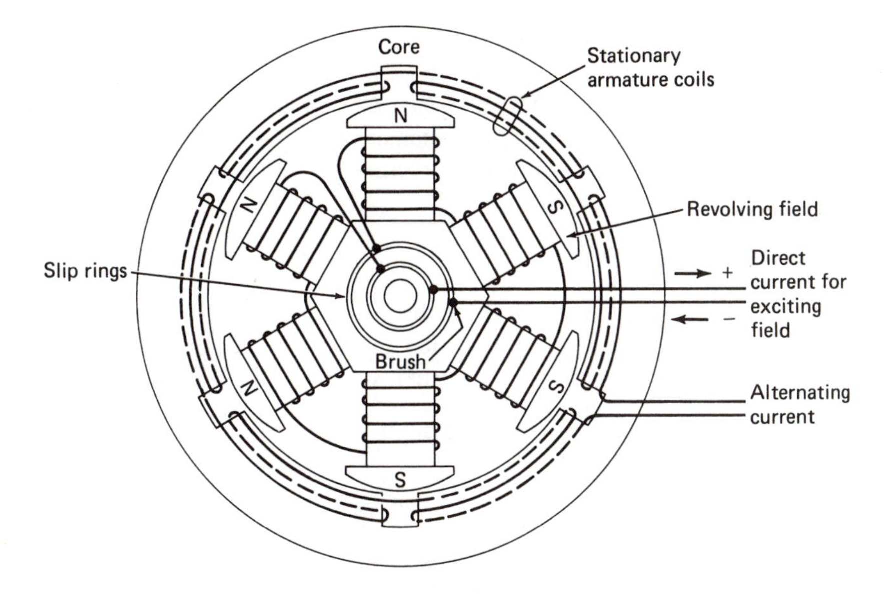

Figure 2-23 Revolving Field Alternating-Current Generator

generation with these machines is that of the moving magnetic field cutting across the fixed conductors. Figure 2-23 diagrammatically shows the revolving field type machine.

Direct-current generators universally use the fixed magnetic field arrangement.

2-23 Measuring Electric Current

To measure the rate of flow of amperes along the conductors, it is necessary to install an instrument called an *ammeter*. A dynamometer-type ammeter is basically composed of a fixed coil and a moving coil through which the line current (or some portion of the line current) flows. Magnetism develops in the fixed coil causing the moving coil to rotate. A pointer is attached to the shaft

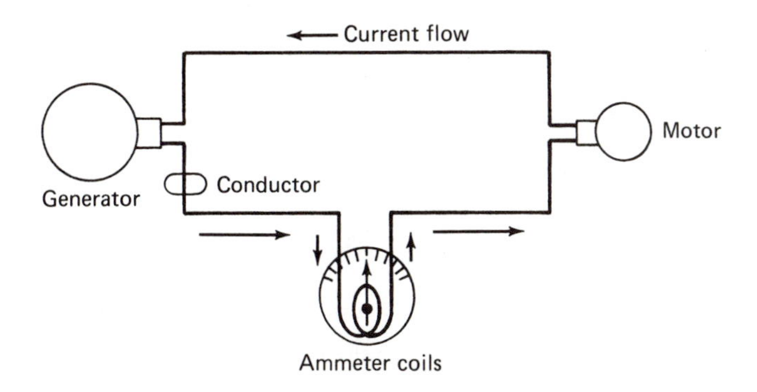

Figure 2-24 Connection of Ammeter in Simple Electric Circuit

of the moving coil and the appropriate scale is provided to indicate current in amperes or some multiple of amperes.

Ammeters are connected *into* the line in order to measure current flowing, as shown in Fig. 2-24. Ammeters for alternating current and direct current are usually different in construction but perform the same function.

2-24 Voltage Measurement

Voltmeters measure the electrical pressure, or voltage, on a system. They are similar in construction to the ammeter described above except that they normally use smaller current flows and have either internal or external resistor elements. They are connected *across* the two lines where a voltage measurement is required, as shown in Fig. 2-25.

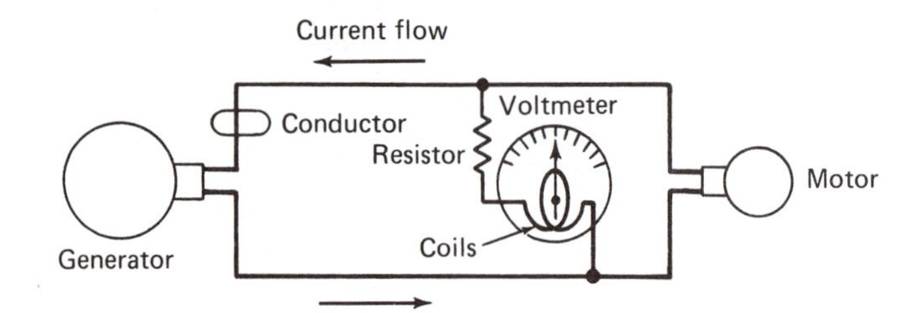

Figure 2-25 Connection of Voltmeter in Simple Electric Circuit

Voltmeters too vary for alternating- and direct-current service so the right type must be used.

2-25 Relationship between Voltage and Current

The voltage required to make current flow depends on the resistance of the circuit. A direct-current voltage of *one volt* will make *one ampere* flow through a resistance of *one ohm*. This relationship is expressed by the formula

$$I = \frac{E}{R}$$

where I = current in *amperes*
E = voltage in *volts*
R = resistance in *ohms*

Other useful formulas are easily derived through transposition as follows:

Known factors	*To find*	*Use*
Current (I) and resistance (R)	Voltage (E)	$E = I \times R$
Voltage (E) and current (I)	Resistance (R)	$R = \frac{E}{I}$

Note that the letters used in the formulas to signify voltage and current are not V for volts and A for amperes as might be expected. The letters used are universally accepted although in other expressions of electrical quantities the letters V and A do appear. This is explained later. The relationships indicated above are commonly called *Ohm's law.* (This is the electrical statement for the general law that governs all physical phenomena; i.e., *the result produced is directly proportional to the effort or cause and inversely proportional to the opposition.* In the electric circuit, the current established is the result; the voltage that establishes it is the effort; the resistance is the opposition.)

This relationship always holds good for direct-current circuits. In alternating-current circuits, the current and voltage are continually changing but at any instant the current flowing will depend on the value of the voltage at exactly that same instant and, provided the circuit has only resistance to effect its current, the formulas stated above will apply. Most alternating-current circuits have other properties besides resistance that have an important influence on the current. These factors are explained in the following sections.

2-26 Inductance, Capacitance, and Impedance

Many alternating-current circuits have coils or windings in some form that are required to produce useful magnetic effects. These magnetic effects however react in turn upon the current. They retard the current and cause it to lag behind the voltage. This is shown diagrammatically in Fig. 2-26.

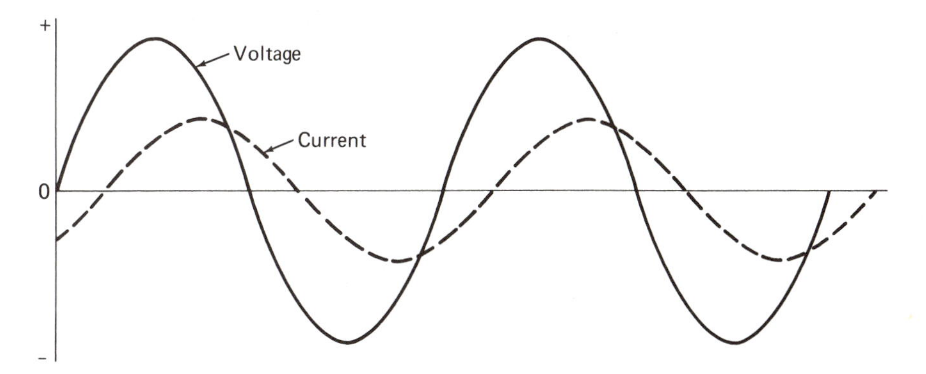

Figure 2-26 Curves Showing Current-Lagging Voltage

From examination of the curves of Fig. 2-26, it can be seen that the voltage has reached its maximum value and started to fall sometime before the current has reached its maximum value. Actually, some current will be flowing in the circuit at the instant that the voltage is zero. This magnetic reaction is called *inductance* in the circuit. The amount of inductance depends on the number of turns in the coil and the presence of iron around or within the coils. Transformers, motors, solenoid coils, and similar devices are inductive.

Devices that have the opposite effect on the voltage–current relationship in a circuit are called *condensers* or *capacitors*. Presence of these in a circuit will cause the current to *lead* the voltage. Condensers are, in a way, the "batteries" in alternating-current circuits because they have the ability to store an electrical charge momentarily. They are not in anyway comparable to the direct-current battery in performance and are not used for the same purpose. They do, however, become charged with electricity when the current is in one direction and will discharge when the current is in the opposite direction. The condensers are composed of two conducting "plates" separated by insulation. They are constructed by laying the plates and insulation together and then rolling both together much like making a jelly roll. Typical construction is shown in Fig. 2-27.

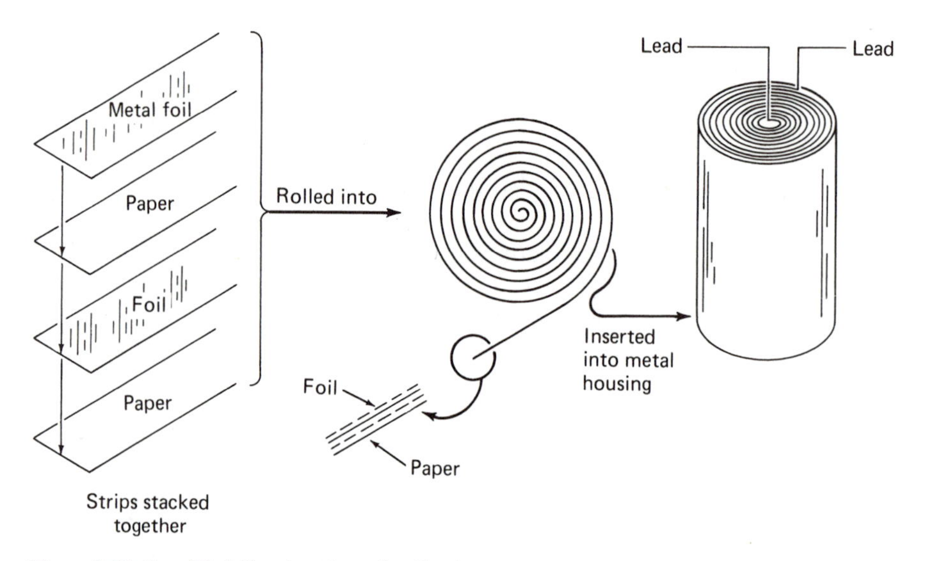

Figure 2-27 Simplified Construction of a Condenser

The condenser/capacitor tends to counteract the inductance in the circuit and is useful in overcoming the inductive lag in current that is inherent in most alternating-current motors. Therefore, in the alternating-current circuit, there may be present resistance, inductance, and capacitance that affect the current–voltage relationship. The combination of all these effects is called the *impedance* of the circuit. Impedance is measured in ohms, just as resistance, but it is invariably greater than the resistance whenever one, or both, of the other factors is also present. Impedance is determined as follows:

$$Z = \sqrt{R^2 + X^2}$$

where Z = impedance in ohms
R = resistance in ohms
X = inductive or capacitive reactance in ohms

Therefore, for alternating-current circuits, Ohm's law is modified by substituting impedance for resistance as follows:

To find current	*To find voltage*	*To find impedance*
$I = \frac{E}{Z}$	$E = IZ$	$Z = \frac{E}{I}$

2-27 Effective Values of Current or Voltage in Alternating Current

Alternating-current circuits present the difficult problem of thinking of voltages and currents that are continually changing. To overcome this problem, the term *ampere*, when applied to an alternating current, means the *effective* value of the current rather than the current value at any instant. The word *effective* is used because the result or *effect* of 1 A flow of alternating current is equal to the *effect* of 1 A flow of direct current. To put it another way, if an electric light attains a given brightness when 1 A of direct current flows through the lamp filament, it will attain the same brightness when 1 *A* of alternating current flows through the filament; the effective current is the same. Extending this logic further, if the filament resistance is 1 Ω, 1 V will be required to cause 1 A to flow through it whether it is a direct voltage with a steady state of 1 V or an alternating voltage with an *effective* value of 1 V. Lamp filaments are strictly a resistance element without inductive or capacitive effects.

2-28 Power—Definition and Measurement

Ultimately electricity is required to do work. Power is defined as the *rate of doing work*. In direct-current circuits, work is done in proportion to current strength and to the voltage level at which the circuit operates. Therefore, power is equal to the product of the current and the voltage. Power is measured in *watts* (W). Mechanical power is measured in *horsepower* (hp) and one horsepower very nearly equals 746 watts. Because 1 W is a very small measure of power, it is necessary in many instances to use a larger unit. Consequently, 1,000 W, called a *kilowatt* (kW), is a common unit of measurement (*kilo* means 1,000).

As an example of the foregoing definitions, if it is desired to find the power in watts being delivered to a direct-current motor, the formula is

$$P = E \times I \quad \text{or} \quad \text{watts} = \text{volts} \times \text{amperes}$$

If the motor is rated 25 hp operating at 230 V and drawing 92 A, its power consumption is approximately

$$P = E \times I$$

$$P = 230 \times 92$$

$$P = 21{,}160 \text{ W} \quad \text{or} \quad \frac{21{,}160}{1{,}000} = 21.16 \text{ kW}$$

The measurement of power in an alternating-current circuit cannot be determined by simply multiplying effective amperes by effective volts unless the circuit contains no inductance or capacitance. A circuit containing only incandescent lamps or electric heating devices, with purely resistive loads, would be of this nature. The current and voltage in such a circuit would rise and fall together and power would be the product of the volts times amperes at any given instant. This is shown in Fig. 2-28.

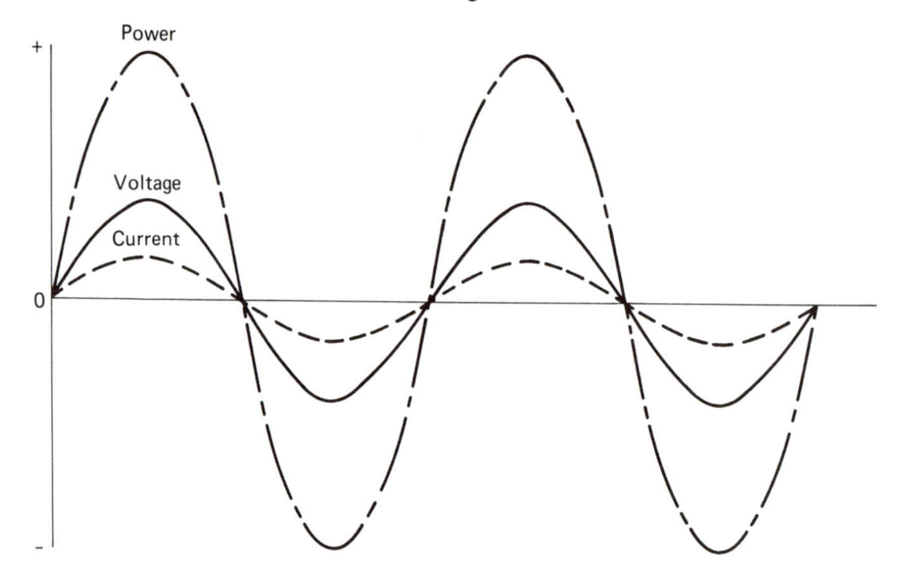

Figure 2-28 Curves Showing Current, Voltage, and Power Relationships in a Purely Resistive Circuit

Of course, power to an alternating-current motor cannot be determined by multiplying effective values of voltage and current together because the motor has windings, which means that magnetic flux is present and therefore the motor is an inductive load. The current will lag the voltage, as shown in Fig. 2-26.

If the effective current and voltage values were multiplied together, the value would be greater than the real power used and therefore is called *apparent power*. Apparent power is measured in voltamperes or VA. It is convenient to use units of 1,000 times the voltampere product, called *kilovoltamperes* or kVA. The reason for this apparent power condition can be more easily understood by studying Fig. 2-29.

The power curve of Fig. 2-29 has been plotted at every instant throughout the cycle as the product of volts times amperes at the same instant. Most of the time, the power is positive; however, when either the current or the voltage is zero, then volts or current times zero must result in power being zero. Also, when the voltage is positive and the current is negative, or vice versa, then the product of the multiplication is also negative and therefore power is a negative

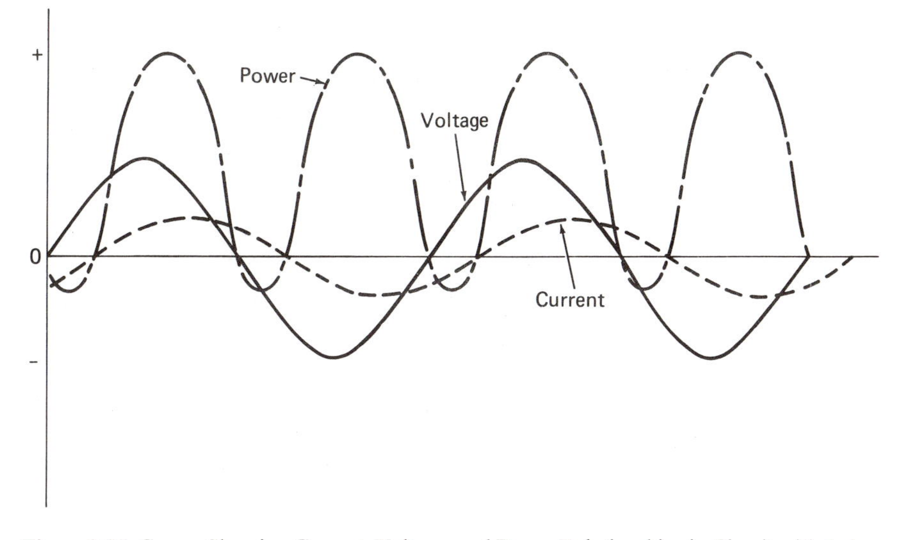

Figure 2-29 Curves Showing Current, Voltage, and Power Relationships in Circuit with Inductive Loads

value. If this curve was representing power taken by an induction motor, the motor will draw power most of the time from the line feeding the motor *but will be returning power to the line where power is a negative value*! The *real power* then is only that power that the motor actually takes *from* the line that it never sends back. Real power is expressed in watts or kilowatts (kW). This cannot be determined from ammeter and voltmeter readings except in rare cases where there is no inductance or capacitance in the circuit. A device called a *wattmeter* or *kilowattmeter* is used to measure real power. This instrument contains two coils. One coil measures voltage and the other measures current. Just like the ammeter, the current coil is connected *into* the line and the potential (voltage) coil is connected across the lines. A simple circuit with a wattmeter connection is shown in Fig. 2-30.

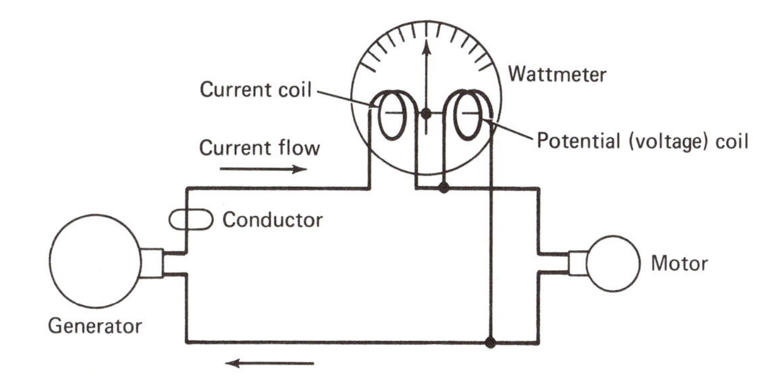

Figure 2-30 Connection of Wattmeter in a Simple Electric Circuit

2-29 Power Factor

Previous explanation in foregoing sections have shown how the current lags behind the voltage in circuits having inductance. The actual current drawn by the motor or other inductive device can be considered as having two components. One component is called the *magnetizing current* and it lags 90 electrical degrees behind the voltage (1 cycle is 360 electrical degrees). The value of this current is zero when the voltage has reached its maximum value. This current is also called the *reactive current*.

The other component is known as the *active current* and it is in phase with the voltage; i.e., the active current and the voltage reach maximum values simultaneously.

The *actual line current* is therefore the resultant of the reactive and active currents; one is lagging and therefore out of phase with the voltage and one is in phase with the voltage. These values are more readily understood if they are shown in a diagram. Figure 2-31 shows a step-by-step representation.

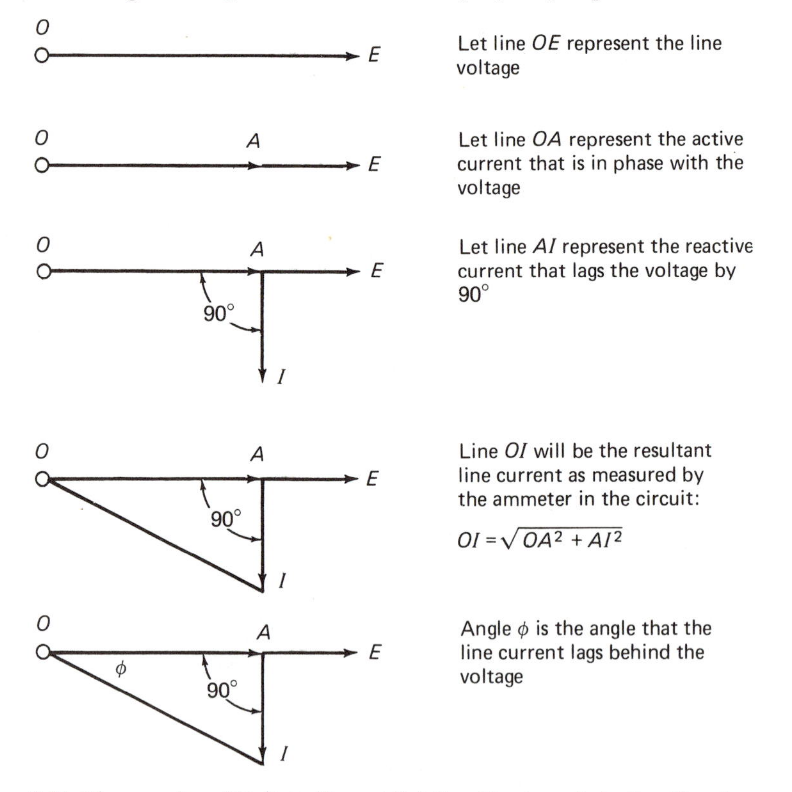

Figure 2-31 Diagramming of Voltage-Current Relationships in an Inductive Circuit

The *power factor* is the ratio of the active current component to the actual current as follows:

$$\text{Power factor} = \frac{OA}{OI} \quad \text{and} \quad \frac{OA}{OI} = \text{cosine of the angle } \phi$$

(The cosine value of the angle ϕ times 100 equals power factor in percent.) If OI is the actual current reading from an ammeter in one phase of a three-phase circuit, then

$$\text{kVA} = \frac{\sqrt{3} \times OE \times OI}{1{,}000} \quad \text{or} \quad \text{kVA} = \frac{\sqrt{3} \times \text{volts} \times \text{amperes}}{1{,}000}$$

The calculations above will provide values of *apparent power*. To determine real or true power, the power factor of the circuit must be factored into the equation as follows: The cosine value of the angle is a decimal figure such as 0.85 and is therefore the power factor value used in the formulas

$$\text{kW} = \frac{\sqrt{3} \times OE \times OI \times \text{p.f.}}{1{,}000} \quad \text{or} \quad \text{kW} = \frac{\sqrt{3} \times \text{volts} \times \text{amperes} \times \text{p.f.}}{1{,}000}$$

Comparing these equations will show that if the power factor is 1 (such as in purely resistive loads), then kVA and kW are exactly the same. If the power factor is less than 1, say 85% (a multiplier of 0.85), then kW will be considerably less than kVA.

Power factor is therefore the ratio of real power to apparent power. Real power is measured by a wattmeter and apparent power is the product of voltmeter and ammeter readings. Power factor is equal to kW/kVA.

2-30 Common Connections of Apparatus Coils

The windings of generators, transformers, and other apparatus can be connected in several ways, for polyphase circuits; however, by far the two most common are delta and wye connections. The wye connection is often grounded at the center point. These two connection arrangements are shown in Fig. 2-32.

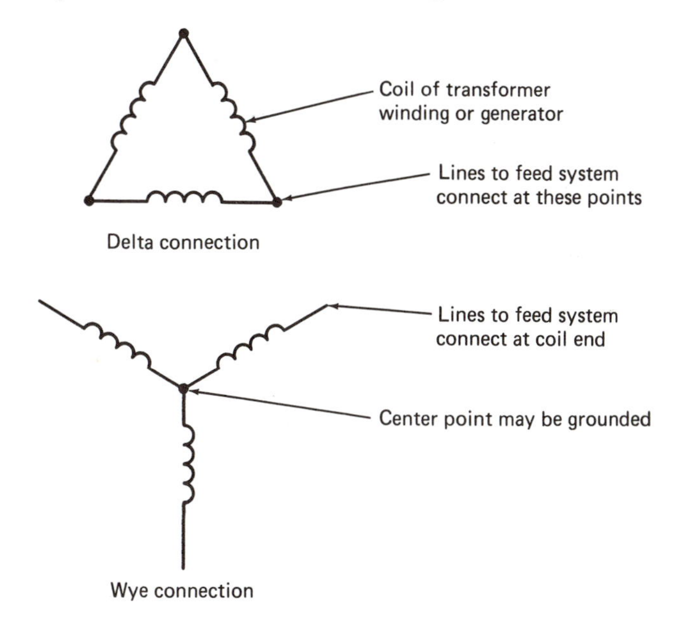

Figure 2-32 Delta and Wye Connections

These connections are discussed further in sec. 2-31 wherein the basic systems for utilization of power are described.

2-31 Basic Systems for Utilization of Power

Modern electrical systems use four basic voltage systems for utilization of power. These are usually secondary distribution from transformers, although generators may have their windings connected to give these same system voltages. Considering transformers only, the primary voltage and connections will vary in voltage levels depending on the system feeding the transformer, but the secondary system will generally be one of the following:

Single-Phase, Three-Wire System

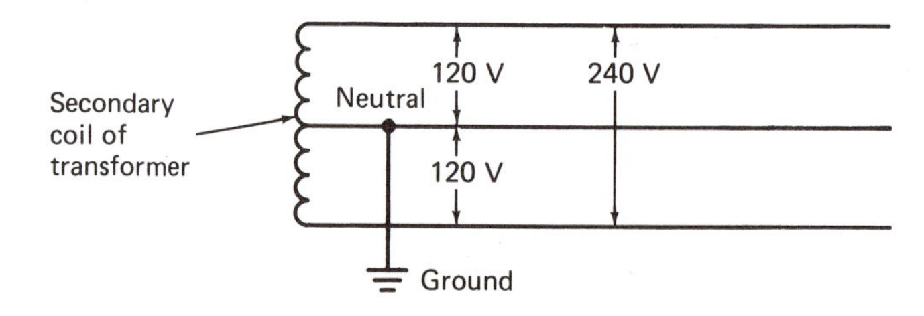

This is the system most commonly used to supply power to individual residences, small apartments, and small commercial buildings. Both lighting and single-phase motor loads can be fed from this system. The motor loads can be either 120 or 240 V. Lighting is usually 120 V. Electric cooking ranges are usually three wire and use both 120 and 240 V. Receptacles for miscellaneous loads can also be either 120 V for general use or 240 V for specific service such as required for large room air conditioners.

Three-Phase, Four-Wire Wye System—208/120 V

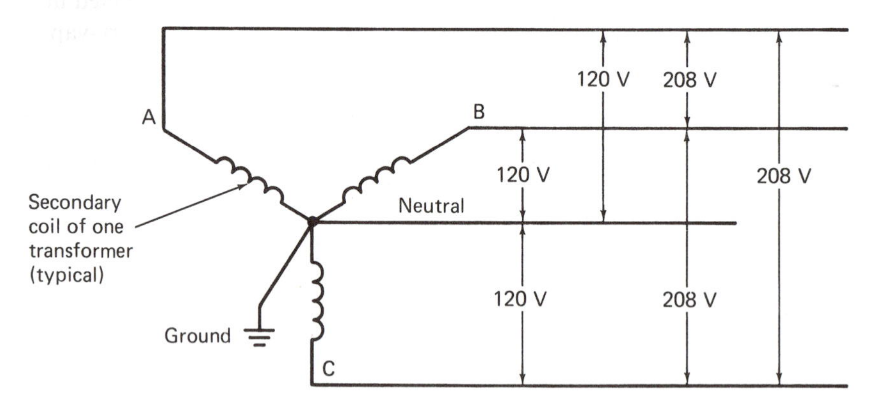

This is one of the most commonly used three-phase secondary distribution systems. The various voltages available are

1. 120 V, single phase
2. 208 V, single phase

3. 208 V, three phase

This system is used extensively in cities as the voltage system for their networks, supplying power to large apartment buildings, stores, theaters, etc., and it is used as the distribution system in large office buildings, shopping center complexes, and similar loads where there is a large demand for low-voltage power.

Three-Phase, Four-Wire Wye System—480/277 V

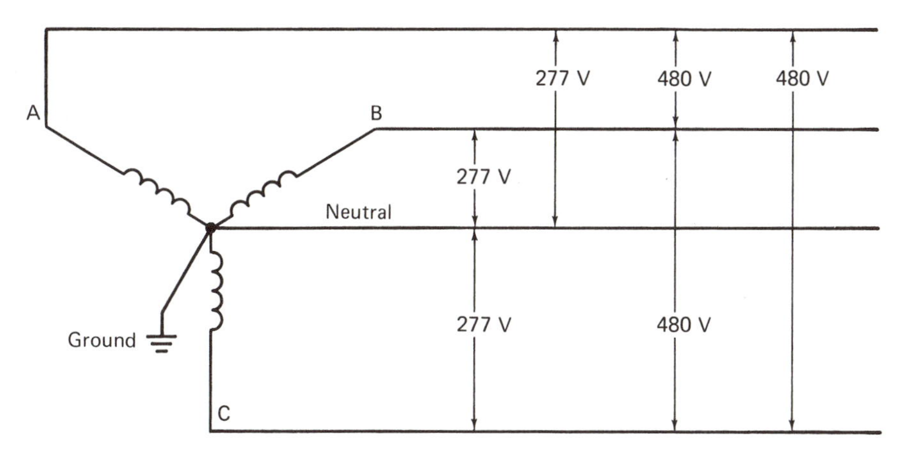

The transformer secondary winding connections for this system are exactly the same as for the 208/120 V system except the number of turns in the windings are different (assuming the same primary voltage) and therefore the voltages are as shown in the diagram above.

This system is widely used in industry where the motor loads are significant and require three-phase 480 V. Large areas to be lighted utilize the single-phase 277 V connections for both fluorescent and mercury-vapor lighting. Voltages available are

1. 277 V, single phase
2. 480 V, single phase
3. 480 V, three phase

Three-Phase, Three-Wire System

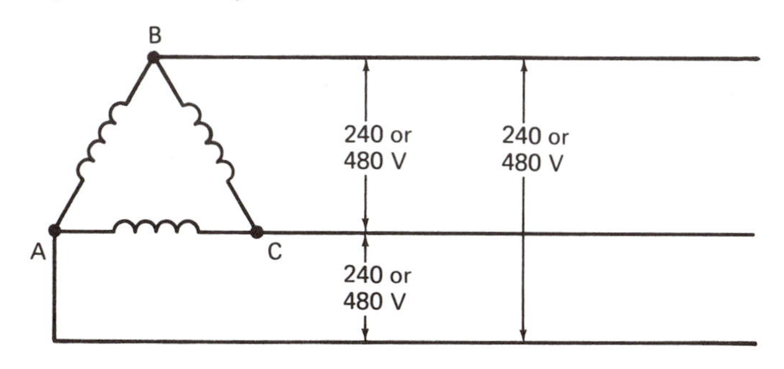

The delta connection will provide phase-to-phase voltages of either 240 or 480 depending on the voltage level of the transformer winding; i.e., either the transformer is wound for 240 or 480 V—not both. This system is used where the motor loads represent a major part of the total load.

2-32 Conversion Formulas

Most calculations required for design problems can be made through the use of a fairly limited set of formulas.

Conversion Formulas

To find	*Direct current*	*Alternating current*	
		Single phase	*Three phase*†
Amperes when horsepower (input) is known	$\frac{\text{hp} \times 746}{\text{V} \times \text{efficiency}}$	$\frac{\text{hp} \times 746}{\text{V} \times \text{efficiency} \times \text{p.f.}}$	$\frac{\text{hp} \times 746}{\text{V} \times 1.73 \times \text{efficiency} \times \text{p.f.}}$
Amperes when kilowatts is known	$\frac{\text{kW} \times 1{,}000}{\text{V}}$	$\frac{\text{kW} \times 1{,}000}{\text{V} \times \text{p.f.}}$	$\frac{\text{KW} \times 1{,}000}{\text{V} \times 1.73 \times \text{p.f.}}$
Amperes when kVA is known		$\frac{\text{kVA} \times 1{,}000}{\text{V}}$	$\frac{\text{kVA} \times 1{,}000}{\text{V} \times 1.73}$
Kilowatts	$\frac{\text{A} \times \text{V}}{1{,}000}$	$\frac{\text{A} \times \text{V} \times \text{p.f.}}{1{,}000}$	$\frac{\text{A} \times \text{V} \times 1.73 \times \text{p.f.}}{1{,}000}$
kVA		$\frac{\text{A} \times \text{V}}{1{,}000}$	$\frac{\text{A} \times \text{V} \times 1.73}{1{,}000}$
Power factor		$\frac{\text{kW} \times 1{,}000}{\text{A} \times \text{V}}$ or $\frac{\text{kW}}{\text{kVA}}$	$\frac{\text{kW} \times 1{,}000}{\text{A} \times \text{V} \times 1.73}$ or $\frac{\text{kW}}{\text{kVA}}$
Horsepower (output)	$\frac{\text{A} \times \text{V} \times \text{efficiency}}{746}$	$\frac{\text{A} \times \text{V} \times \text{efficiency} \times \text{p.f.}}{746}$	$\frac{\text{A} \times \text{V} \times 1.73 \times \text{efficiency} \times \text{p.f.}}{746}$

Power factor and efficiency when used in formulas above should be expressed as decimals.
†For two-phase, four-wire, substitute 2 instead of 1.73.
For two-phase, three-wire, substitute 1.41 instead of 1.73.

In some of these same formulas, used in other sections of this text, current in amperes may appear as I and voltage may appear as E.

Chapter Review Topics and Exercises

2-1 What does the electron theory assume?

2-2 What are electrons, protons, and neutrons?

2-3 How do electrons relate to the protons?

2-4 How is electrical energy put to work as regards electrons?

2-5 What is electromotive force?

2-6 What does the word *potential* mean?

2-7 What are the principal sources of EMF?

2-8 How does electric current relate to electrons?

2-9 How is the term *ampere* used in electricity?

2-10 Name six or more materials classified as dielectrics.

2-11 What is a microhm? a megohm?

2-12 Describe how heat may be generated in an electrical conductor.

2-13 A pump in a water system is analogous to a generator in an electric circuit because they both do essentially what same thing?

2-14 Why is an understanding of a loadstone's characteristic important?

2-15 What is magnetic flux?

2-16 What action must be made to cause electrons to flow in a conductor? Can another action bring the same result?

2-17 Sketch a representation of direct current in graphic form.

2-18 Sketch alternating current in graphic form.

2-19 Describe a single-phase generator.

2-20 In electrical language, what does the word *phase* signify?

2-21 How does the term *hertz* apply to electricity?

2-22 Draw a sketch of generator coils connected in delta. Show the coils connected in wye.

2-23 Describe how a transformer works.

2-24 Why are most electrical systems alternating current?

2-25 Direct current has several advantages. Name three useful applications.

2-26 If a generator is designed to run at 300 rpm and generate electricity at 60 Hz, how many poles will the machine have?

2-27 Why do large generators have revolving fields?

2-28 What instrument is used to measure current flow?

2-29 How is a voltmeter connected to measure voltage in a single-phase, alternating-current circuit? Draw a sketch to show the connections.

2-30 If a lamp has a resistance of 240 Ω and .5 A is flowing through it, what voltage is causing this current flow?

2-31 What causes a reactive current?

2-32 Describe a capacitor. What effect does capacitance in a circuit have in regards to the voltage and current relationship?

2-33 What power is represented in Exercise 2-30?

2-34 When electrical loads are inductive in nature, power is sometimes returned to the line rather than taking power from the line. Why does this happen?

2-35 What measurements are combined in a wattmeter?

2-36 If kVA is known and real power in kW is required, what factor must be applied to make the conversion?

2-37 If the power factor of a load is 100%, how do kVA and kW compare?

2-38 What is the phase to neutral voltage on a 120/208 V, three-phase, four-wire distribution system?

2-39 Why is the 480/277 V, three-phase, four-wire distribution system used extensively?

2-40 If a transformer is rated 1,000 kVA, three-phase, 60 Htz, 13,800 V, 480/277 V, three-phase, four-wire, what is the full load primary current? the full load secondary current?

3

ONE-LINE DIAGRAMS

3-1 Definition

In Chapter 2, the electric *circuit* was defined and types of circuits or systems were discussed. From these explanations, it was learned that systems can be composed of two, three, or four conductors and therefore they can become complicated. It is cumbersome to have to work with the complete circuit in order to understand the system design so an abbreviated form is universally used that is called the *one-line diagram*. This type of drawing uses a single line to represent all the conductors and other elements of the system. Through the use of appropriate notes and graphic symbols, the system design can be illustrated and understood. One-line diagrams are defined as *a diagram that indicates by means of single lines and simplified symbols the interconnection and component parts of an electric circuit or system of circuits.* As an example of this simplification, Fig. 3-1 shows both a full three-line diagram and the much more simplified one-line diagram of the same portion of a system circuit.

Note that the separate parts of the three-line diagram have been bracketed to indicate the same element as represented in one-line form.

3-2 Purpose of the One-Line Diagram

As the electrical engineer or designer develops the system design, he needs a simple means of recording his thoughts and the development of system ideas. He also needs a means of communicating this information to others for comments and approvals. Ultimately, the system design will progress from the thinking and development stages, where it is generally used in rough form, to a finished drawing for use in construction and later reference in system operation. This diagram is generally the basis for all electrical design and therefore is a very important drawing. Its purpose can be summed up as *the means for formulating system design and recording design concepts and is the basis for total design development.*

3-3 Basic System Selection

Before the one-line diagram can be started, the engineer must consider the several ways in which power can be distributed and controlled. Distribution of

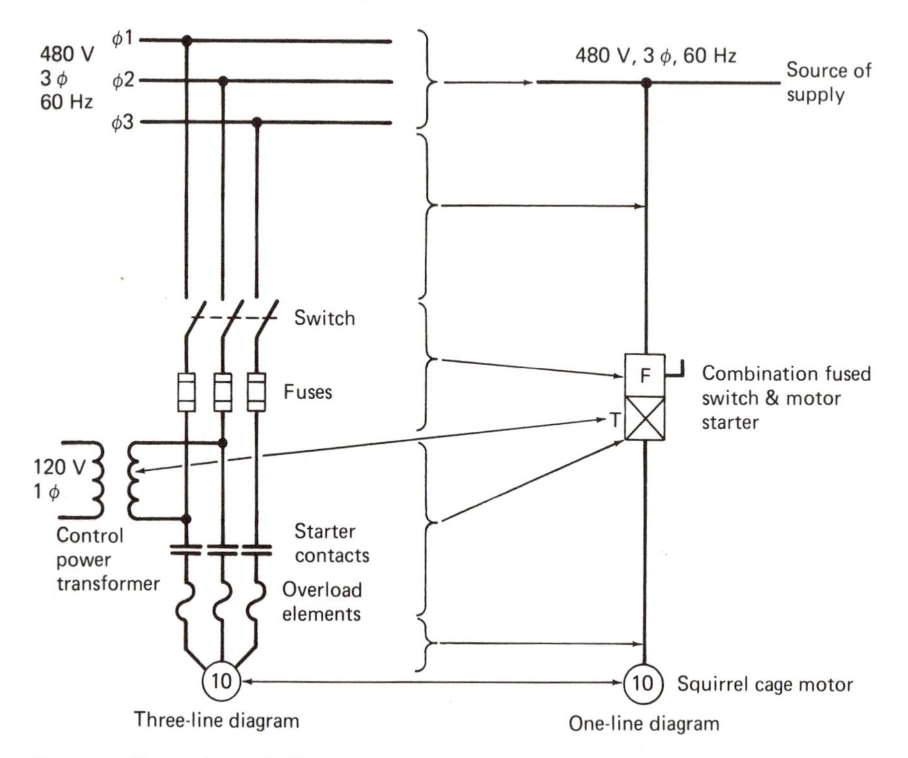

Figure 3-1 Comparison of Three-Line to One-Line Diagram Illustrating Simplification of Equal Elements through Use of the One-Line Diagram

power generally starts for industrial- or commercial-type projects at the point of power supply to the project from the utility company. There are at least four basic systems for power distribution that have developed from the many possible arrangements that might be used. The electrical engineer will take all factors into consideration and choose a system that will meet the demands and criteria needs of the particular project, bearing in mind reliability of service, cost, etc. The four systems are as follows:

Simple Radial System

The simple radial system is the most economical for direct distribution of power-to-load centers where power is further distributed for utilization. It is an adequate system for many applications where reliability is not a major factor. Figure 3-2 shows the simple radial system from the incoming power source to the load centers. It can be seen that the system uses radial-type feeders from a single source point similar to the radii of a circle.

The obvious disadvantage to this system is that loss of source power means loss of power at all the load centers.

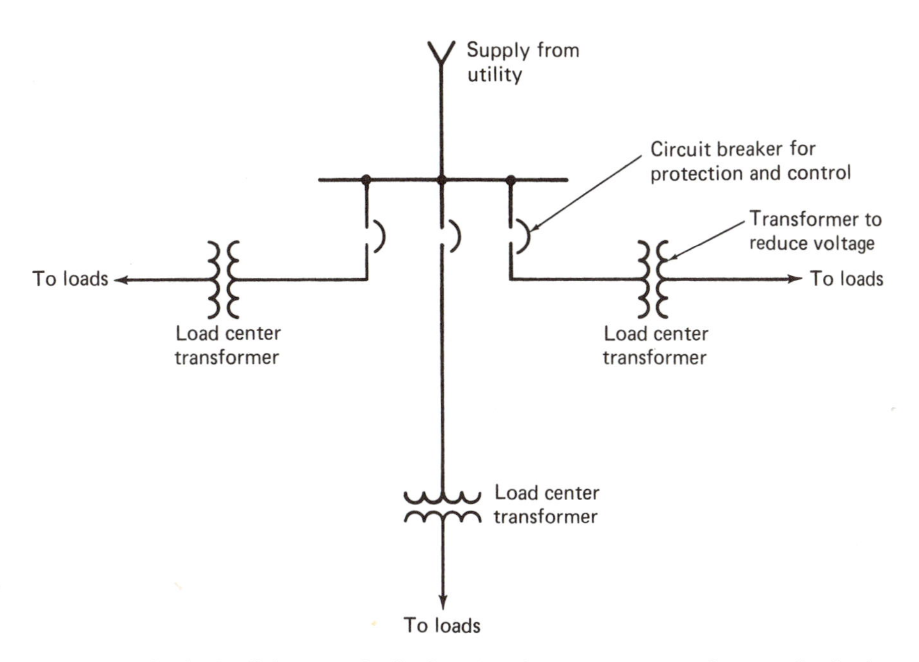

Figure 3-2 Simple Radial-Type Distribution (Load centers are transformer—distribution units that are generally located near the center of the electrical load)

Primary Selective System

The primary selective system provides an alternate supply to each load center. Here two lines are run to each load center unit. If a line fails (a fault occurs), then only some of the load centers lose their power and through simple switching they are put back on the line while the faulted feeder is being repaired. Figure 3-3 shows this system.

This system, of course, costs more because of the additional feeders and switching equipment. On the other hand, it may very well pay for itself in keeping power on the line if service is vital.

Secondary Selective System

The secondary selective system shown in Fig. 3-4 is actually a primary radial system with secondary ties between the busses.

This system, like the simple radial system, has the disadvantage of the single source. It is possible, of course, to have more than one radial source that improves reliability. Through the use of secondary bus ties, it is possible to isolate any radial feeder and feed the secondary bus by closing the bus tie breakers. For example, if the No. 1 transformer had to be taken out of service or if there were a fault on the feeder, the A and D circuit breakers could be opened to isolate the feeder circuit and the transformer. Bus tie breakers E and I could be closed and all loads on the transformer No. 1 secondary bus would be temporarily fed from transformer No. 2. This system requires careful analysis because if the

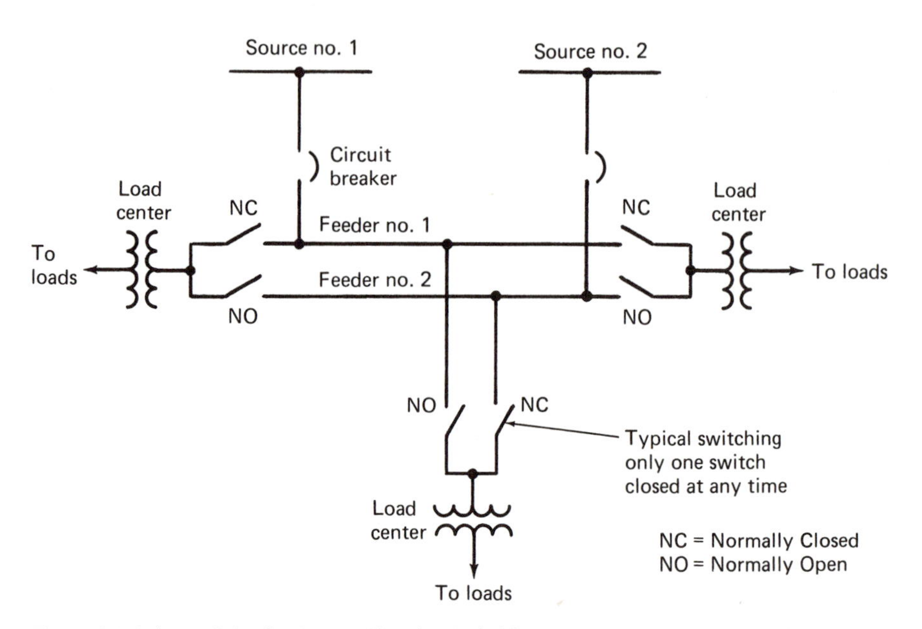

Figure 3-3 Primary Selective System Showing Switching at Each Load Center to Allow for Use of Either Feeder

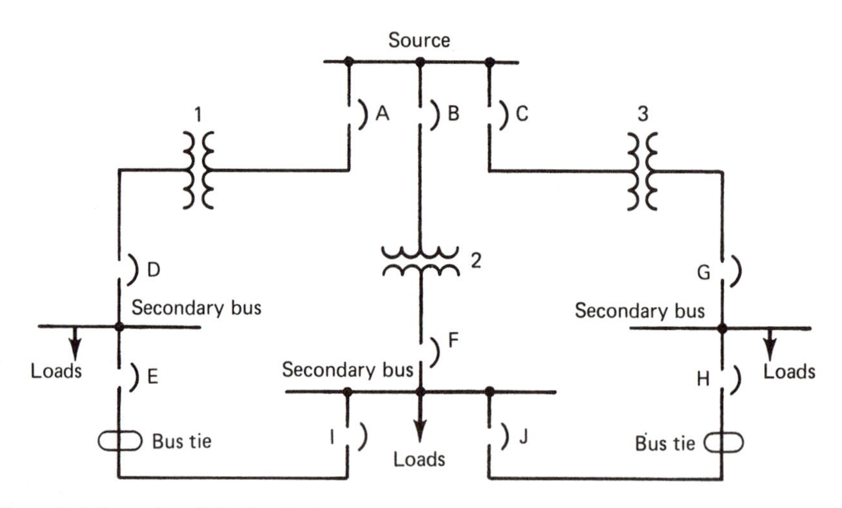

Figure 3-4 Secondary Selective System

loads on one bus plus the loads on a second bus are greater than the capability of the one transformer, it may cause further outages. Some systems using secondary selective system are designed so that each transformer is only half loaded in order to allow for picking up additional load on bus tie closing. The bus tie closing scheme is often arranged to operate automatically. In the example cited, tripping of circuit breakers A or D or both would initiate closing of circuit breakers E and I to complete the secondary tie. If the secondary tie, under normal operating conditions, has one breaker normally closed and the other normally open, then only the normally open breaker would close. Control wiring would be arranged, however, to assure that both breakers were closed when required.

Primary Selective, Secondary Network System

The primary selective, secondary network system is a logical extension of the other systems described as a means of increasing reliability of the system. This is also called a *spot network* and is shown in Fig. 3-5.

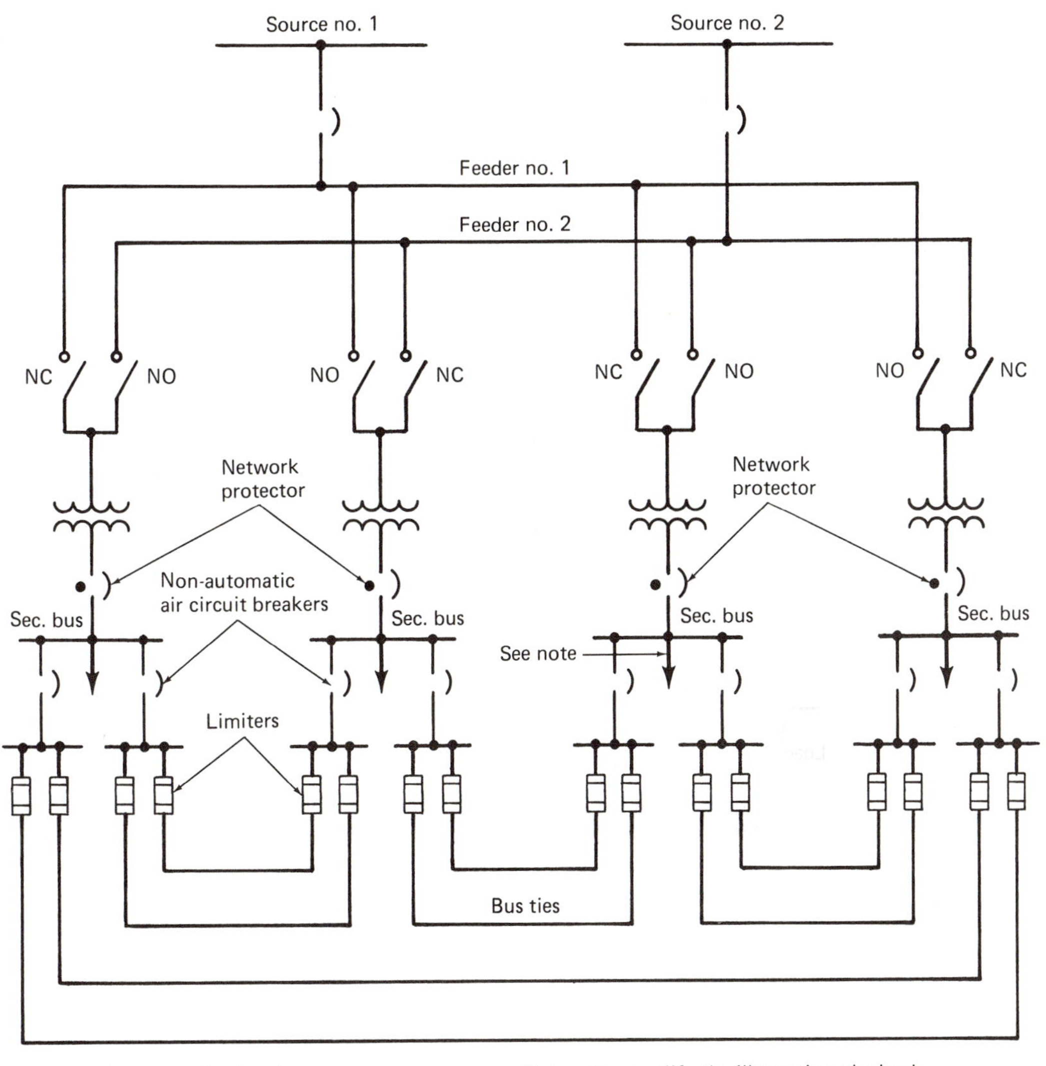

NC = normally closed
NO = normally open

Note: To simplify the illustration, the load circuit breakers are not shown on the secondary bus. ⟶ indicates load circuit breakers are connected here.

Figure 3-5 The Primary Selective, Secondary Network System

Following the diagram of Fig. 3-5 from the two sources, it can be seen that half the transformers are energized from feeder No. 1 and the other half from feeder No. 2. If a primary feeder fault occurs, the feeder is isolated from the system by the automatic tripping of the source feeder circuit breaker and all the network protectors that are associated with the fault. Because the secondary busses are connected together through the bus ties, the entire load is fed from the feeder still on the line and the other half of the transformers. All the transformers can be restored to service by switching over to the live feeder. After the fault has been cleared, the switches are thrown over to their normal operating position and the feeder breaker is closed. The network protectors will automatically close to restore the entire network back to its original condition. The beauty of this arrangement is that there has been no interruption of service to the loads.

The *network protector* is required in order to allow the two primary feeders to operate in parallel (both energized at the same time and feeding onto a common bus). It is basically an electrically operated air circuit breaker controlled by a directional power relay that senses power flow in one direction only and operates to open the breaker if the power flow is from the secondary bus toward the primary feeders. If a fault occurs in the feeder, power will flow toward the fault from the other feeder by way of the network secondary bus tie. The fault current flows from the secondary bus *toward* the source and the relay (directional power) senses this flow is in the *wrong* direction and trips the breaker. There are other relays required for reclosing functions.

The *limiter* is actually a fuse with a long time delay characteristic. Because it may need replacing, it is necessary to put nonautomatic circuit breakers in the circuit to allow for de-energizing the fuse clips or connections. They are installed as protection devices for the bus tie cables. Other devices can be used for this protection, including automatic circuit breakers.

3-4 Information Sources for Preparation of the One-Line Diagram

These sources are relatively simple for the draftsman or designer in the large engineering firm because the rough one lines are usually developed by the project engineer. He furnishes the rough sketches in preliminary form to the designer or draftsman to develop into a finished diagram. In the smaller engineering offices, the engineer and designer or draftsman may have to work closely to develop the one line so it is important that the designer and draftsman consider several basic items of information that should be shown. Some of these items are as follows:

1. The amount of short-circuit current that will come from the utility company supplying power: This is usually called the *primary available short-circuit current.*

2. The characteristics of the cables feeding the system from the utility: This data should include the number of conductors, size and type (copper,

aluminum, or other) conductor insulation, and the jacket covering the insulation if there is any.

3. Rating of all major equipment: Examples of some items follow.

Power Transformer

Power transformer data include the number of transformers, the kVA rating, primary and secondary voltages, whether the transformer is single phase or three phase (or other), and the impedance in percent. Symbols called *vector diagrams* noting how the transformer windings are connected must also be shown if the system is other than single phase. A typical transformer rating is shown in Fig. 3-6. The notes indicate that there is one 1,000-kVA, oil-immersed, self-cooled (OA) transformer connected in delta (Δ) on the primary side at 13,800 V (13.8 kV) and connected in wye (Y) on the secondary side at 480/277 V. The neutral of the secondary connection is grounded (⏚) so secondary voltages will be 480 V between phases and 277 V from any phase to neutral or ground. The impedance of the transformer is 5.75%, based on its own rating. It can be seen that these notations tell a great deal about the transformer.

Rating data is usually shown in the right-hand side of the symbol and the vector diagrams are shown on the left.

In some cases it is desirable to increase transformer capacity above its self-cooled rating. This is accomplished by providing means of cooling the oil or other coolant into which the coils are immersed or by cooling the enclosure of dry-type transformers. The transformer windings in the transformers in Fig. 3-6 are immersed in oil and the oil is self-cooled by the air surrounding the tank. Because transformer capacity relates to the effectiveness of keeping coil temperature controlled, several means are available for cooling the coils. The types of transformer cooling methods and their identifying-type letters are as follows:

Type letter	*Method of cooling*
OA	Oil-immersed, self-cooled
OW	Oil-immersed, water-cooled
OW/A	Oil-immersed, water-cooled/self-cooled
OA/FA	Oil-immersed, self-cooled/forced air-cooled (above 500 kVA)
OA/FA/FA	Oil-immersed, self-cooled (forced air-cooled) forced air-cooled (10,000 kVA and larger)
OA/FOA/FOA	Oil-immersed, self-cooled/forced air–forced oil-cooled/forced air–forced oil–cooled (10,000 kVA and larger)
FOA	Oil-immersed, forced oil-cooled with forced air cooler (10,000 kVA and larger)
FOW	Oil-immersed, forced oil-cooled with forced water cooler (10,000 kVA and larger)
AA	Dry type, self-cooled
AFA	Dry type, forced air-cooled
AA/FA	Dry type, self-cooled/forced air-cooled (above 500 kVA)

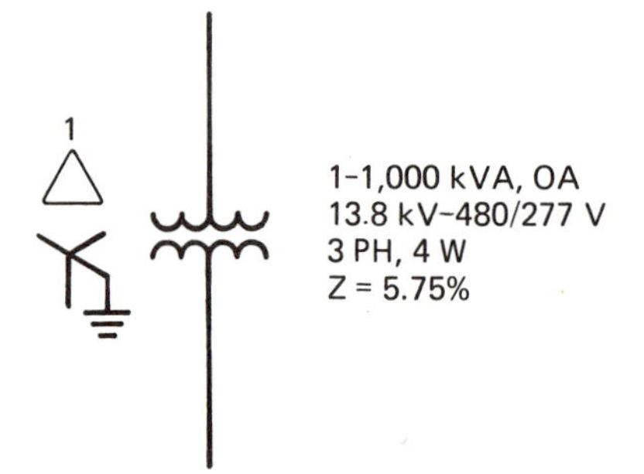

Figure 3-6
Typical Notations for Power Transformers

From the foregoing data, it can be seen that transformers are cooled in several ways. The increase in capability is published for the methods indicated in the American National Standards Institute (ANSI) and National Electrical Manufacturers Association (NEMA) standards as well as in catalogs published by manufacturers. As an example of the increase in capability for the transformer shown in Fig. 3-6, if fans were added on the transformer radiators so that the transformer were rated OA/FA, then its capability would be increased by 15% to 1,150 kVA. Usually the decision to increase transformer capability through use of auxiliary cooling is made when the transformer is specified so that the radiators or other cooling means are properly sized when the unit is purchased. Some transformers can be bought without the special cooling facilities but with provision to add them later. Fans, for example, are relatively simple to add if the controls and wiring are provided when the transformer is purchased.

Circuit Breakers

Circuit Breakers divide into several types and therefore rating data vary. Figure 3-7 shows several types with the notations relating to their ratings that should generally appear on one-line diagrams.

Some further clarification of these notations is warranted. The *continuous current* rating is the maximum current that the breaker trip coil carries under *normal* conditions without tripping. The *interrupting rating* is the power level that the breaker must be capable of interrupting without injury to itself. For the high-voltage breakers, the voltage rating is sometimes given because it may need to be rated higher than the operating voltage.

Low-voltage (under 600 V) breakers generally have their *frame size* indicated. The frame of a circuit breaker relates to its physical size and overall dimensions and the term applies to a group of breakers that are physically interchangeable with each other. Frame sizes are expressed in amperes and correspond to the largest ampere rating available in the group. The same frame size designation may be applied to more than one group of circuit breakers. These groups may or may not be physically interchangeable with each other, whether furnished by one manufacturer or various manufacturers. By noting frame size on the one line, the breaker is given a physical grouping characteristic. Further, it aids in later purchasing information as well as to tell if the space allowed in a particular piece of equipment will accommodate a change in breaker

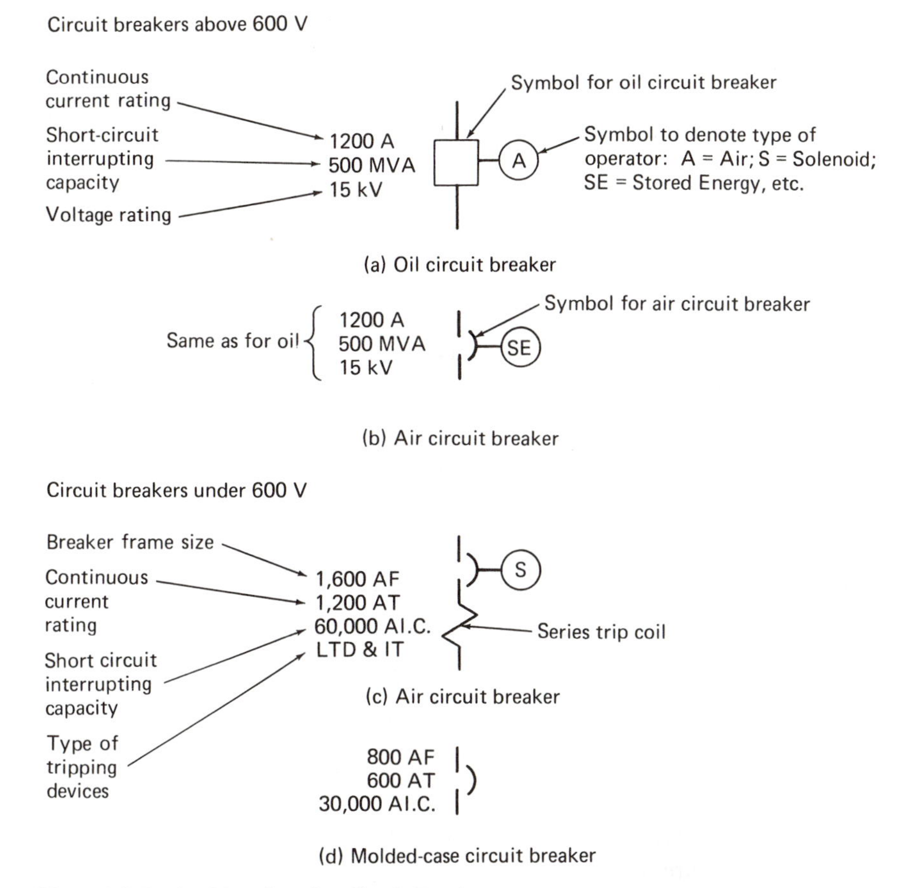

Figure 3-7 Rating Notations for Circuit Breakers

continuous current or interrupting current rating. Figure 3-8 indicates representative basic data relating to the type air circuit breakers indicated in Fig. 3-7(c)

Examination of the table in Fig. 3-8 will show that the frame size data must be included in the breaker rating information. For example, a breaker with a standard 600-A trip coil is made in either the 600-A or the 1,600-A frame size but they may not be physically interchangeable so the frame size must be noted.

The type of tripping device is also an essential part of the rating data. The three types generally specified are

1. Long time delay—LTD
2. Short time delay—STD
3. Instantaneous trip—IT

The type of device or combination of types to be used are determined by the engineer. He must study the system and determine how the breakers and other devices must react one to another.

Frame size (A)	*AC (V)*	*Continuous current (A)*	*Interrupting amperes— asymmetrical*	*Standard trip coil ratings (A)*
225	600 480 240	15–225 25–225 35–225	15,000 25,000 30,000	15, 20, 30, 40, 50 70, 90, 100, 125 150, 175, 200, 225
600	600 480 240	40–600 100–600 150–600	25,000 35,000 50,000	40, 50, 70, 90, 100 125, 150, 175, 200 225, 250, 300, 350 400, 500, 600
1600	600 480 240	200–1,600 400–1,600 600–1,600	50,000 60,000 75,000	200, 225, 250, 300 350, 400, 500, 600 800, 1,000, 1,200, 1,600
3000	600 480 240	2,000–3,000 2,000–3,000 2,000–3,000	75,000 75,000 100,000	2,000, 2,500, 3,000
4000	600 480 240	4,000 4,000 4,000	100,000 100,000 150,000	4,000

Figure 3-8 Representative Rating Data for Air Circuit Breakers

Molded-case circuit breakers are rated in much the same way as the air circuit breakers. Their frame size and continuous and interrupting rating should be indicated. Frame sizes for these breakers generally are 50, 100, 125, 150, 200, 225, 400, 600, 800, 1,000, 1,200, 1,600, 2,000, and 2,500 A. Interrupting ratings range in general from 5,000 rms A symmetrical at 120/240 V ac to 65,000 A at 240 V. They vary with the particular type of breaker. Manufacturers furnish this data in their catalogs. The continuous current ratings range from 15 to 2,500 A. Standard sizes are 15, 20, 25, 30, 35, 40, 45, 50, 60, 70, 80, 90, 100, 110, 125, 150, 175, 200, 225, 250, 300, 350, 400, 450, 500, 600, 700, 800, 1,000, 1,200, 1,600, 2,000, and 2500.

Fuses

Fuses are generally noted with their continuous current rating and as to the type of fuse that must be used. This kind of data provides for the normal (continuous) as well as the abnormal (interrupting) rating of the fuse. In the case of circuit breakers, the interrupting rating itself was shown but in the case of fuses the type covers this requirement. Figure 3-9 indicates several types of fuse notations for one-line diagrams.

Figure 3-9(a) shows a fixed fuse with a continuous rating of 30 A. STD indicates a standard "one-time" fuse. It may have an interrupting rating as high as 10,000 A (some newer types go as high as 50,000 A). The exact type to be used would be specified.

Figure 3-9(b) shows a 100-A, time delay (T.D.) fuse in a fused safety switch. Time delay fuses have interrupting ratings up to as high as 100,000 A depending on the type specified.

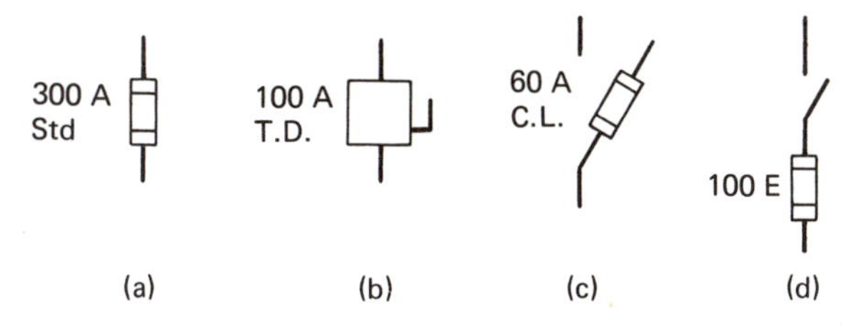

Figure 3-9 Rating Notations for Fuses

Figure 3-9(c) is a fused disconnect switch. This would probably be on a circuit above 600 V where this type switch is commonly used (sometimes this symbol is used for fused cutouts on high-voltage systems). The C.L. indicates a current-limiting–type fuse. These have interrupting ratings up to 200,000 A.

Figure 3-9(d) shows a high-voltage fuse in a fixed holder connected directly to a disconnect switch. The 100 E designation is NEMA for a high-voltage fuse with a continuous rating of 100 A.

Connected Loads

The connected loads and their characteristics must be clearly shown. Examples of notations for a variety of loads are shown in Fig. 3-10.

Each of the notations of Fig. 3-10 indicate to the person reading the one-line diagram just how much electrical power is needed for each device or piece of equipment. For the motor in (a) there is a load of 20 hp. Tables in the appendix would quickly show how much the full load current will be. The heater in (b) with a rating of 30 kW is readily converted to amperes by the formulas in Chap. 2. With a small transformer such as that shown in (c), the full-load transformer current would be assumed to be the demand for small power when figuring load. This analysis should also be made for the bus in (d); i.e., it would be possible to load the plug-in bus to 400 A if necessary.

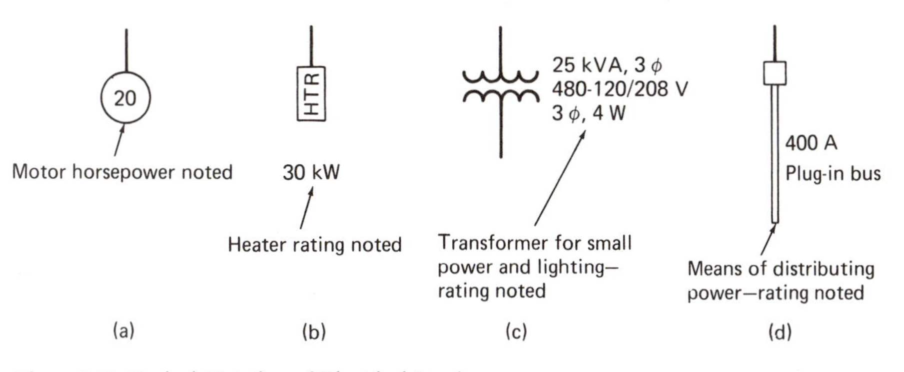

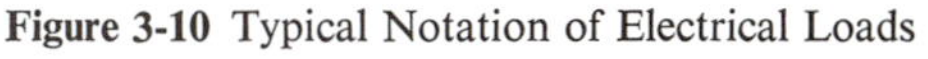
Figure 3-10 Typical Notation of Electrical Loads

Metering Devices

Metering devices must be shown and properly noted because they are important to monitoring system operation. Typical instrument transformer notations are shown in Fig. 3-11.

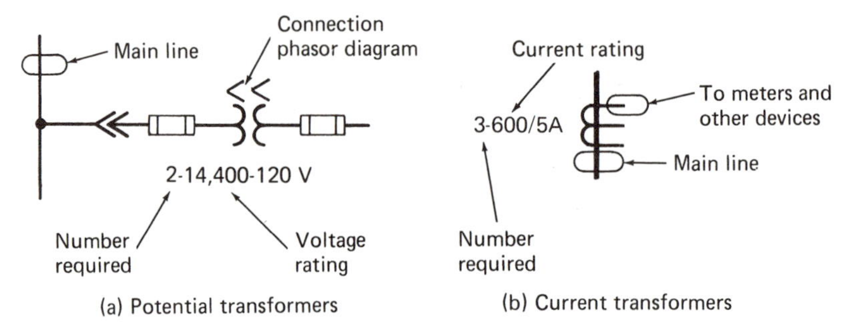

Figure 3-11 Notations for Instrument Transformers

The potential transformer in Fig. 3-11(a) is used to reduce the main line voltage from 14,400 V down to 120 V for meters and other devices. The current transformers in (b) reduce main line current for use at meters; i.e., 600 A in a main line would cause only 5 A to flow to meters.

Meters

Meters are readily identified by their symbols; however, meter scale should be indicated. Examples are shown in Fig. 3-12.

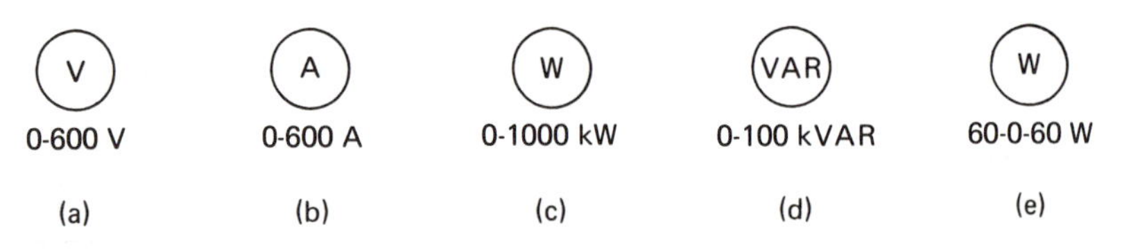

Figure 3-12 Typical Notations for Electrical Meters

Miscellaneous Devices

Miscellaneous other devices such as lightning arresters, interlocking devices, and protective relays are also identified by their won unique symbol. Where size is a consideration, it should be properly noted.

Geographical Considerations

Geographical considerations should be noted. For example, some elements of the system may be indoors and others outdoors. This division can be indicated by a distinguishing line such as a dashed line.

3-5 Typical One-Line Diagram

A typical one-line diagram showing major equipment and devices with their proper notations is shown in Fig. 3-13. Note that the layout of the diagram starts with the incoming lines at the higher voltage at the top of the sheet and power flow is down the sheet with loads shown near the bottom.

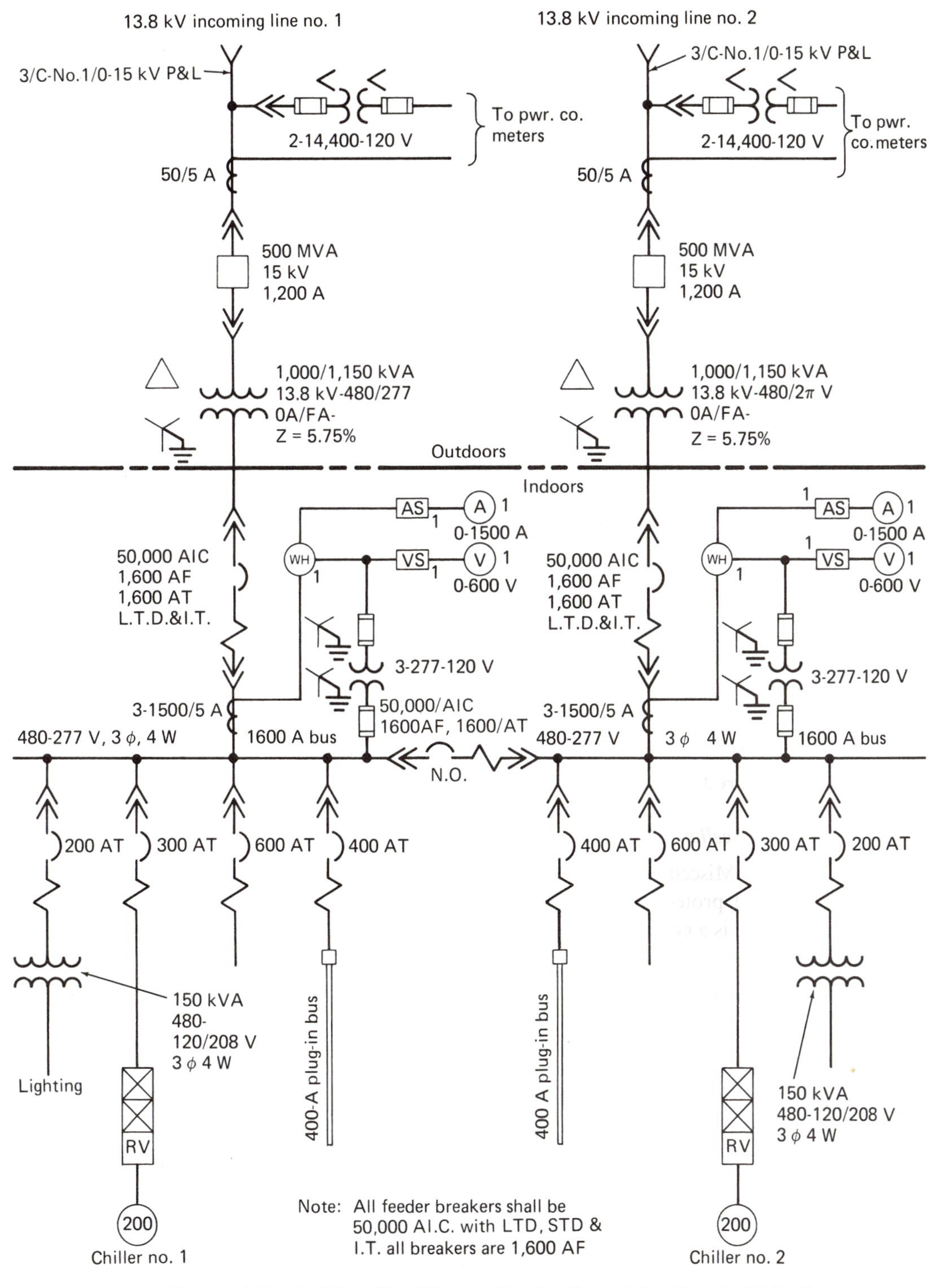

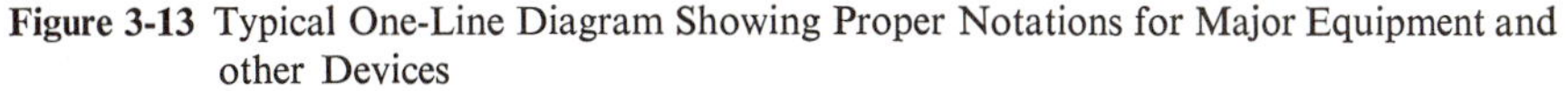

Figure 3-13 Typical One-Line Diagram Showing Proper Notations for Major Equipment and other Devices

3-6 Standard Symbols

The use of standard symbols and other standard notations is encouraged because these diagrams are reviewed and read by many different people. Standard American National Standards Institute symbols are shown in the text. These should be used wherever possible. If they are modified to cover a unique situation, the symbol should be shown and explained on the drawing. Many engineering firms use a prepared legend and symbol sheet as the lead sheet in a set of drawings. This technique overcomes many problems in reading the drawings for those who do not immediately recognize the symbols. If this method is not used, then it is always good practice to include a symbol legend in the upper right-hand corner of the one-line drawing sheet.

3-7 Simple and Straightforward Diagrams

Simple and straightforward diagrams are much easier to read. Keeping the line work straight up and down without offsets and straight across without offsets is to be desired if possible. This often requires rough drafting of the one line several times to keep the line work simple. One lines, during the conceptual stage, are continually changing so several rough drafts are common, however, once the layouts are ready for final drafting, they should be examined for areas of simplification. Physical relationships between elements can generally be ignored so placing a symbol or line to the right or left or on top or underneath another symbol is not important. The physical relationships are shown on other types of drawings. As an example, by turning the current transformer around in Fig. 3-14, it is possible to overcome a line crossing and a crowding of the relay symbol. The circuit is exactly the same but is clearer.

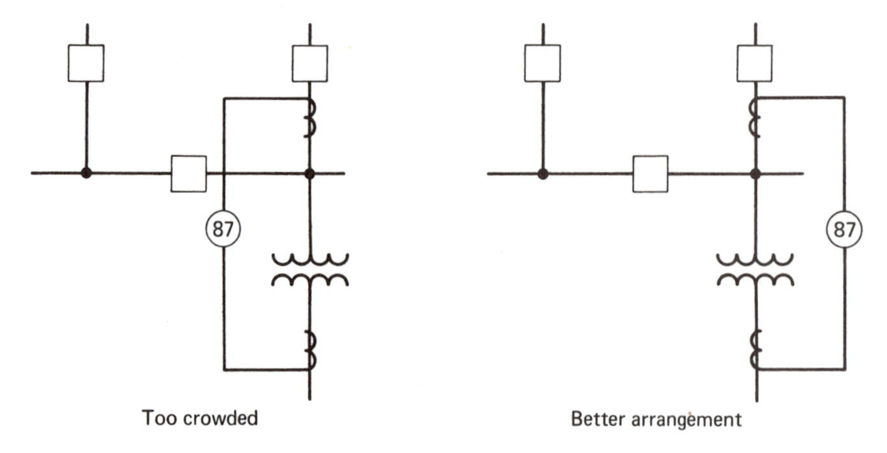

Figure 3-14 Example of Simple and Straightforward Diagram

3-8 Duplication of Data

Duplication of data is to be avoided. It is poor practice in any set of drawings to put the same information in two or more places. Invariably if the data are changed, they may be corrected in one place only because the person making the change may not be aware that they exist in other locations as well. Symbols usually do not need further identification. For example, it is not necessary to label current or potential transformers or any other standard device because the connection shown and the rating given to the device clearly tells what it is. Notes are often used to say that all devices throughout a drawing are of a certain size, etc., *unless otherwise noted.* This is a common practice and eliminates much duplication from the body of a drawing but *care must be used* to be sure the *otherwise noted* devices are identified.

3-9 Known Data

All known data should be shown rather than assuming that showing a device or a circuit is irrelevant and unimportant. Because most of the conceptual design ideas are initially recorded on the one-line diagram, it is better to include all the elements rather than leaving them to guesswork later. An old rule that applies well here is *when in doubt, show it.*

3-10 Future Planning

Show future planning using dotted lines or some other appropriate symbol because the future work can have a very direct relationship to size of cables, transformers, etc. Identifying equipment to be installed now but to be used later for future needs is very important because the equipment may not appear to have any purpose unless explained. As an example, the first two isolating disconnect switches of the four shown in series in Fig. 3-15(a) do not seem to

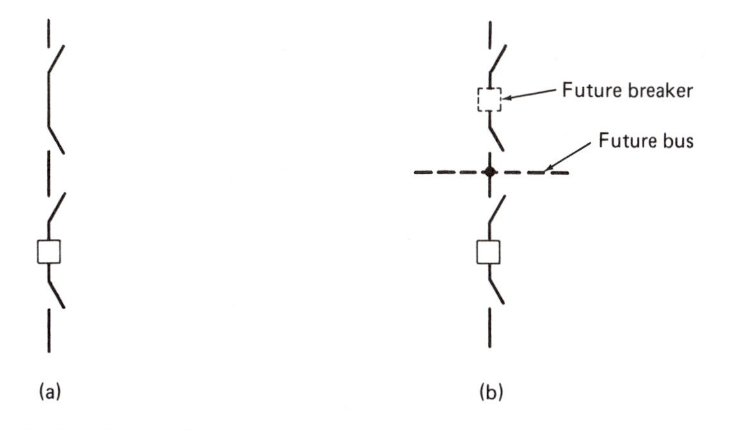

Figure 3-15 Show Future Planning in Diagrams

perform any function but if the circuit were drawn as shown in Fig. 3-15(b), the need for the first two switches is clear.

3-11 Draftsman or Designer Check

The draftsman or designer should check his own work after finishing the drawing and before releasing it to the engineer for approval. Some items to check are as follows:

1. Ratings of all devices
2. Phasor diagrams
3. Ratios of instrument transformers
4. Fuse ratings
5. Circuit breaker ratings
6. Solid "dots" at points in the diagram where a connection is required
7. Geographical data (indoor–outdoor, etc.)
8. Ratings of cables and wires if this data is to be included
9. Title of drawing and correct names of substations, busses, generators, etc.
10. Neatness of drawing
11. Minimum line crossings
12. Crowding
13. Spelling
14. Proper abbreviations if they must be used—always spell out words if possible
15. Uniformity of lettering
16. Instructions by notes are conclusive and can be readily understood
17. *Do not take anything for granted*

3-12 Typical One-Line Diagram for Office Building

A typical one-line diagram for an office building is shown in Fig. 3-38 at the end of this chapter. To understand the drawing, some background information on the building itself will be helpful.

The building is called a *high rise* because it towers above surrounding buildings with its 16 stories above the street level. At the lobby level, there is an upper mezzanine; below the lobby level, there is a lower mezzanine that runs above the bottom level of the structure and is called the *concourse*. This level joins to other concourse levels for adjoining buildings and to the subway systems. Altogether, there are 20 levels plus the roof. The sixteenth-story level is called the *penthouse*. The building is shown in block form in Fig. 3-16.

The entire building is to be air-conditioned. There will be six elevators to service all floors. Electrical service will be furnished by the local utility from

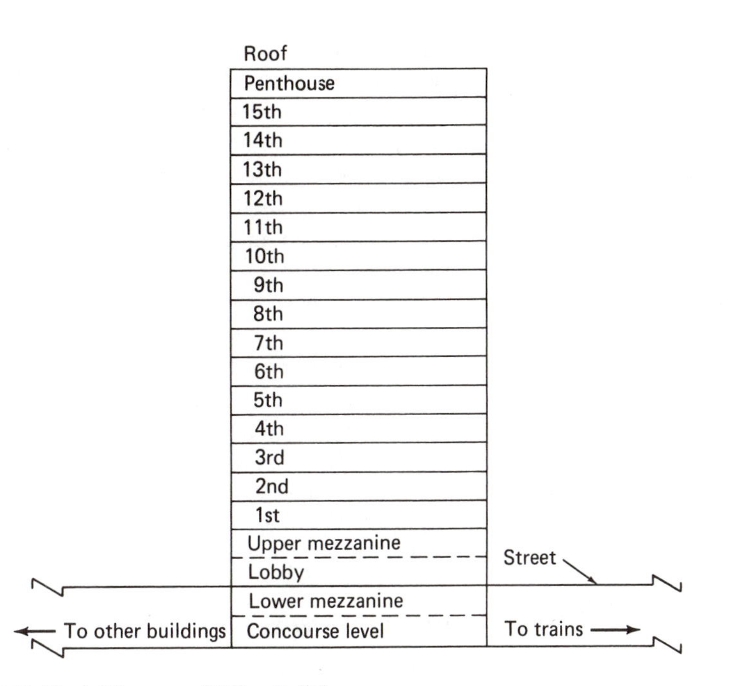

Figure 3-16 Block Diagram of Office Building

its underground system. The primary system is 13,800 V, three phase, 60 Hz.

The engineer elected to use a primary selective system [see Sec. 3-3(d)] because of its reliability of service. The 15-kV service entrance switchgear will provide for the service to the building and will be located on the lower mezzanine. It is a totally enclosed metal structure called *metal-clad* switchgear, which is purchased as a custom-made unit to receive the incoming feeders, house relays, and meters and to provide for distribution to the load center substations. The switchgear arrangement is shown in block diagram form in Fig. 3-17.

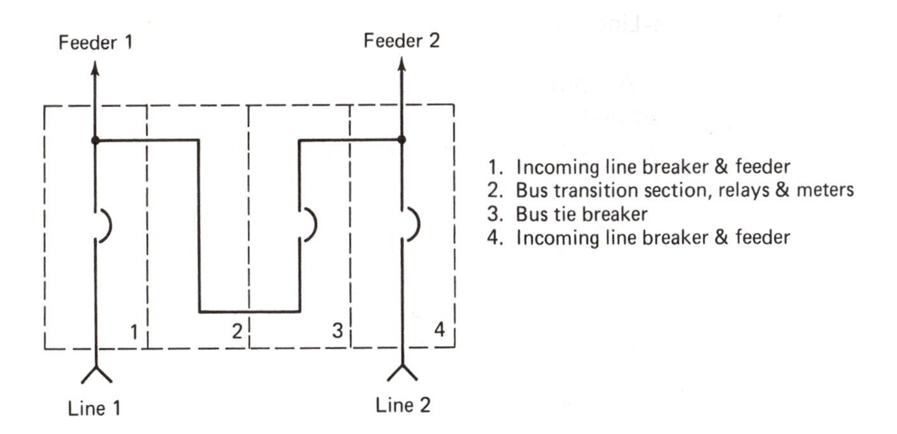

Figure 3-17 A 15-kV Metal-Clad Service Entrance Switchgear

The primary incoming lines from the utility company are fed from two different substations to improve reliability further. These lines are to be furnished and installed by the utility company. This is common practice because utility companies prefer to run and test cables that are connected to their systems. The cable selected is to be a 250-MCM (see Chap. 14, Sec. 14-3) copper conductor that is paper insulated, lead encased, and rated 15 kV. Three single conductors will be installed in conduit and will terminate in three single-conductor potheads in the 15-kV switchgear. The physical arrangement of this primary service entrance is shown to portray the simplification of this part of the circuit. A typical pothead is detailed to show how the cables are terminated in order to connect into the switchgear busses (Fig. 3-18).

Potheads are not always used. Many times the cables are properly terminated with stress cones only and connected directly to the terminals provided. Where lead-encased cable has been used, as in this case, the lead sheath must be properly terminated. This is done by wiping the lead directly to the wiping sleeve of the pothead. This technique requires special training and can only be done by qualified people.

Primary Relaying and Metering

Primary relaying and metering is provided in the 15-kV switchgear. These relays and meters are usually mounted in the doors of each cubicle. Figure 3-19 shows the metering for one of the incoming line and feeder sections.

The relaying and metering shown for the primary line/feeder section is relatively simple. Basically, relaying is limited to undervoltage and overcurrent protection and metering is limited to voltage indication and current flow on each phase. Power company metering is provided in the form of a standard kilowatthour meter with a demand register. This latter device measures the peak demand for electricity made during any set period of time and is used to bill for power at a separate demand rate. The standard register on the meter measures the energy used and is calibrated in kilowatthours.

Primary Feeders

The primary feeders from the 15-kV service entrance switchgear go to three separate unit substations located in electric equipment rooms. These rooms are provided in strategic locations in order to serve best the loads connected to them. Very often they are located at the *center of the load* and are therefore referred to as *load center unit substations*. In this building, one unit substation is located in a room on the lower mezzanine and the other two units are located in the penthouse equipment room. This arrangement provides for power distribution from the lower floors *up* through the building and from the top floor *down* through the building. In a way, it is similar to two smaller buildings with one sitting on top of the other. The diagrammatic arrangement is shown in Fig. 3-20.

The primary feeder cables are single-conductor, copper, oil-base rubber-insulated cables with a polyvinyl chloride (plastic) jacket. This is a tough-

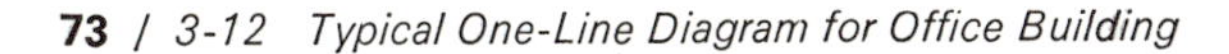

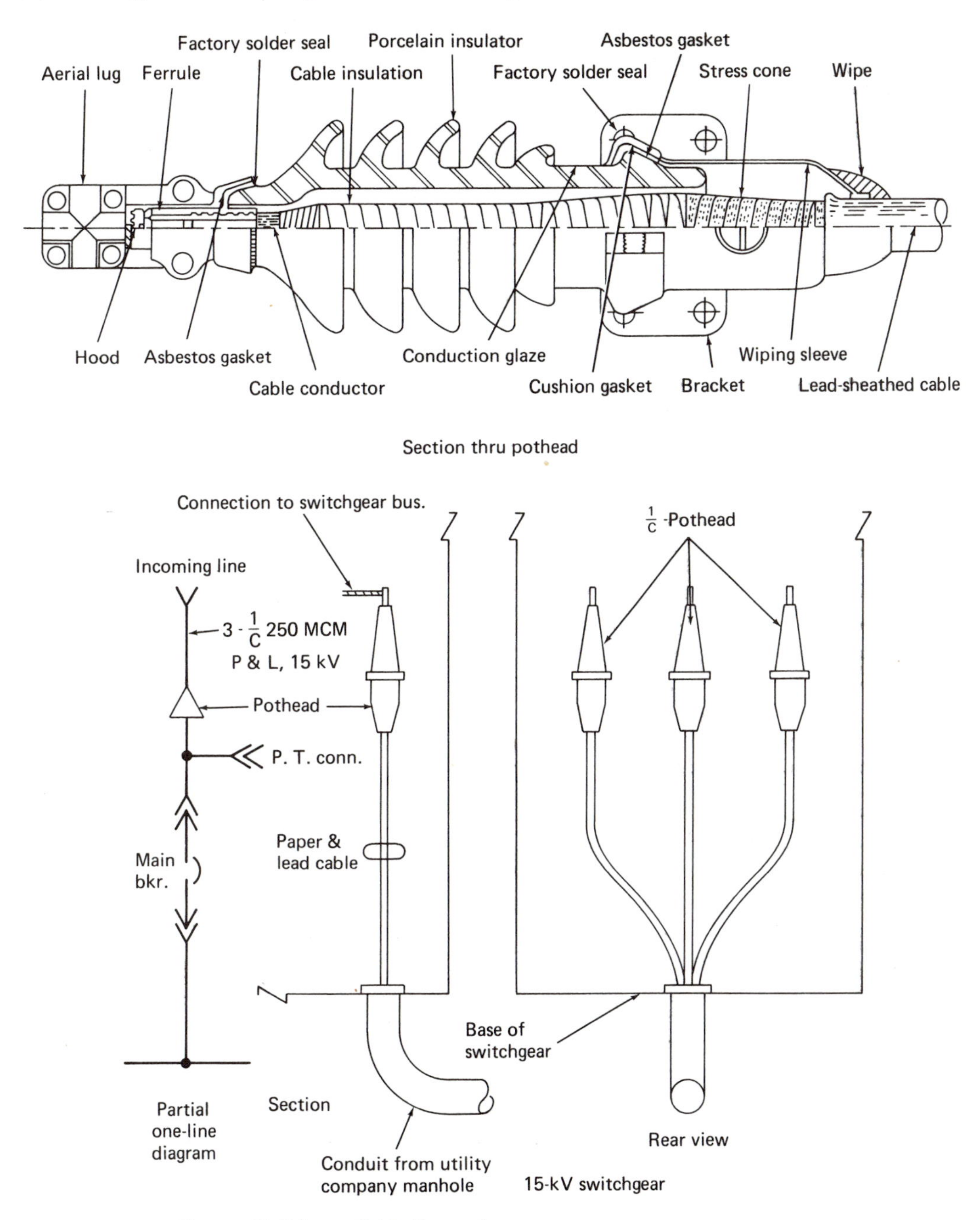

Figure 3-18 Primary Cable Connections

jacketed cable that will take considerable abuse. Unlike the primary incoming line cables, these are not lead encased and no potheads will be used. (Potheads are often used with this type of cable but the engineer has decided that they are not required for this installation.)

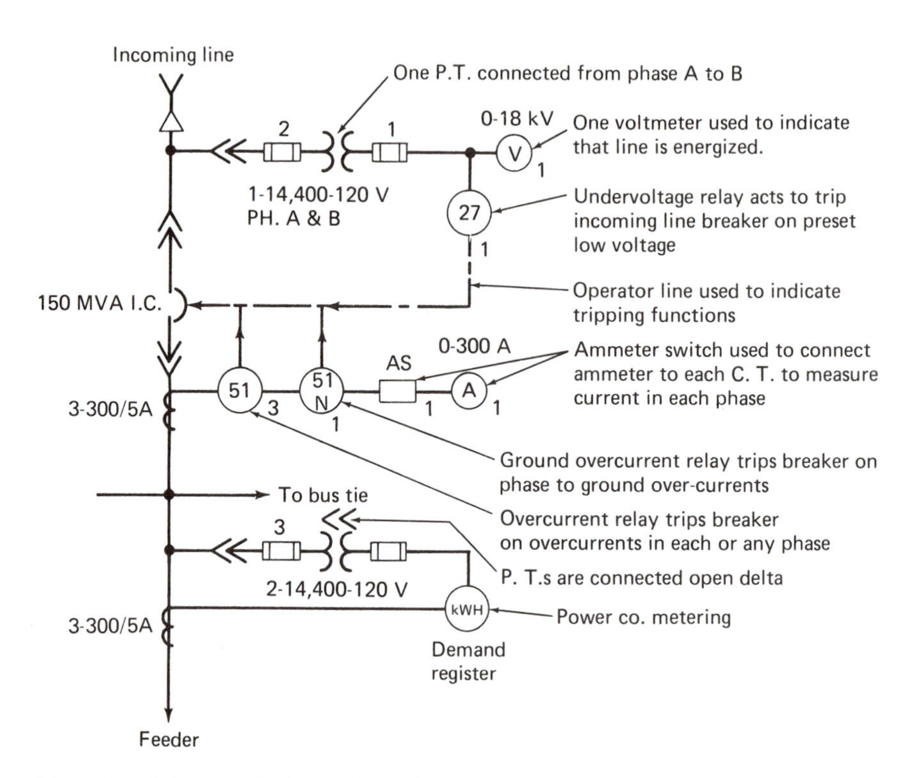

Figure 3-19 Primary Relaying and Metering

The cables run from the 15-kV service entrance switchgear in conduit and terminate in the switch enclosure on the unit substation. A block diagram arrangement of this section of the switchgear together with the corresponding one-line diagram is shown in Fig. 3-21.

The switch arrangement, in this case, is of the primary selective break type as manufactured by suppliers of major electrical equipment. Because there are two incoming feeders, this switch arrangement provides for maximum service continuity by allowing the switching from one incoming line to the other in case of failure or to the middle Open point for planned maintenance. The dotted line in the one-line diagram with the Ⓚ symbol indicates two things. First, the dotted line indicates that the two switches are mechanically joined together so that both incoming feeder switches cannot be closed at the same time. If the operator throws the switch on the feeder, No. 2 cubicle to Line 1, the blades in the feeder No. 1 cubicle will also move to the Open position in the middle and then over to the Feeder No. 1 position that accomplishes the transfer by closing onto Feeder No. 1 and simultaneously opens from Feeder No. 2. Second, because the symbol was shown, it is indicated that a key must first be secured from some other device in the system before these switches can be operated. This feature is incorporated in the design because these interrupter switches are capable of breaking transformer magnetizing current only as opposed to *load*-interrupter

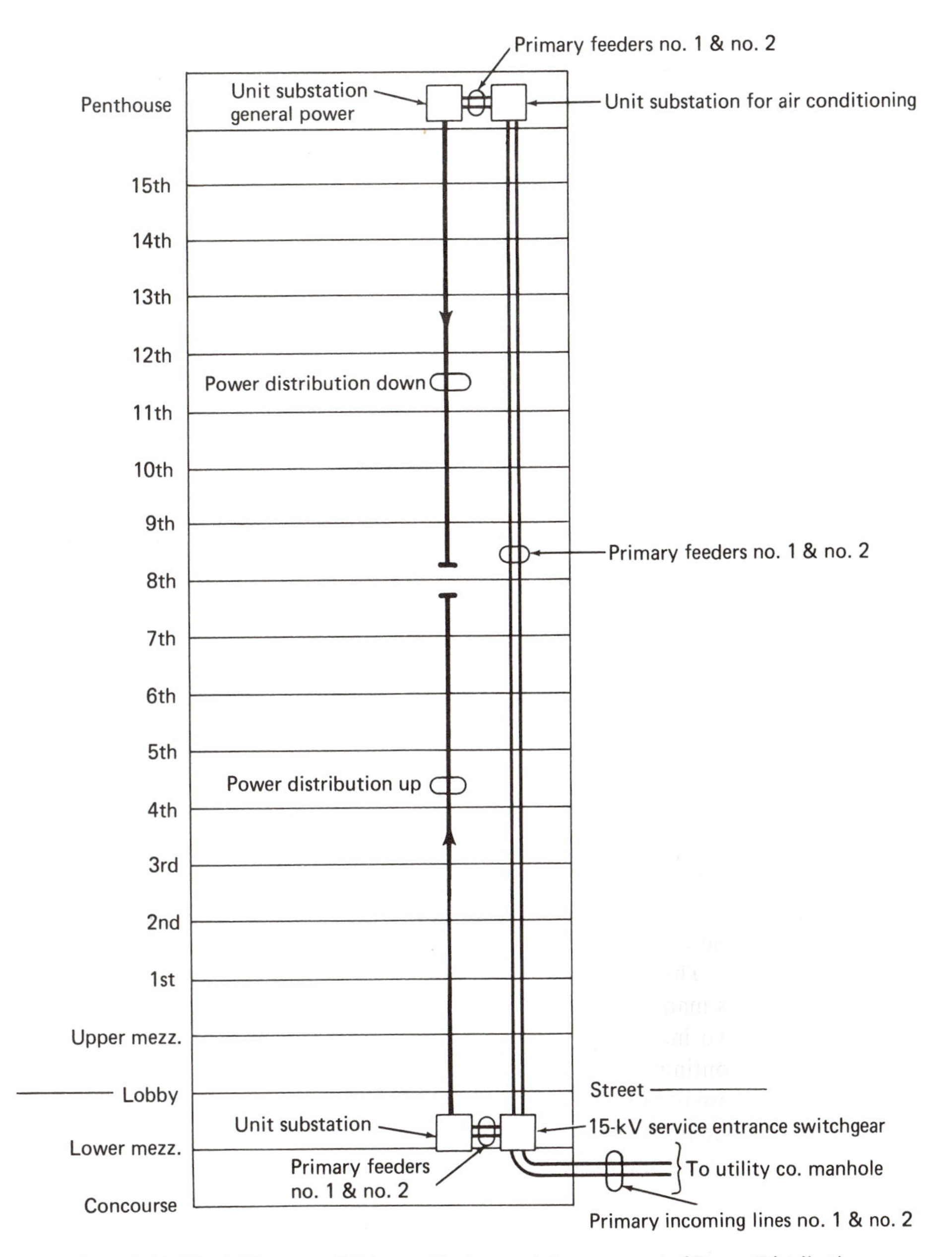

Figure 3-20 Block Diagram of Primary Feeders and Arrangement of Power Distribution

types and must therefore never be operated if the feeders are carrying load. To provide the protection, the design requires tripping the main circuit breaker in the secondary side of the transformer, which releases a key in the cubicle housing that breaker. Once the key is released, it can be taken to the feeder switches and inserted in the lock, which will allow for unlocking the switches

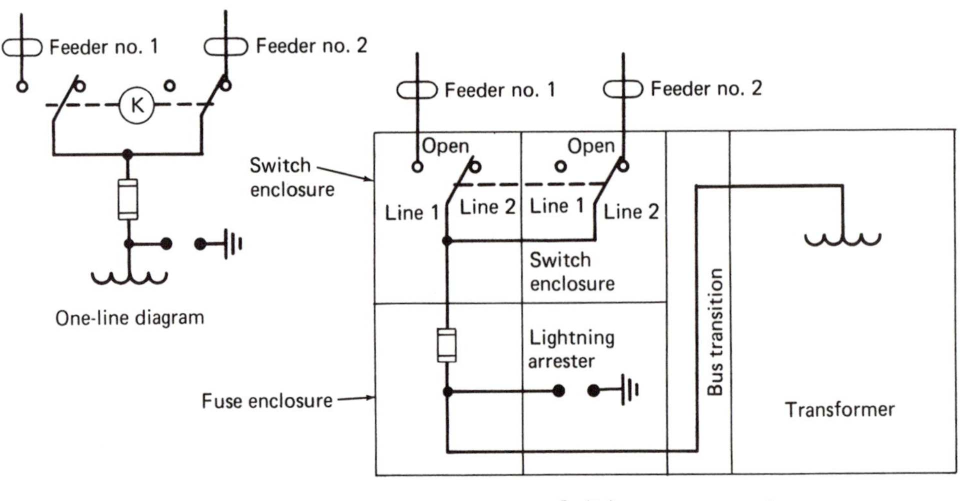

Figure 3-21 Arrangement of Feeder Entrance to Unit Substation and Transformer Connection

so that they can be operated without carrying any load. Variations on the key interlocking scheme are extensive and provide an almost foolproof method for preventing switching and other operating errors.

There are other switching arrangements from that used in the foregoing example. Some of these are shown in Figs. 3-22 and 3-23.

This air-interrupter switch is a three-pole, two-position (open–closed) switch with all three poles operated simultaneously. Obviously, this can only be used with a single feeder.

Where there are two separate incoming lines, this three-position (Line 1/ Open/Line 2) switch provides for service continuity by allowing the operator to switch from one incoming line to the other in case of failure of the feeder or to the Open position for maintenance.

This arrangement consists of a two-position (open–closed) air-interrupter switch in series with a two-position (Line 1/Line 2) selector switch. The selector switch cannot be operated unless the interrupter switch is Open.

There are other means available for providing the switching function and transformer protection. Some switches are load interrupting and overcome the

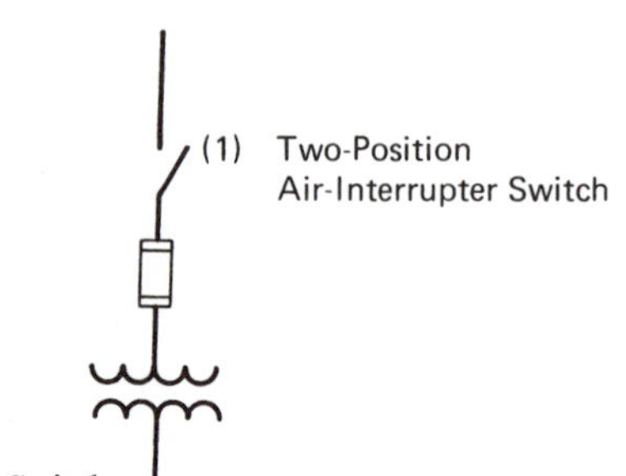

Figure 3-22 Two-Position Air-Interrupter Switch

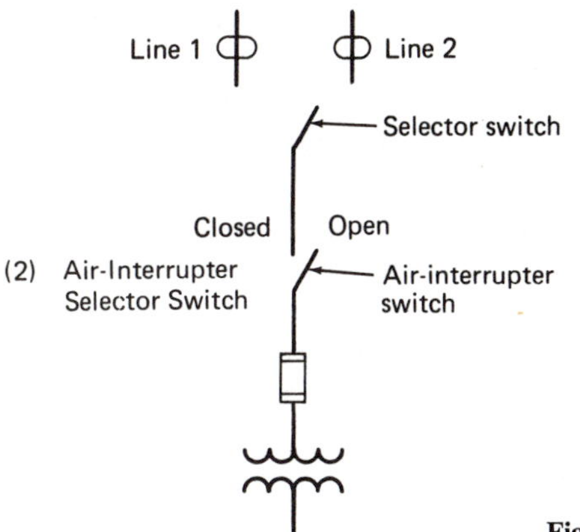

Figure 3-23 Air-Interrupter Selector Switch

key-interlock requirement. These switches can be of either the air-load-breaker interrupter type or a switch utilizing oil or nonflammable liquid as a means of controlling the arc that forms when the load currents are interrupted. The various means of switching control are evaluated by the engineer as to value in operation and cost before a decision is made. Manufacturers' catalogs show the various types available and explain to the potential user the value of each.

Figure 3-21 also showed a requirement for lightning arresters on the load side of the fuses. Protection of incoming line switches and transformers against lightning surges is generally recommended for lines connected to exposed circuits. A circuit is considered exposed if it connects to any open line (line on overhead poles, etc.) including aerial cable either directly or through a cable reactor or regulator. A circuit connected to open-line wires through a power transformer or a metal-sheathed cable is not considered exposed if adequate protection is provided on the line side of the transformer *or* at the junctions of the cable and the open-line wires. Circuits confined entirely to the interior of a building, such as an industrial plant, are not considered exposed and ordinarily require no lightning protection. In the example used, the incoming lines and feeders are not protected by any other arresters and therefore it was considered a prudent addition to provide arresters to protect the transformers even though service to the building is underground because farther out along the utility line these lines were run overhead in the open.

Unit Substation

The unit substation for low-voltage (0–600 V) power is a custom-made piece of equipment that includes the incoming line and fuse sections, the transformer, the main secondary breaker, and all feeder breakers together with all necessary meters, relays, and switches. These units are made up for each job; however, the individual elements are of repetitive manufacture and many are stock items so the units are produced in somewhat of an assembly line fashion. They are often ordered by using standard specifying forms furnished by the manufacturer wherein the particular requirements are inserted into blank spaces in the form. The front view of the unit together with a one-line diagram is generally required. These unit substations can be purchased as single units or *double-enders* such as the double-enders required for the office building. The floor plan front view and internal main power one-line diagram of the double-ender is shown in Fig. 3-24. Typical dimensions are shown as an indication of the general overall size of this kind of equipment.

It can be seen that this particular double-ender is quite long (nearly 40′-0″) and therefore will require a good-sized room. To overcome the length problem, sometimes the equipment is split into two separate sections and placed on the floor so that a common aisle serves the two. In order to provide the bus tie, an overhead piece of bus duct is supplied to connect the two units. This arrangement is shown in Fig. 3-25.

The floor space requirements for the split unit arrangement is close to a square shape when adequate space is provided behind the units for installation and maintenance requirements. The clearance distance from the back to any wall or other fixed barrier should be checked through data provided by the manufacturer and also rules governing these installations in the latest edition of the National Electrical Code.

Unit Substation Transformer

The unit substation transformer is generally rated at one of the following primary voltages:

2,400
4,160
4,800
6,900
7,200
12,000
13,200
13,800
All are delta-connected ratings.

Standard kilovoltampere ratings generally available for the primary voltages listed above are as follows:

Secondary Voltages			
208 Y/120	*240Δ*	*480 Y/277*	*480Δ*
112.5	112.5	112.5	112.5
150	150	150	150
225	225	225	225
300	300	300	300
500	500	500	500
750	750	750	750
1,000	1,000	1,000	1,000
1,500	1,500	1,500	1,500
		2,000	2,000
		2,500	2,500
		3,000	3,000

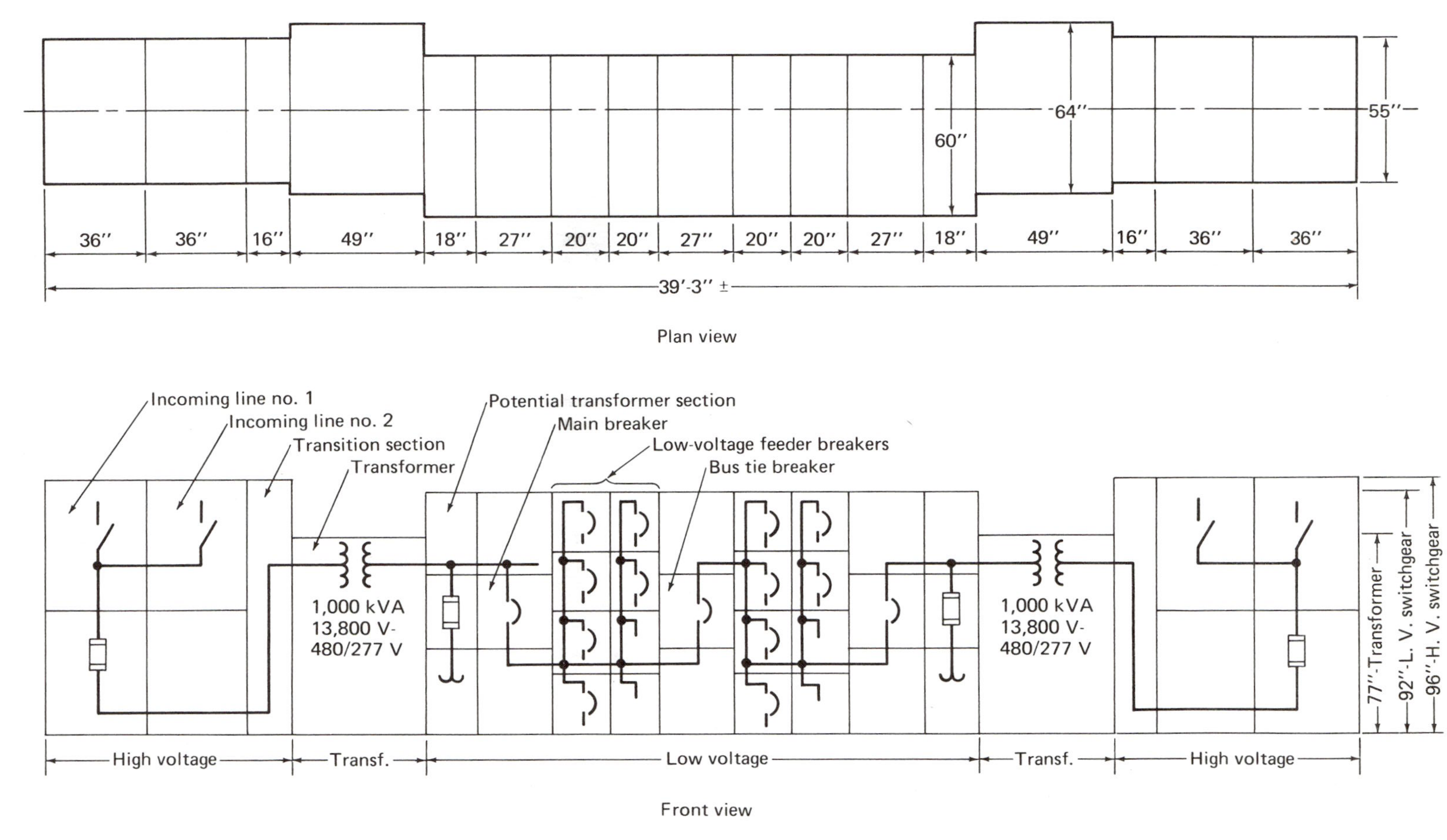

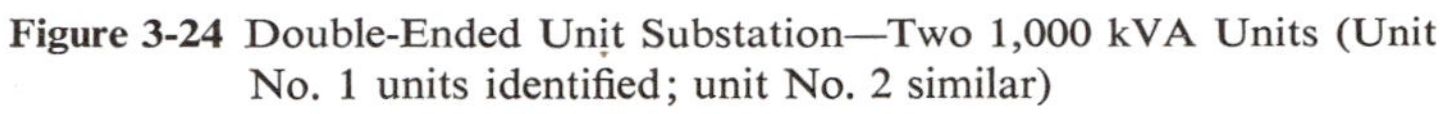

Figure 3-24 Double-Ended Unit Substation—Two 1,000 kVA Units (Unit No. 1 units identified; unit No. 2 similar)

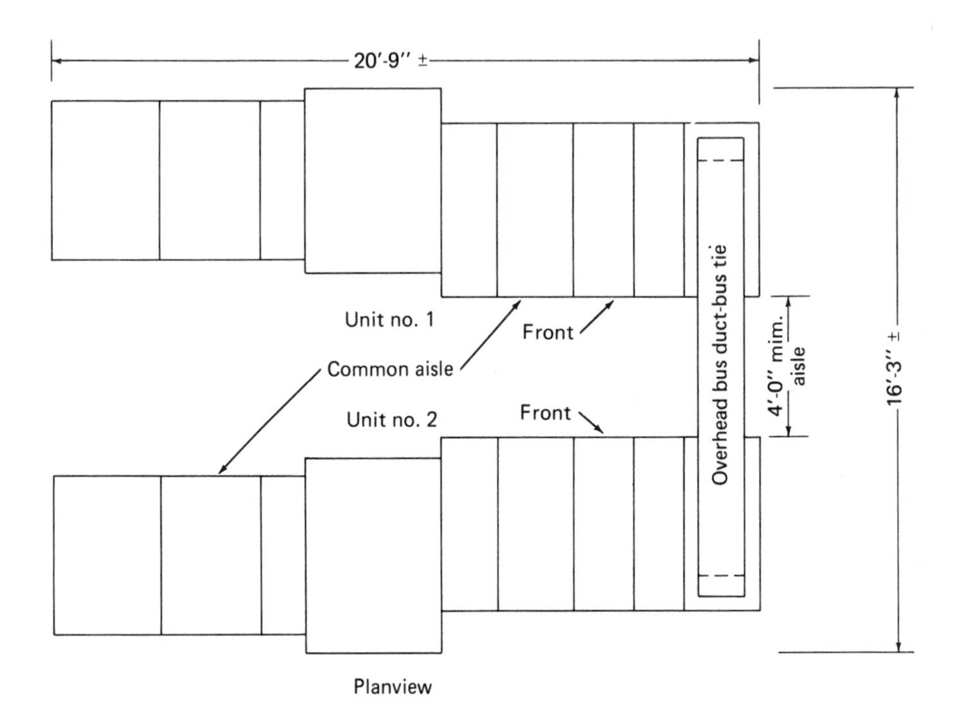

Figure 3-25 Unit Substation Showing Units Arranged on a Common Aisle with a Bus Duct Used for the Bus Tie

Sizing the transformers requires a great deal of engineering study. There are lighting loads to serve as well as power required for air conditioning fans, pumps, etc. All these data are accumulated and compiled into areas of load and a picture begins to form as to where the unit substations are needed and some idea of size is established. Because the building would have substantial lighting as well as power loads, the lighting could be supplied at 277 V, single phase from the same power supply provided for power. This overcomes the need for separate lighting transformers. Power is generally supplied at 480 V, three phase for motors $\frac{1}{2}$ hp and larger up to about 200 hp. Motors smaller than $\frac{1}{2}$ hp are usually supplied from 120-V, single-phase sources. *These are by no means fixed rules*; however, they are a guide.

In the case of the office building, the 480Y/277-V system was ideal and the 1,000-kVA transformers suited the load requirements with capacity allowed for growth. Lower voltages for smaller motors could be obtained by installing 480 V to 208Y/120-V dry transformers in locations throughout the building as required.

Instrumentation in the Unit Substation

Instrumentation in the unit substation is generally held to a minimum because these are unattended pieces of equipment and therefore the metering needs are limited to those required to check on the unit's operation only. The meter-

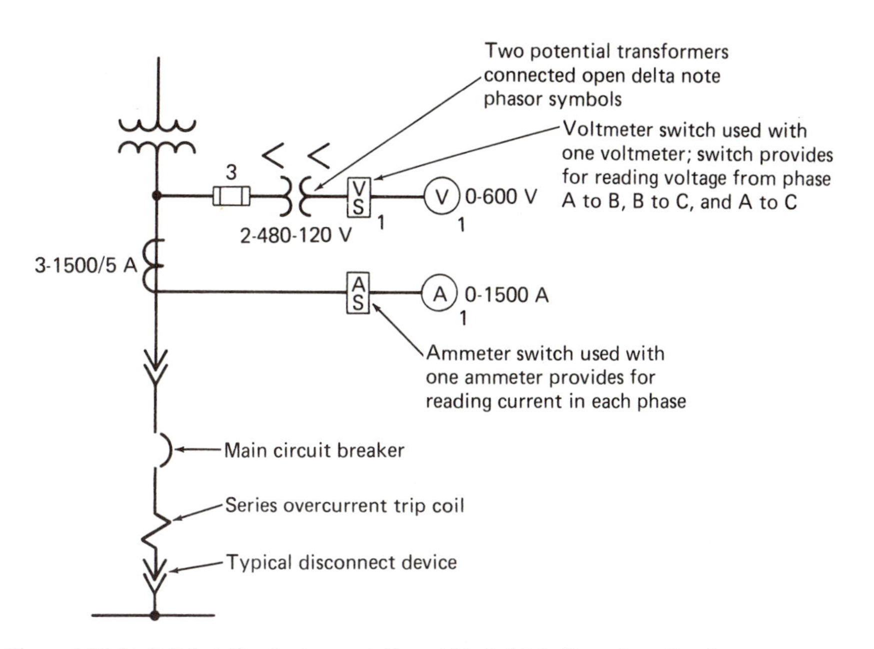

Figure 3-26 Unit Substation Instrumentation at Each Main Secondary Breaker

ing in this case is in the main secondary circuit and used to check voltage and loads only. Figure 3-26 shows the devices used.

For detailed connection of the devices shown in Fig. 3-26, refer to Chapter 4, p. 124 and 125. Of course, additional instrumentation can be provided; however, the cost of such devices must be weighed against the need for the information that would be acquired.

Feeders from the Unit Substation

Feeders from the unit substation are protected by circuit breakers connected to the main bus. The size of these breakers is dependent on the loads they serve plus other considerations. The engineer will size the brakers considering many things such as

1. Present connected load
2. Future growth
3. Length of feeder—(Is a larger size justified in the initial installation because the feeder is long and it will cost less now to install the additional capacity than to have to make a second feeder run later?)
4. Cost of several smaller breakers as opposed to larger units
5. Short-circuit requirements

In this case it was decided that all breakers would be 600-A frame size with an interrupting rating of 35,000 A asymmetrical at 480 V. A typical feeder with its notations is shown in Fig. 3-27. 250 A indicates the normal current

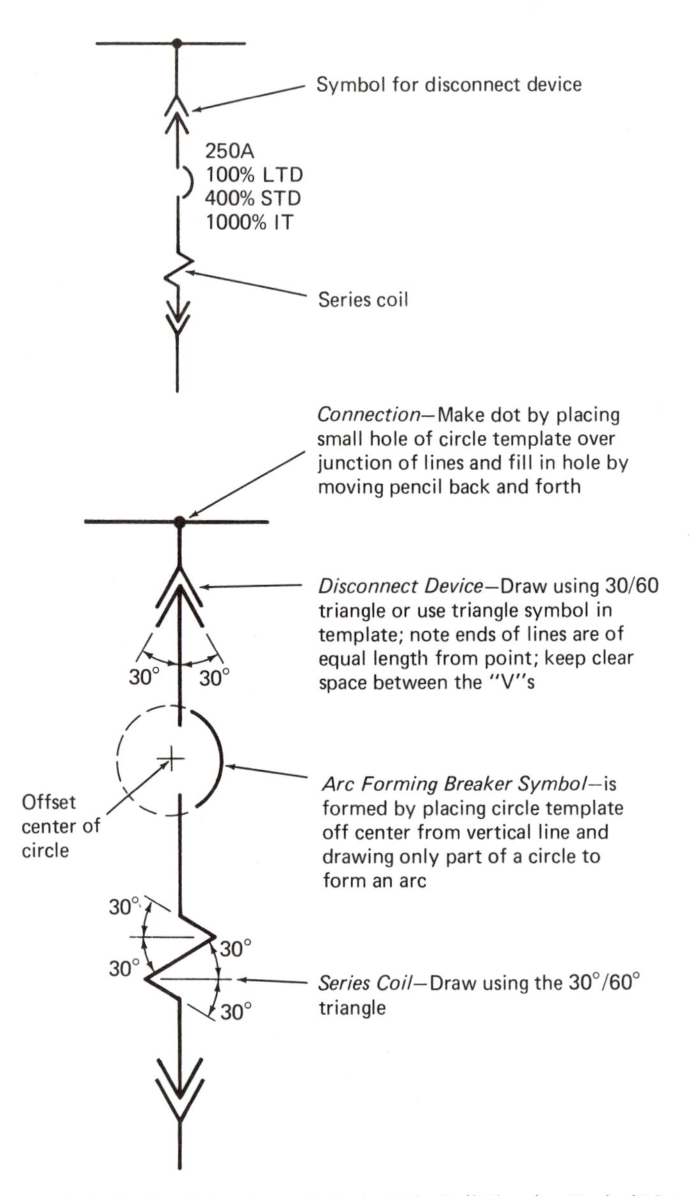

Figure 3-27 Typical Air Circuit Breaker with Series Trip Coil Showing Typical Notations and Method for Drafting the Symbols

trip rating. This is more correctly called the *coil size* because it relates to the rating of the series coil in the tripping devices. It is the Z-shaped symbol part of the breaker symbol (see Fig. 3-27). The tripping characteristics are also identified beside each breaker:

100% LTD indicates the pickup adjustment as a percent of tap setting for the *long time delay* pickup.

400% STD indicates the pickup adjustment as a percent of tap setting for the *short time delay* pickup.

1000% IT indicates the pickup adjustment as a percent of tap setting for the *instantaneous* pickup.

Choice of the proper device setting requires extensive engineering study to make certain that these devices coordinate properly with other protective devices in the system (refer to Chapter 13).

These feeders distribute the power to other points of distribution throughout the building. An examination of the distribution from the unit substation on the lower mezzanine will show the extent of these feeders (see Fig. 3-38).

Feeder No. 1: Supplies 480-V, three-phase, three-wire power to a 150-kVA, three-phase, 480-208/120Y, three-phase, four-wire transformer that feeds Power Center No. 1 for some of the power requirements on the mezzanine, in the lobby, and in the concourse areas.

Feeder No. 2: Supplies 480-V, three-phase power to a Motor Control Center in the concourse equipment room.

Feeder No. 3: Supplies 480-V, three-phase, four-wire power to a 600-A plug-in bus duct that runs vertically up through electrical closets on the west side of the building to each floor through the seventh-floor level. Power can be taken from this bus duct at each floor.

Feeder No. 4: This is a spare breaker for future use. The breaker mechanism is in place and ready for use.

Feeder No. 5: Supplies 480-V, three-phase power to Motor Control Center No. LM1 in the lower mezzanine equipment room.

Feeder No. 6: This is one of the two feeders to the fire pump. Two sources are required for reliability. This one is fed from transformer No. 1.

Feeder No. 7: This is the second feeder to the fire pump and it is fed from transformer No. 2.

Feeder No. 8: This feeder is similar to Feeder No. 1 except the transformer is rated 225 kVA and the transformer supplies Power Center No. 2 (one-line diagram is not shown in this text).

Feeder No. 9: Similar to Feeder No. 3 except it is for the east side of the building.

Feeder No. 10: Supplies 480-V, three-phase power to a 300-kVA transformer that will supply 208/120-V, three-phase, four-wire power to a computer (one-line diagram is not shown in this text).

Feeder No. 11: Supplies 480-V, three-phase power to Motor Control Center No. LM2 in the lower mezzanine equipment room (one-line diagram is not shown in this text).

An analysis of the distribution provided by these feeders will indicate that the lower part of the building from the seventh floor down is being supplied power from power center panelboards, motor control centers, and plug-in bus ducts.

"Power Center"

The "power center" distribution point is actually a panelboard containing circuit breakers or other protective devices that provide protection for the sub-feeders being fed from this distribution point. The data on these panelboards are given in a schedule rather than in diagram form. A typical schedule is shown in Fig. 3-28. This particular panelboard distributes 208/120 V, three-phase, four-wire power to several lighting panelboards. Note that the lighting panelboards are given key letter-type designations that help to identify their location. For example, LL-3 is *lighting for the lobby*; and LLM-1 is *lighting, lower mezzanine*.

Power center No. 1 208/120V 3PH 4W					*Lower mezzanine elec. equipt. room*		
					Load (W)		
Bkr. No.	*Poles*	*Frame (A)*	*Trip (A)*	*Designation*	*ϕA*	*ϕB*	*ϕC*
1	3	225	70	Spare			
2	3	225	70	Spare			
3	3	225	70	Panel LL-3	6,000	6,300	6,300
4	3	225	70	13.2 kV switchgear			
5	3	225	150	Panel LC-1	8,925	8,550	8,500
6	3	225	150	Panel LL-1	8,800	800	1,200
7	3	100	90	Panel LLM-1	4,500	4,250	3,100
8	3	100	70	Panel LUM-1	1,200	1,500	1,500

Figure 3-28 Panel Schedule for Power Center No. 1

Lighting Panelboards

The lighting panelboards being fed from the power center panelboard are also shown in table form rather than a one-line diagram because the form provides a simpler means of recording the necessary data. These forms can be preprinted in a reverse-reading, stick-on decal that reads correctly through the face of a drawing when stuck onto the back. A typical form for this type of panelboard is shown in Fig. 3-29.

The internal wiring of the board is different from the power center panelboard because in most cases the circuit breakers are single pole and therefore connect to only one of the main busses; whereas in the power center they are three-pole breakers and therefore all main busses are connected to each breaker. A typical arrangement for a lighting panelboard is shown in Fig. 3-30.

No. LL3 Location: LOBBY WEST CORE

208/120 V, 3 PH., 4 W 100 A Mains, 100 A solid neutral; Grd. bus
10,000 A AIC Bkrs. @ 120 V; Surface mtg. - Flush mtg.

Designation	Load (W)	No outlets	Ckt. bkr. Trip	Ckt. bkr. No.	Phase	Ckt. bkr. No.	Ckt. bkr. Trip	No outlets	Load (W)	Designation
CANOPY LIGHTING	1,200	4	20	1	A	2	20	4	1,200	CANOPY LIGHTING
	1,200	4		3	B	4		4	1,200	
	1,200	4		5	C	6		4	1,200	
	1,200	4		7	A	8		4	1,200	
	1,500	5		9	B	10		4	1,200	
	1,500	5		11	C	12		4	1,200	
	1,200	4		13	A	14				SPARE
	1,200	4		15	B	16				
	1,200	4		17	C	18				
SPARE				19	A	20				
				21	B	22				
				23	C	24				

Lighting Panelboard Schedule

Figure 3-29 Lighting Panelboard Schedule

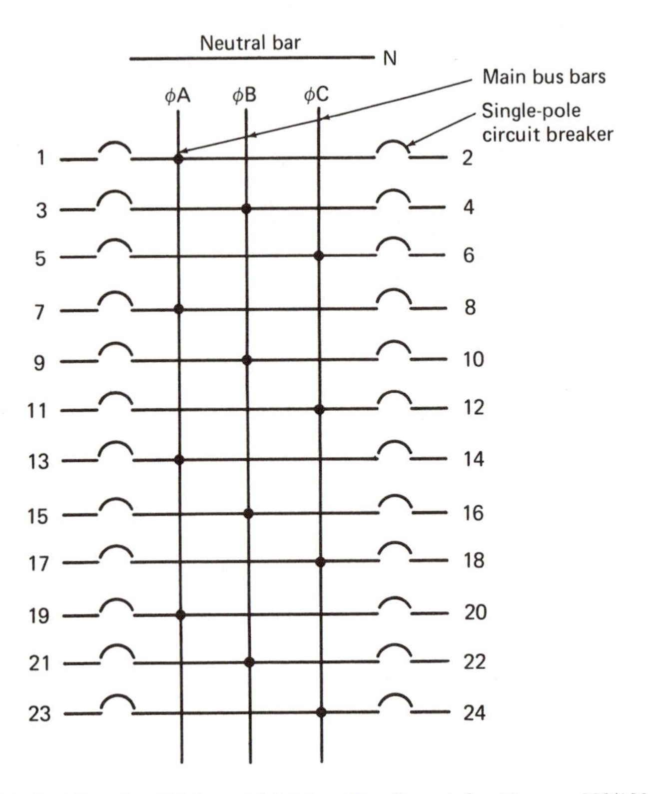

Figure 3-30 Typical Interior Wiring of Lighting Panelboard for Use on 208/120 V, Three-Phase, Four-Wire Service

A comparison with the panelboard schedule shown in Fig. 3-29 and the internal wiring shown in Fig. 3-30 will show that circuits 1 and 2 are connected to phase A, 3 and 4 to phase B, and 5 and 6 to phase C, etc.

Motor Control Centers

Motor control centers are also a prefabricated item of equipment that are similar to the unit substation in that many of the elements are of standard manufacture and the total unit is assembled to the custom requirements of the job. These packaged assemblies were a natural growth development by manufacturers. They recognized that a piece of equipment should be provided that would overcome the need to mount and wire individual motor starters in built-up arrangements. In the latter case, metal wireways were generally required and wiring had to be made from the feeders into the starters with all the attendant problems of connections, taping of joints, etc.

The motor control centers identified as MCC LM-1 and MCC LM-2 for the office building are shown in one-line form in Fig. 3-31. A front view of these units is shown in Fig. 3-32. Note on the one-line diagram that the individual starters and other devices have disconnect device-type connections that

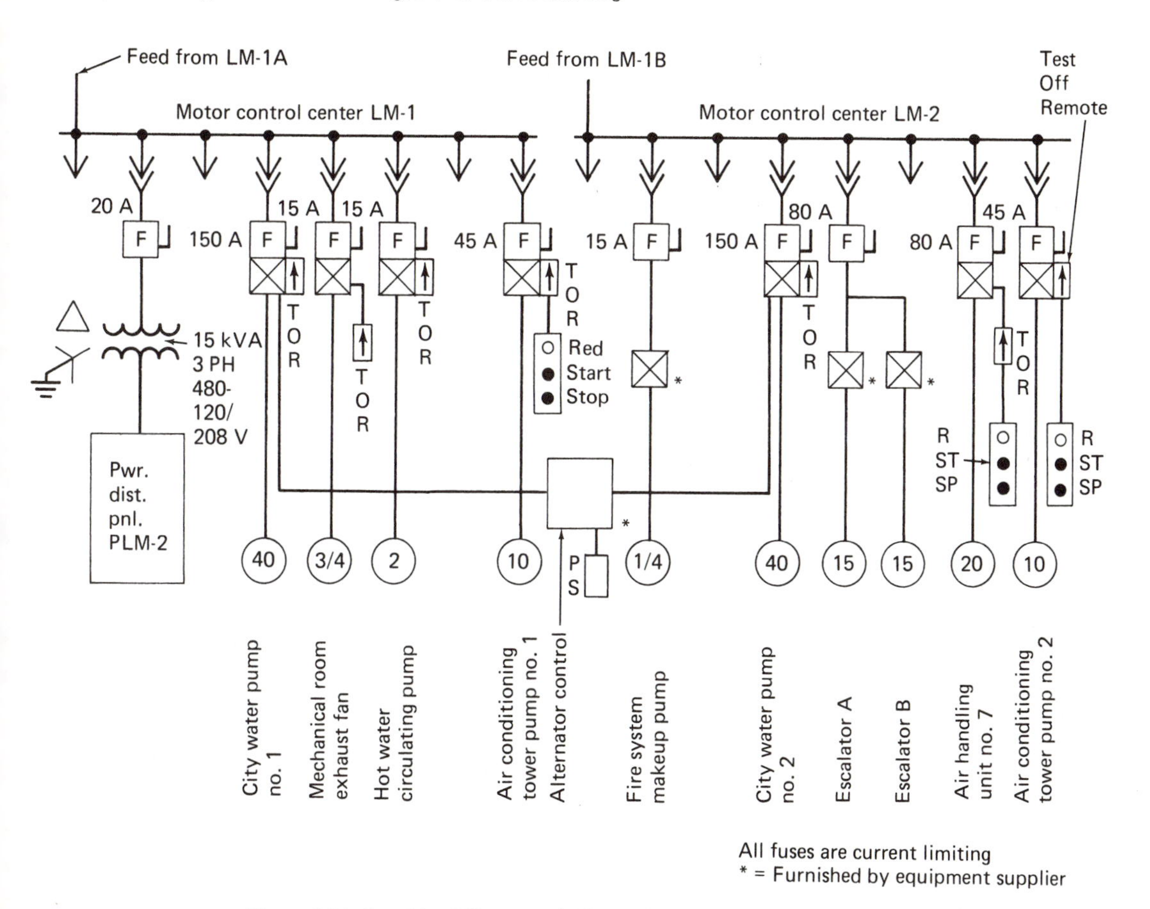

Figure 3-31 One-Line Diagram of Motor Control Centers LM-1 and LM-2 (Portion of Fig. 3-38)

allow for removing the units from the framework for replacement and also provide a means of adding to the motor control center at a later date in spaces provided for future use.

Motor control centers can be located where they best serve the loads connected to them. They do not need to be enclosed in a room; however, they often are because then unauthorized persons cannot operate switches, etc. Branch circuits are run from these centers to the individual motors or other loads making a neat and well-organized installation.

Plug-In Bus Duct

Plug-in bus duct is used as a means of supplying power up (or down) through the building that allows for ready access to a power source on each floor.

Plug-in bus duct is made up of copper or aluminum bus bars supported on insulators enclosed within a metal housing. The bus duct is normally supplied in standard 10-ft sections that bolt together to form a continuous bus of the

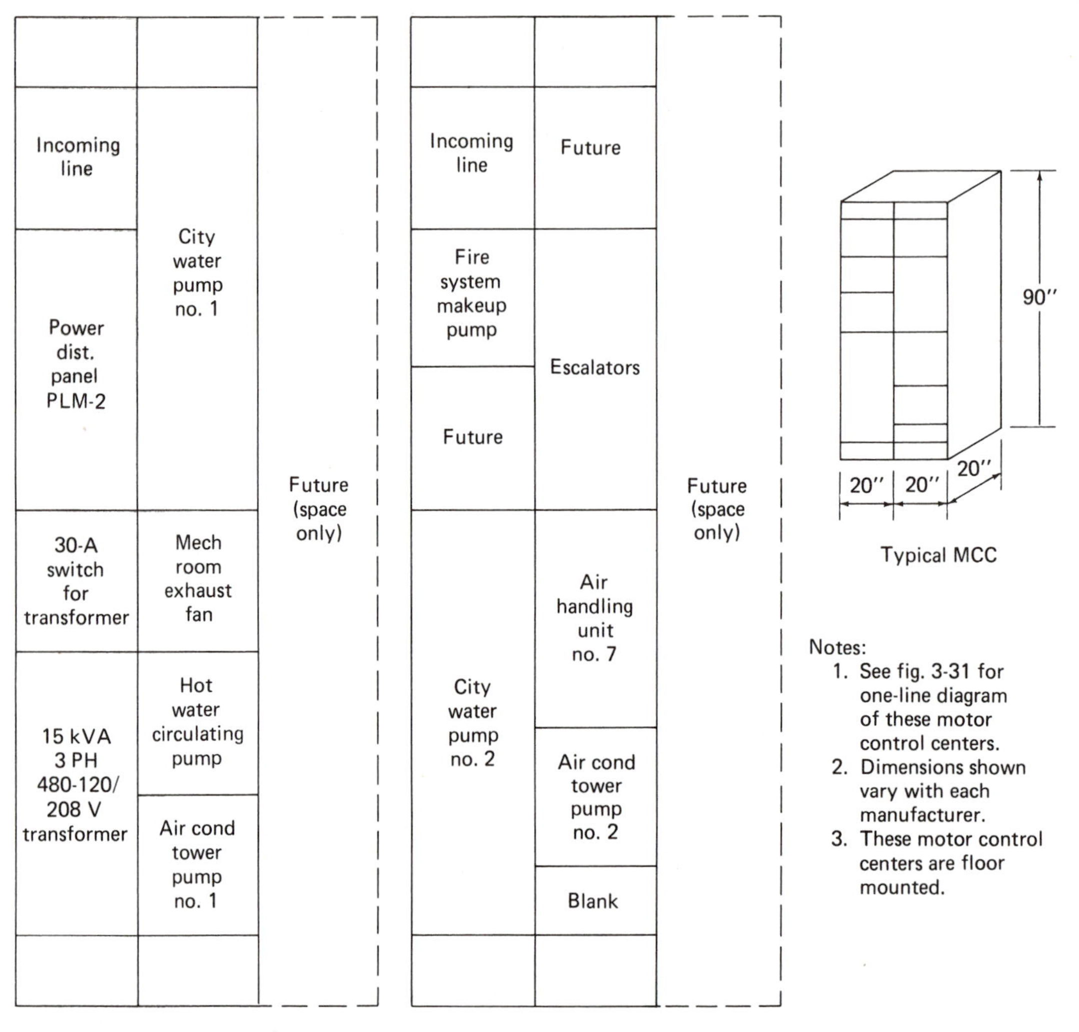

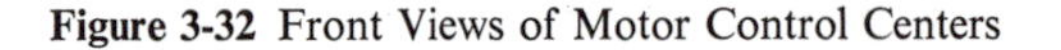

Figure 3-32 Front Views of Motor Control Centers

length required. Busses generally range in rating for plug-in-type applications from 225 to 1,000 A. Fittings are available for making changes in direction such as elbows, tees, and crosses. Tap boxes, end feed boxes, switchboard stubs, and a variety of hangers are manufactured as well. This line of prefabricated bus duct allows the engineer and designer a safe, flexible means of power distribution that can be tapped at set intervals along its length and thus provides for further distribution through panelboards, switches, motor control centers, etc.

The taps are made by plugging in switches with finger-like contacts onto the bus through *doors* in the side of the bus enclosure. The doors are normally closed but can be readily opened by swinging the plate aside or through some other means and *stabbing* the switch in place. Figure 3-33 shows a typical cross section of the duct and a switch ready to plug in onto the bus bars.

This type of plug-in bus duct was run vertically in the office building as indicated in Fig. 3-20. It passed through electrical closets on each floor. At each floor, switches were plugged in to tap the bus and make possible further distribution of power. A typical bus connection at the first-floor closet is shown diagrammatically in Fig. 3-34.

Bus duct is used extensively in industrial plants where machines may be

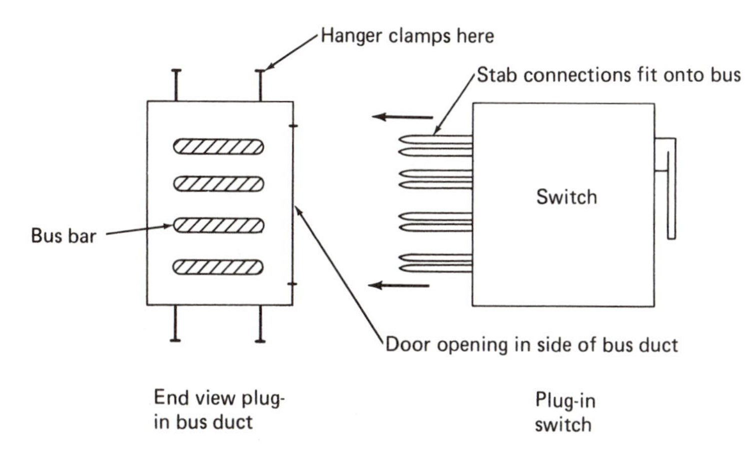

Figure 3-33 Scheme Used to Plug Switches into Plug-In Bus Duct

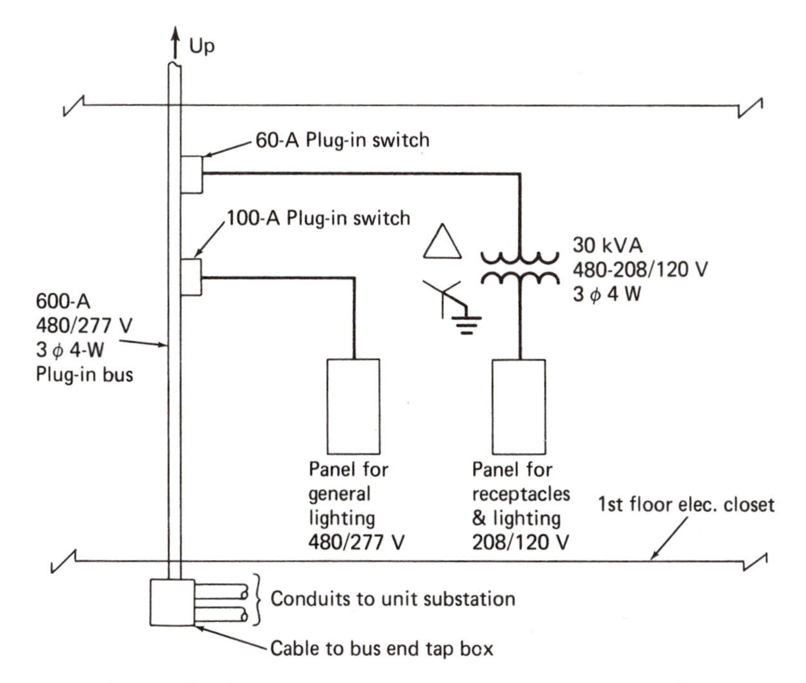

Figure 3-34 Diagram Showing Scheme of Plugging in to Bus Duct at Electrical Closets

relocated frequently and the plugging in or unplugging provides a ready means of securing power along the route of the bus duct.

Unit Substation for Power over 600 V

The unit substation for power over 600 V is sometimes required for large motor loads as is the case for the air conditioning compressors (also called

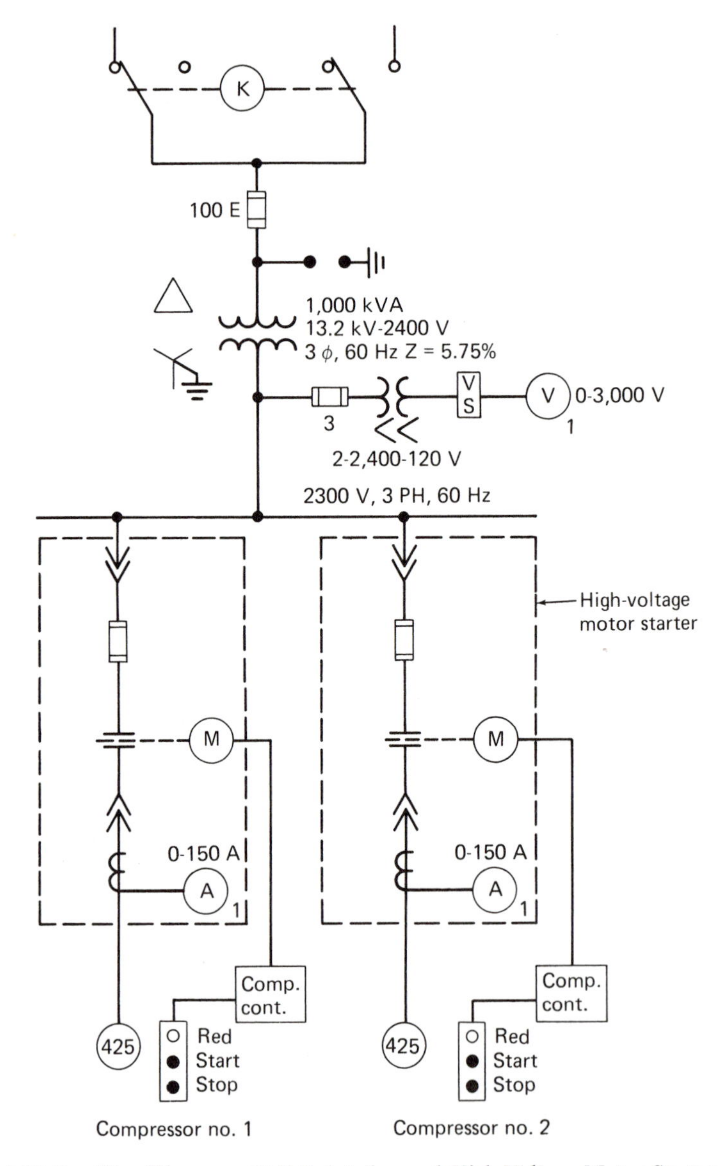

Figure 3-35 One-Line Diagram—Unit Substation and High-Voltage Motor Starters for Air Conditioning Compressors

chillers) in the office building. The motors driving the compressors were rated approximately 425 hp and two compressors were required. An analysis of the costs involved proved that the motor and necessary switchgear could be installed at a lower cost if the motors were rated 2,300 V, three phase rather than 480 V, three phase.

Motor starters were combined with the unit substation to form a package unit that could be installed in the penthouse electrical equipment room. The one-line diagram for this equipment is shown in Fig. 3-35. The front view of the unit substation with the high-voltage motor starters is shown in Fig. 3-36. These starters are provided with high-voltage fuses and a motor starter together with all the necessary controls, housed in a completely enclosed structure. Naturally the arrangement will vary with the particular manufacturer's design. This type of equipment is often made with "drawout" provisions so that the fuses and operating devices can be withdrawn from the enclosure for maintenance purposes.

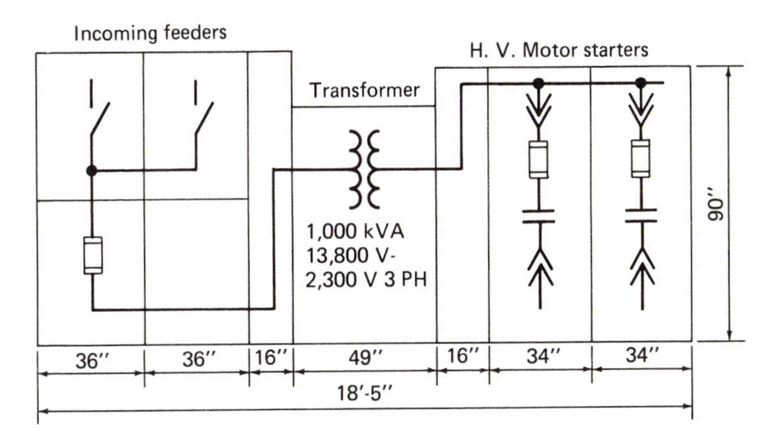

Figure 3-36 Front View and One-Line Diagram of Unit Substation and High Voltage Motor Starters

Legend Showing the Symbols

A Legend showing the symbols used and their identification is required if a standard legend sheet has not been used (see sec. 3-6). For the office building, one-line diagram, the symbols used are shown in Fig. 3-37.

Notes required to clarify further the symbols or to give information that would be too extensive to show on the body of the drawing are generally lettered on the drawing in the lower right-hand corner. Necessary notes for the office building one line are shown in Fig. 3-38.

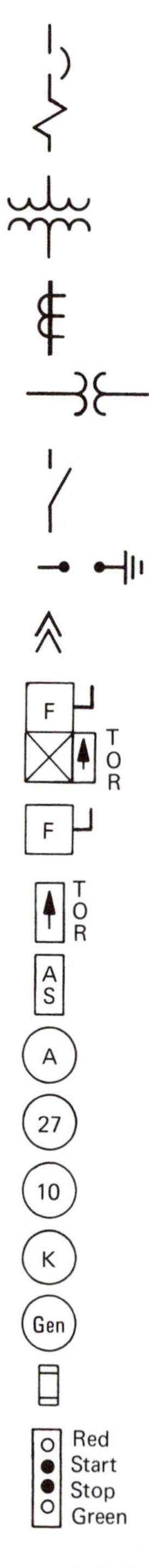

Air circuit breaker with series trip device
250A = coil rating; 100% LTD = long time delay setting; 400% STD = short time delay settings; 1000% inst. = instantaneous setting. See note.

Power transformer

Current transformer

Potential transformer

Disconnect switch

Lightning arrester

Disconnecting device

Combination fused disconnect switch and magnetic motor starter. TOR = cover mounted "test-off-remote" selector switch.

Fused safety switch

Selector switch - "test-off-remove"

AS = ammeter switch; VS = voltmeter switch

A = ammeter; V = voltmeter

Relay - 27 = undervoltage; 51 = phase, time overcurrent
51N = ground, time overcurrent

Motor—numeral indicates horsepower

Key interlock

Generator

Fuse

Pushbutton station with indicating lights

Figure 3-37 Symbols for Office Building One-Line Diagram (Fig. 3-38)

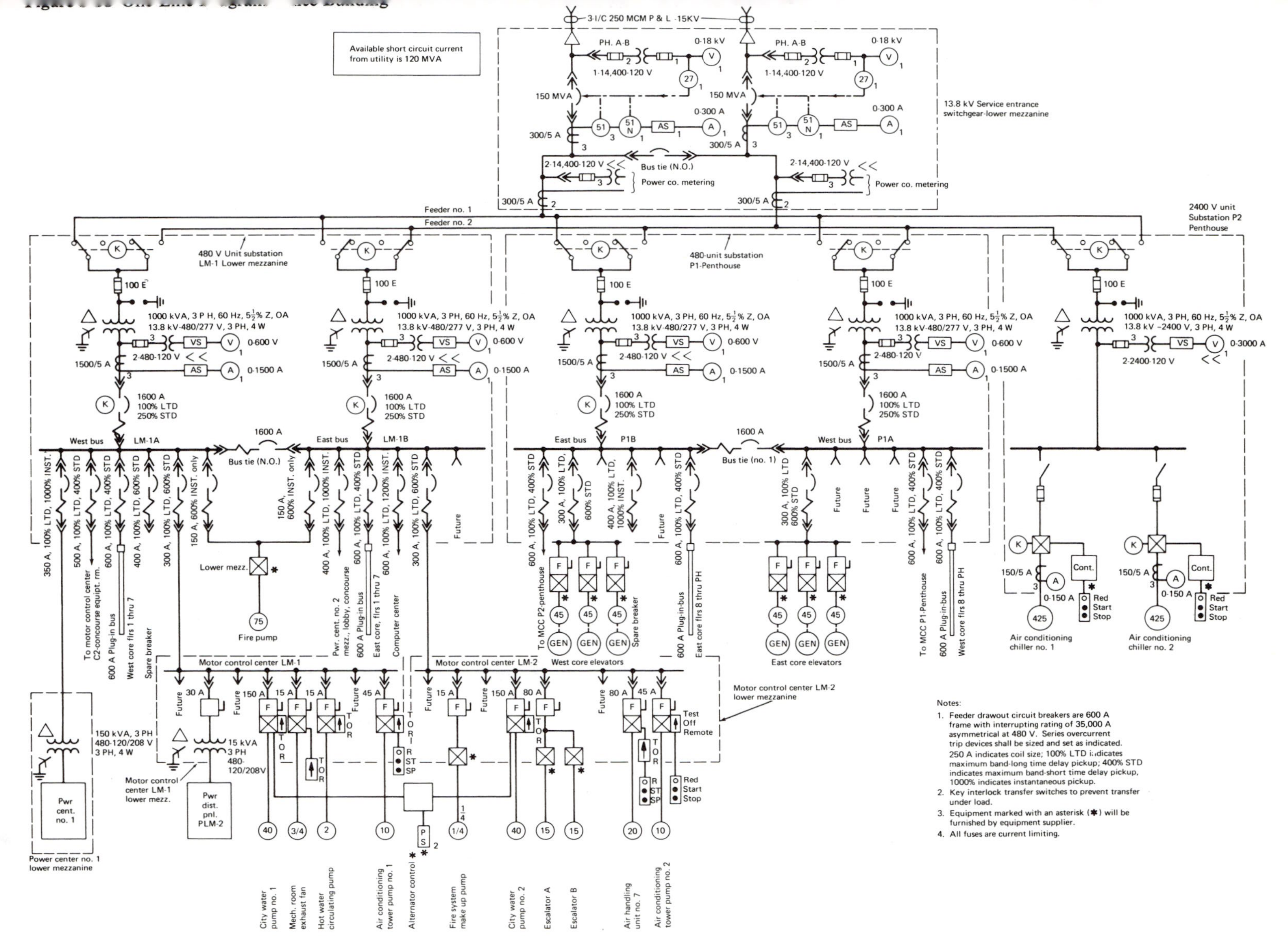

Notes:

1. Feeder drawout circuit breakers are 600 A frame with interrupting rating of 35,000 A asymmetrical at 480 V. Series overcurrent trip devices shall be sized and set as indicated. 250 A indicates coil size; 100% LTD indicates maximum band-long time delay pickup; 400% STD indicates maximum band-short time delay pickup, 1000% indicates instantaneous pickup.
2. Key interlock transfer switches to prevent transfer under load.
3. Equipment marked with an asterisk (✱) will be furnished by equipment supplier.
4. All fuses are current limiting.

Chapter Review Topics and Exercises

3-1 Why are one-line diagrams drawn?

3-2 Describe a radial-type distribution system. What is its principle disadvantage?

3-3 How does the primary selective system vary from the radial distribution system? Is this a better system? Why?

3-4 Describe the secondary selective system and its advantages over the primary radial system.

3-5 How does a spot network system compare with a secondary selective system in terms of reliability of service?

3-6 Referring to Fig. 3-38, what is the available short-circuit current from the utility?

3-7 In Fig. 3-38 the load center transformers are 1,000 kVA, oil immersed, and air cooled. How could their capability be increased?

3-8 Why are frame sizes indicated on some circuit breakers?

3-9 System design requires that a 480-V air circuit breaker have a 600-A continuous current rating, an interrupting capability of 60,000 A, asymmetrical. What frame size is required?

3-10 Fuses vary in their interrupting capability. How is the proper interrupting rating covered on the drawing?

3-11 What device is used to supply a low-voltage signal to meters if the main line voltage is 13.8 kV or something similar?

3-12 What device is used to provide a small current signal to meters, etc., if the line current is, say, in the order of 600 A?

3-13 Why are standard symbols used on drawings?

3-14 Are physical relationships on one-line diagrams important? Explain.

3-15 Why is duplication of data to be avoided?

3-16 How does it help to show future requirements?

3-17 What function does a pothead serve?

3-18 Describe a key interlock.

3-19 Why are lightning arresters sometimes used on indoor switchgear?

3-20 Describe a unit substation.

3-21 If the principal distribution systems in a building is 480/277 V, three phase, four wire, how can 120-V service be supplied for small motors and other 120-V loads?

3-22 Sketch a typical lighting panelboard bus and circuit breaker connection diagram for three-phase supply and single-pole breakers.

3-23 Describe a motor control center.

3-24 Why does plug-in bus make an excellent form of distribution?

3-25 Sketch symbols of the following:

(a) Current transformer rated 300/5 A
(b) Delta phasor diagram
(c) Wye with center point grounded
(d) Key interlock
(e) Pothead
(f) Open delta phasor diagram
(g) Air circuit breaker with disconnect devices and series trip coil
(h) Potential transformer—13.8 kV to 120 V complete with fuses
(i) Fused safety switch with 20-A time delay fuses
(j) Start–stop push button with red indicating light

3-26 Draft the one-line diagram of Fig. 3-38 on a sheet of drafting paper or cloth approximately 17″ × 22″. Use $\frac{1}{8}$-in. grid paper under the sheet if available and make lettering $\frac{1}{8}$-in. high. Plan ahead by working out space requirements before doing final line work. (It will all fit on the sheet nicely.)

4

ONE-LINE AND RELAY DIAGRAM

4-1 Purpose

The one-line and relay diagram, for a power system, shows the principal power elements, the protective relay requirements, and the major metering devices. Figure 4-1 is an example of such a diagram for generation of power from a large turbine generator. This diagram has been abbreviated to some degree due to space limitations; however, the essential elements for understanding its purpose are shown.

4-2 The Main Turbine Generator

To understand the diagram better, sections can be considered individually. The main turbine generator is the source of the power generation and is shown in Fig. 4-2.

This symbol and related data indicate several things. The coils of the machine are to be connected in wye as indicated by the upside down Y in the center of the circle. With rotation in a counterclockwise direction, the electromotive force will rise first in phase 1, second in phase 2, and third in phase 3. Phase rotation information is important in determining connections into other systems for the power generated by this machine.

The turbine generator is number one in the station so it may be assumed that installation of this unit is the first at a new station site. *Turbo* is an abbreviation for the word *turbine* and indicates that the generator is probably driven by a steam turbine. (Water-driven turbines are generally not so large in rating.) Nameplate output rating is 511.4 MVA, which at 90% power factor is 460 MW. Thinking of this value in other ways,

$$460 \text{ megawatts} = 460{,}000 \text{ kilowatts}$$

$$460{,}000 \text{ kilowatts} = 460{,}000{,}000 \text{ watts}$$

Voltage generated at the terminals of the machine is 22 kV (22,000 V). The machine is wound for three-phase generation at 60 Hz (60 cps). Speed is 3,600 rpm.

Figure 4-1 One-Line and Relay Diagram

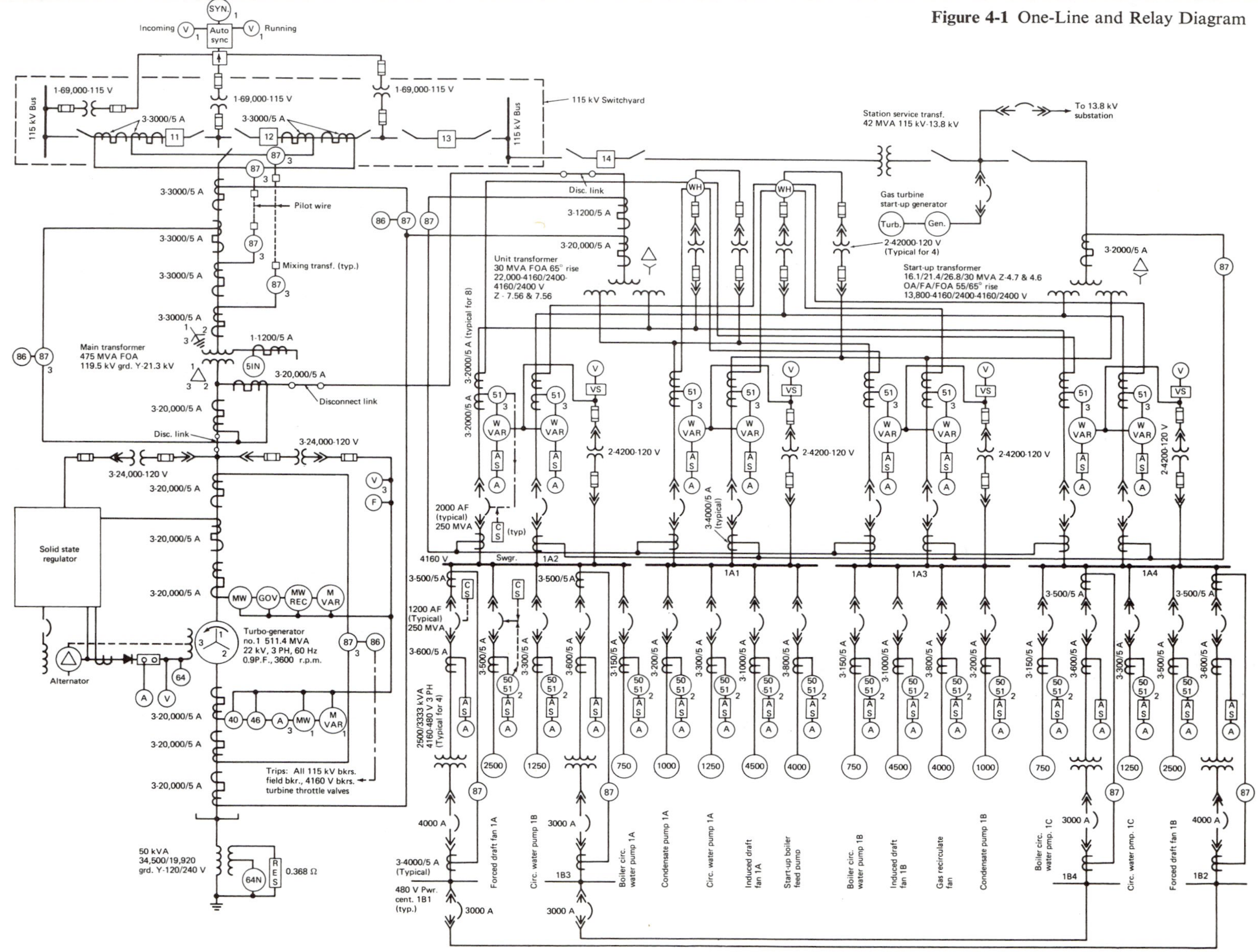

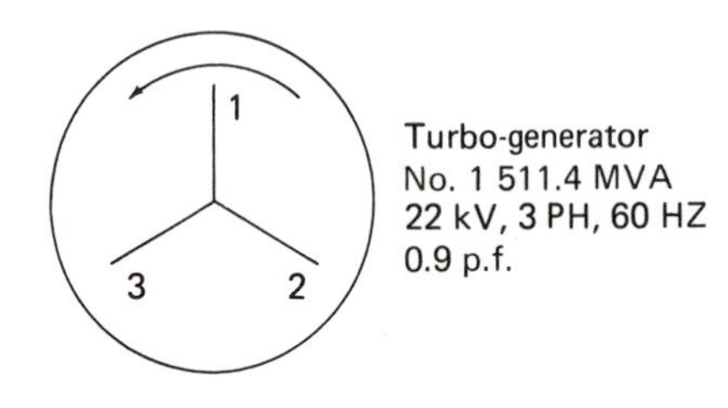

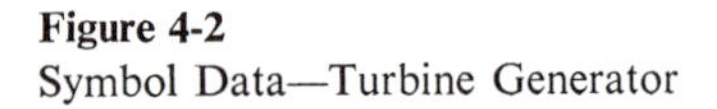
Figure 4-2
Symbol Data—Turbine Generator

4-3 The Generator Neutral Connection

With a wye-connected machine, the midpoint of the wye is generally grounded in some manner. To understand how this is accomplished, it is first necessary to have some knowledge of how the ends of the machine windings are terminated outside the machine casing. Figure 4-3 portrays location of the terminals as viewed both from the end and from the side of the underside of the machine. In addition, a diagrammatic view is shown to illustrate the coil connections.

With the machine windings terminating, as shown in Fig. 4-3, the ends of all coils can be joined in one common connection by bolting a conductor in place across either all three front or all three back terminals. Figure 4-4 shows the results of this arrangement using the back terminals.

The neutral connection is shown in one-line diagram form in Fig. 4-5. Proceeding toward the ground connection ( ) from the generator symbol, the first main line element is the common connection joining the coil ends, as described above ( ). Beyond this point only one conductor is required so the one-line diagram, in this case, is a true representation of the actual condition. One side of the primary winding of a distribution transformer is connected to this line and the other side of the transformer is connected to a grounding terminal in the station.

If the insulation on a machine winding were to fail and a live conductor came in contact with the casing of the machine, which is grounded, current would flow in the neutral conductor through the transformer primary winding. The resistor connected to the transformer secondary winding, as shown, will help to limit the amount of current that will flow to ground. Some machine neutral grounding connections are made using only a resistor in series with the neutral lead instead of the distribution transformer. The purpose of the resistor is the same in either case.

Three sets of bushing-type current transformers are shown in the lines between the machine and the common connection. Because these transformers are located between the machine casing and the common connection, they must be made to fit around the insulator bushings. For that reason, they are called *bushing-type* current transformers and are often called donut CT's. Figure 4-6 shows a typical transformer and its location on the bushings of the machine.

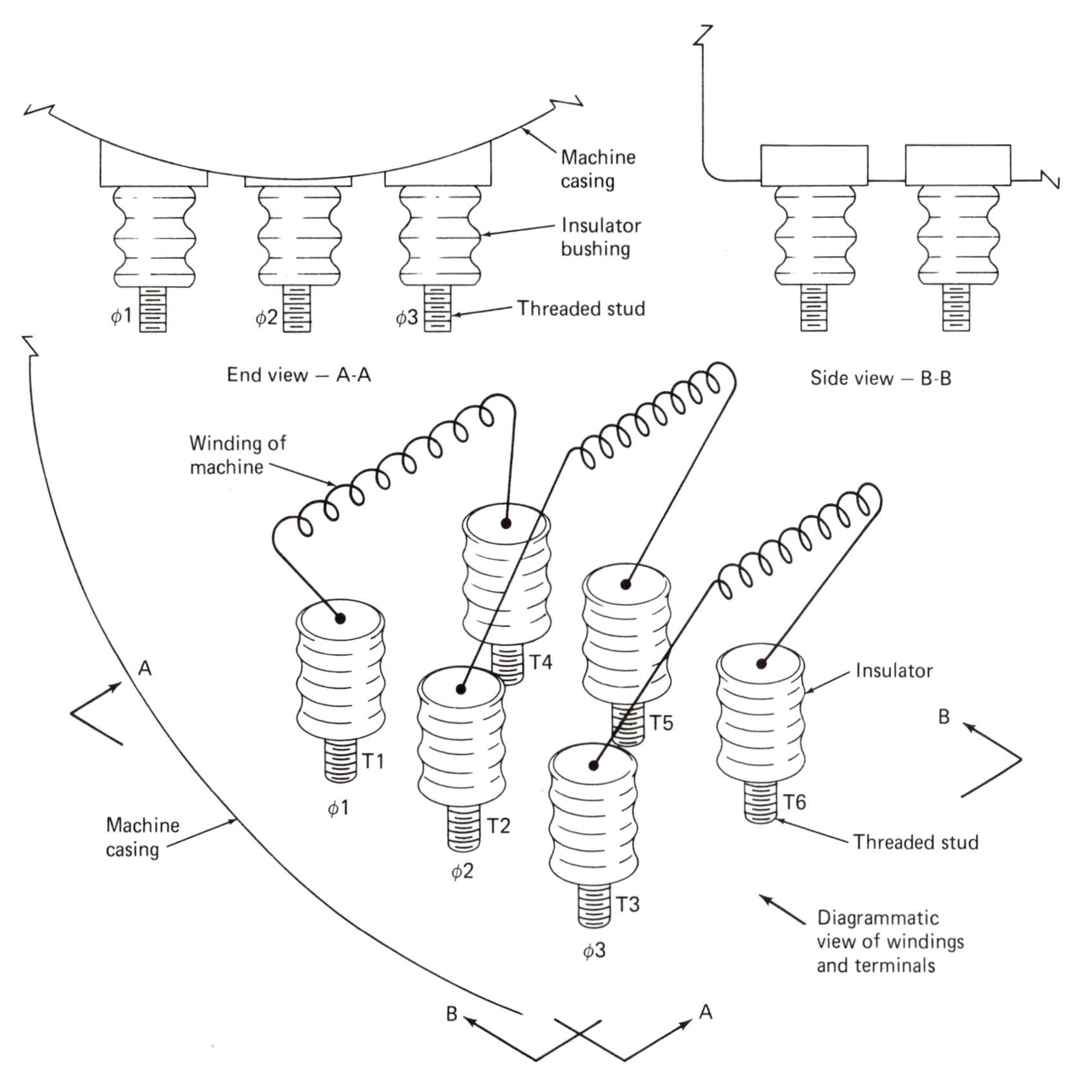

Figure 4-3 Termination of Machine Windings on Outside of Machine Casing

These particular current transformers are rated 20,000/5 A, which means that 20,000 A flowing in the main power circuit will produce 5 A flowing in the transformer winding. There is no direct connection between the main line and the winding of the transformer; therefore, the induced current in the winding is caused by the magnetic flux surrounding the main conductor. As the flux rises and falls, with the 60-Hz current generation, the magnetic lines cut across the winding of the transformer to induce the necessary low-level current for instrumentation.

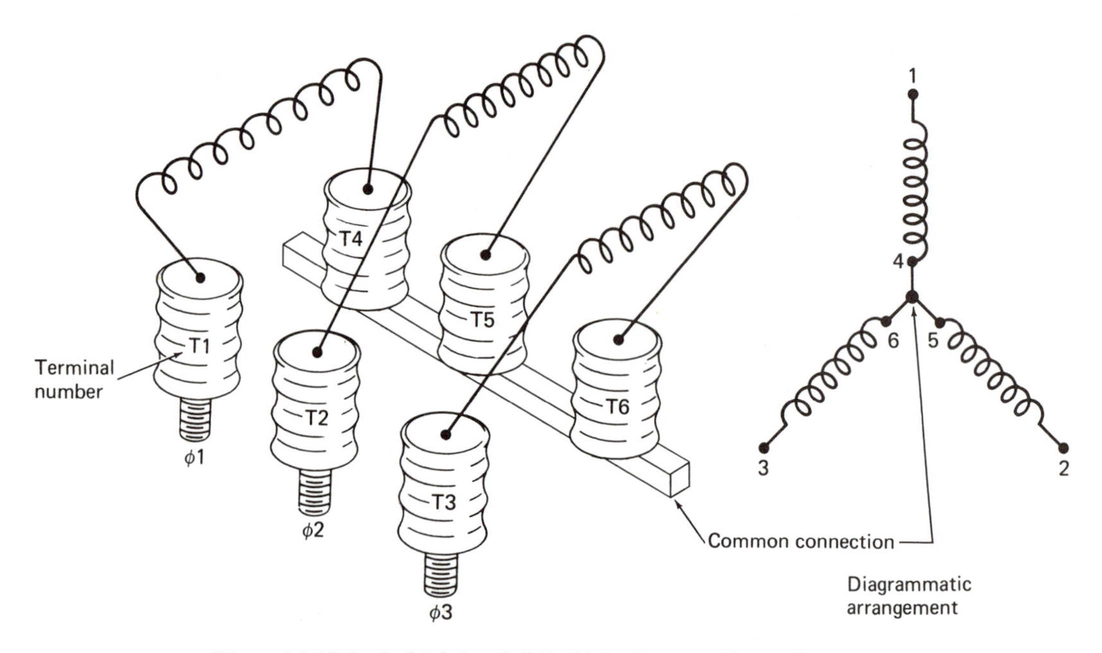

Figure 4-4 Method of Joining Coil End into Common Connection

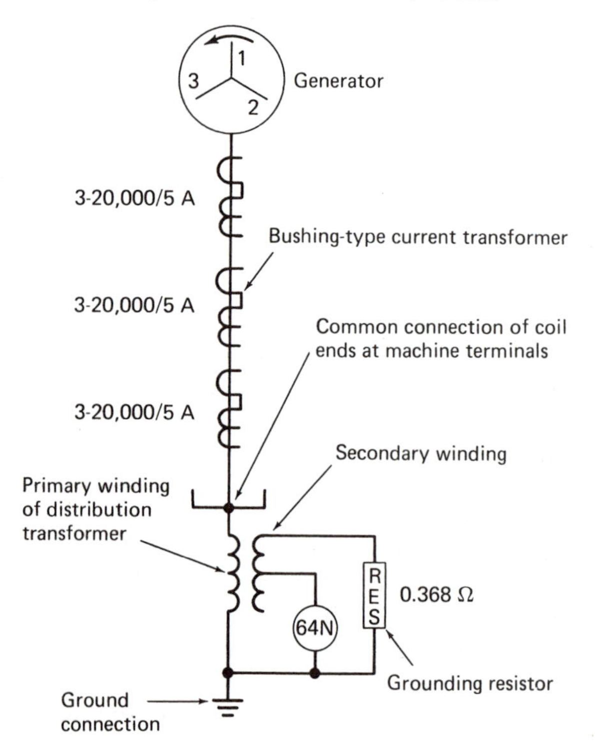

Figure 4-5 One-Line Diagram of Generator Neutral Connection

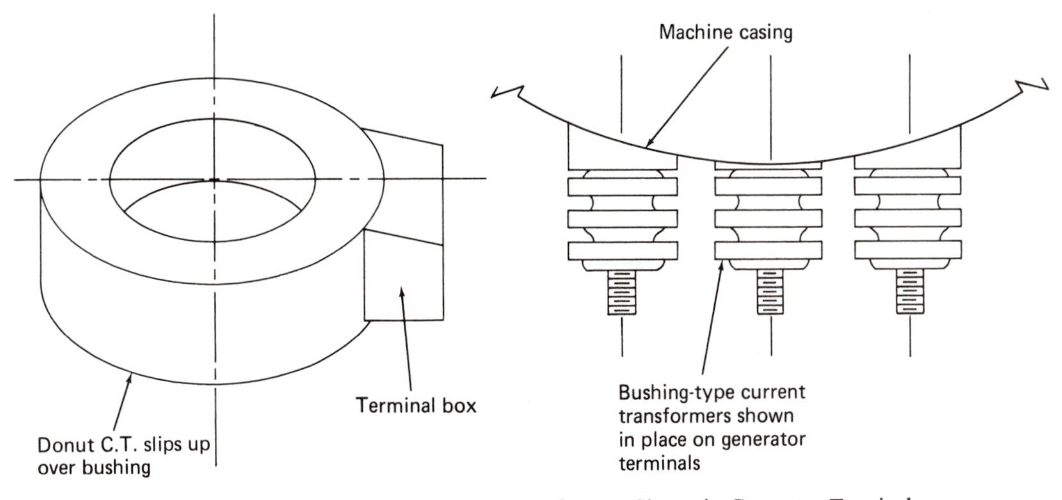

Figure 4-6 Bushing-Type Current Transformers Shown in Generator Terminals

4-4 The Main Power Circuit

The main power circuit of the diagram shown in Fig. 4-1 extends from the generator through the main transformer to the 115-kV switchyard. Circuit breakers in the switchyard control power distribution into the rest of the system. This portion of the one-line diagram is shown in Fig. 4-7.

Bushing-type current transformers are installed on the main line bushings of the machine in the same way that they were put on the bushings for the neutral connection. Bushing-type transformers are also installed at the transformer and in the switchyard.

The neutral tap of the 115-kV winding of the transformer is grounded directly with a bushing-type current transformer installed at the transformer for relaying purposes.

Connection between the generator and the main transformer is made using isolated phase bus. A pictorial view of this connection is shown in Fig. 4-8.

The isolated phase bus is furnished with a short section of bus, which can be unbolted and removed. This is called a *disconnect link* and is furnished so the generator can be disconnected from the transformer.

Also included in this portion of the one-line diagram are the phasor symbols for indicating how the transformer coils are to be connected. The 22-kV coils are connected delta (Δ) and the 115-kV coils are connected wye (Y). Figure 4-9 shows the complete line and coil connections and voltage levels.

Circuit breakers in the 115-kV switchyard are connected in what is called a *breaker-and-a-half* scheme. This provides a great deal of flexibility in control

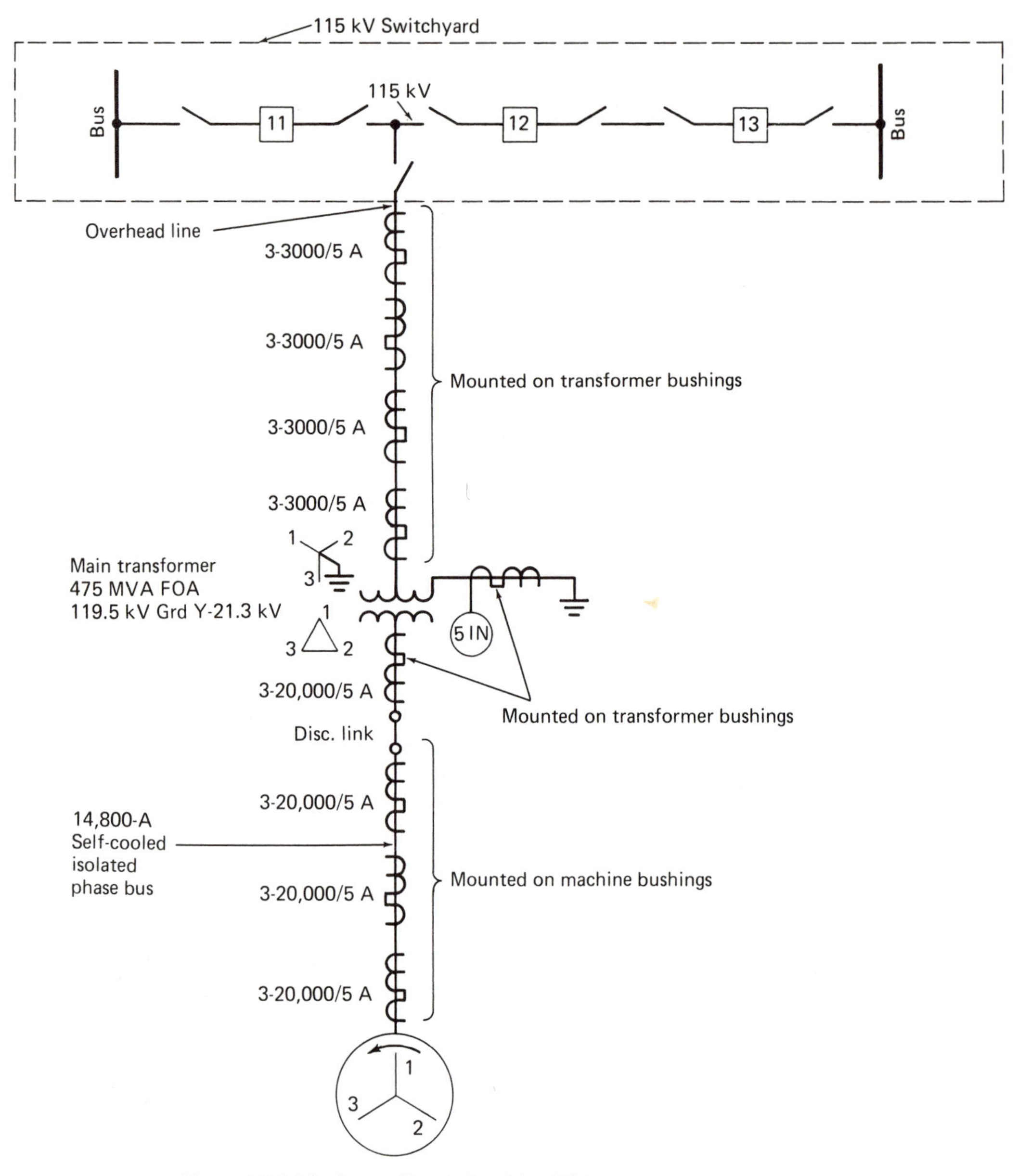

Figure 4-7 Main Power Circuit One-Line Diagram

of the lines feeding out from the switchyard so breakers can be isolated out of service for maintenance. The basic arrangement is shown in Fig. 4-10.

By switching of circuit breakers, any one breaker can be taken out of the circuit without disrupting service on the system. Note that there are disconnect switches on either side of each breaker to allow for isolating the breaker.

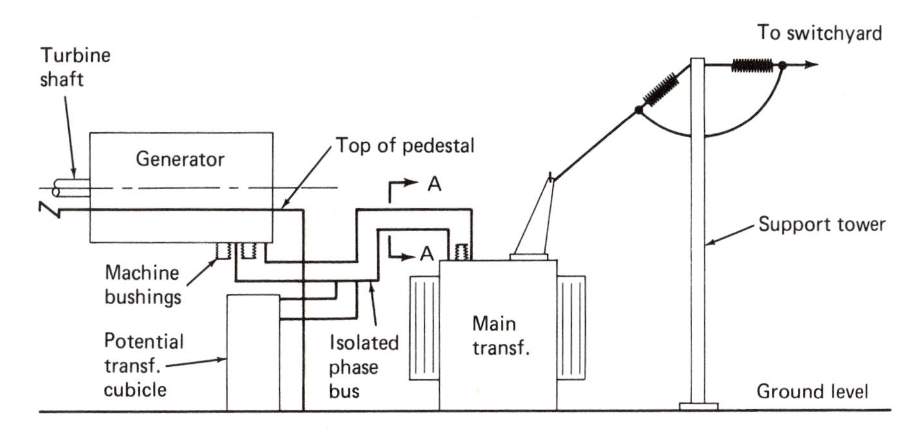

Connection of generator to main transformer

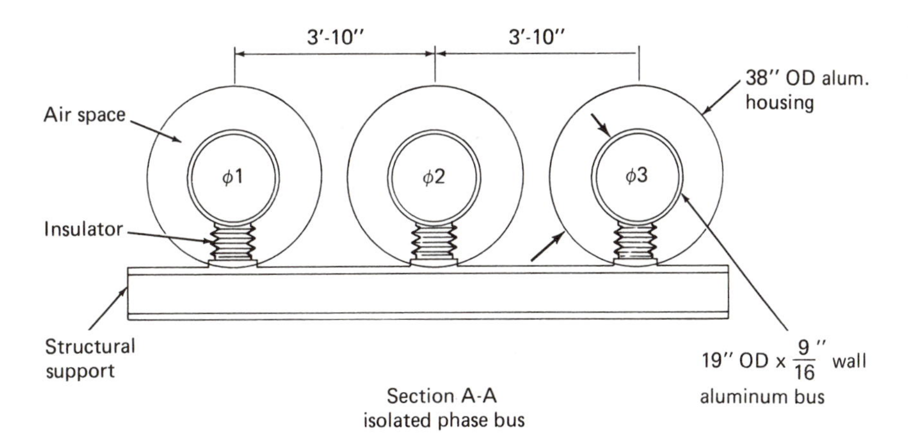

Section A-A
isolated phase bus

Figure 4-8 Isolated Phase Bus Connection between Generator and Transformer

4-5 The Start-Up Circuit

In order to start the turbine generator rolling, it is necessary to get steam up in the boiler and from the boiler to the turbine section of the machine. The boiler cannot produce steam until fuel is pumped to the burners, fans are started, pumps are started, etc. To provide power for these loads, an outside source of power must be brought to the station. This can take several forms. In the one-line diagram of Fig. 4-1, power can be provided from three different sources. First, there is a start-up gas turbine generator provided. The generator in this unit is driven by a gas turbine similar to a jet airplane engine. It starts on batteries and will supply sufficient power to get station auxiliaries running. There is also a tie to a remote 13.8-kV substation, which will provide start-up power, and there is a tie back into the 115-kV switchyard so that start-up power can be taken from this source. If the generator in Fig. 4-1 were not the first

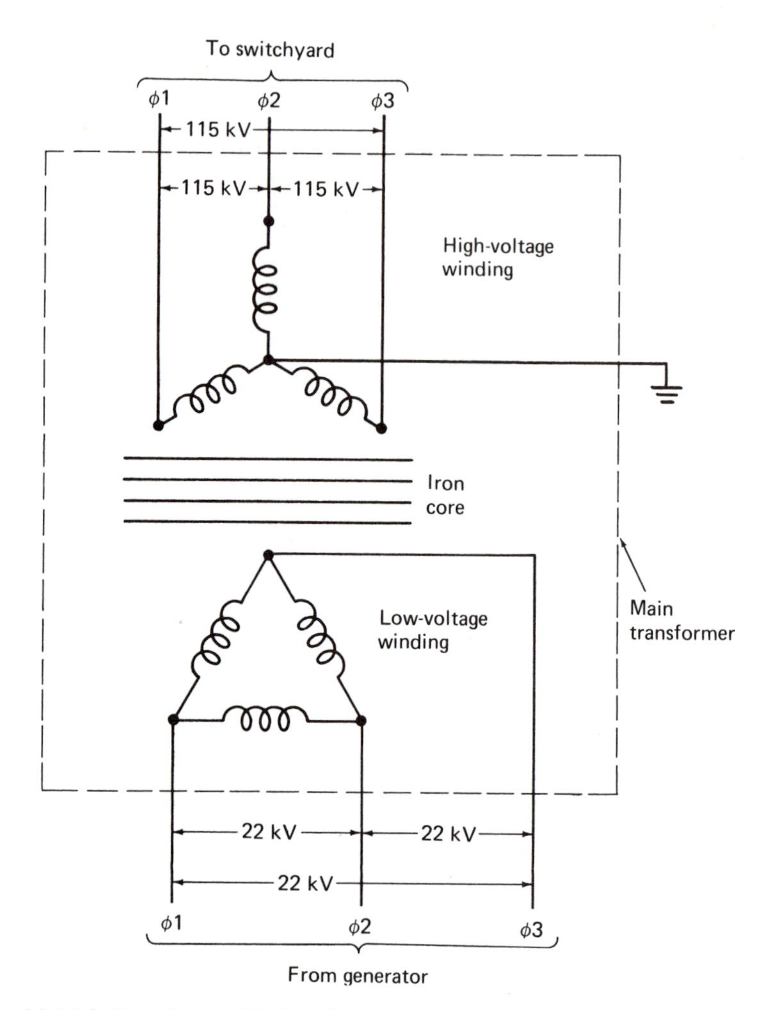

Figure 4-9 Main Transformer Winding Connections

unit in the station, power could be taken from the other units; but in any arrangement for start-up some outside source must be provided to get the first unit going. Figure 4-11 shows the start-up power supply section of Fig. 4-1.

Outside power comes to the start-up transformer at 13,800 V, three phase. A transformer is provided in the circuit from the 115-kV bus to reduce the voltage to 13.8 kV from that source. The start-up generator is wound for 13.8 kV and power from the remote substation is at 13.8 kV. The start-up transformer reduces the voltage to 4,160 V, three phase in two separate secondary windings. Feeders from the transformer supply power to all four station service busses, which in turn are the source of power for the several large motors shown in Fig. 4-1. These busses also feed to the primary side of the bank of transformers to reduce the voltage further to 480 V, three phase. Station loads rated at this voltage are fed from the four separate busses provided for this service. Secondary

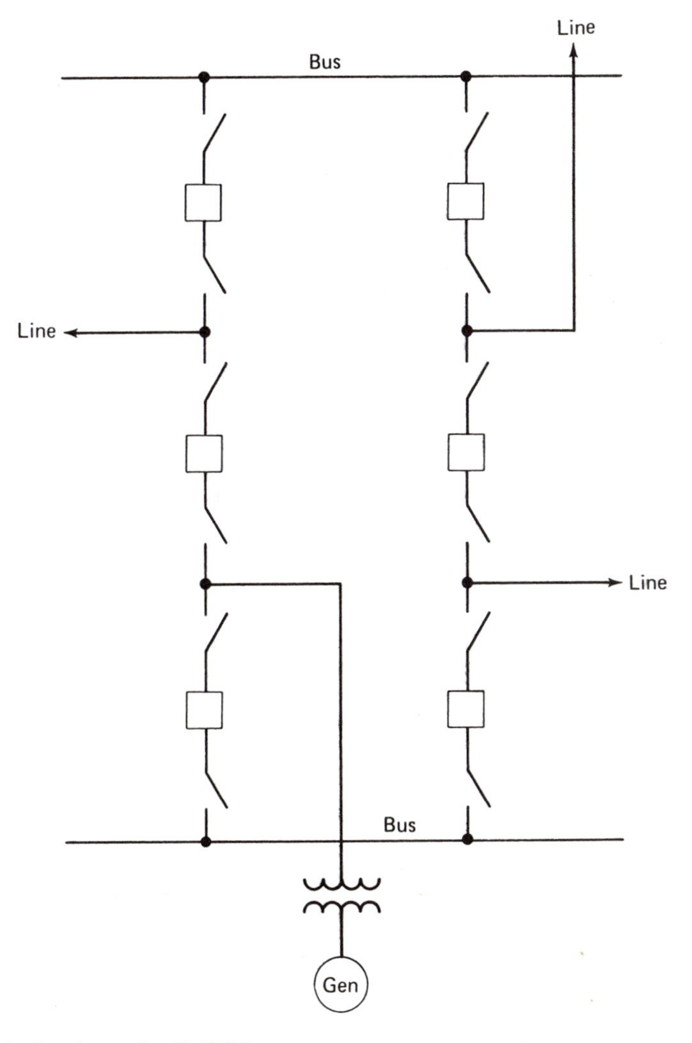

Figure 4-10 Breaker-and-a-Half Scheme

bus ties interconnect the 480-V busses to provide further reliability. This arrangement provides a source of power to all the station loads so the main generator can be started and put on the line.

4-6 The Unit Station Service Circuit

Once the generator is producing power, it can supply its own power to the station service loads necessary to keep the unit running. This is called the *unit station service circuit.* Transfer of the station loads from the start-up power supply to the unit supply is made after the machine is connected into the total system at the 115-kV switchyard (in the diagram of Fig. 4-1). The unit station service circuit is shown in Fig. 4-12.

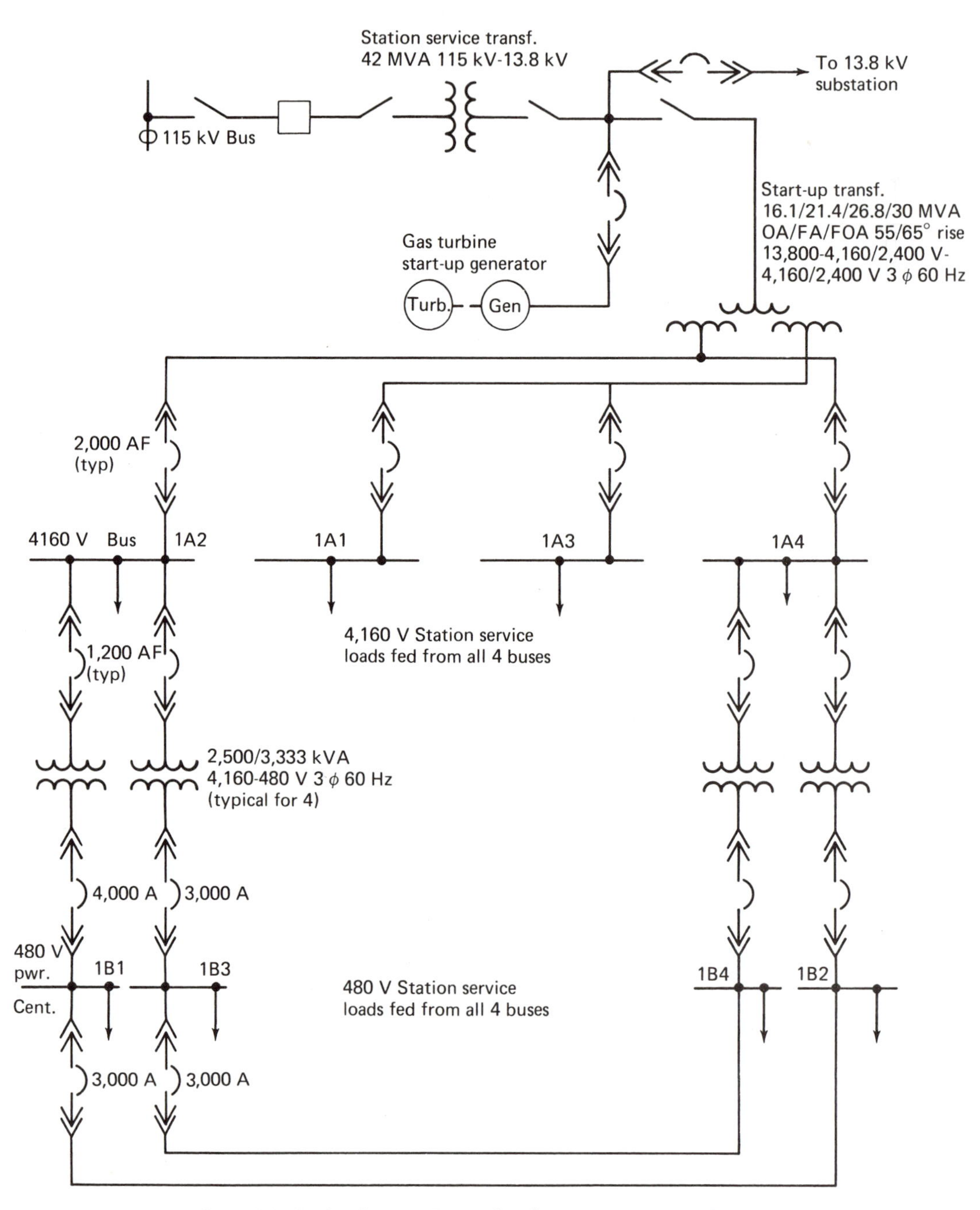

Figure 4-11 Station Start-up Power Supply

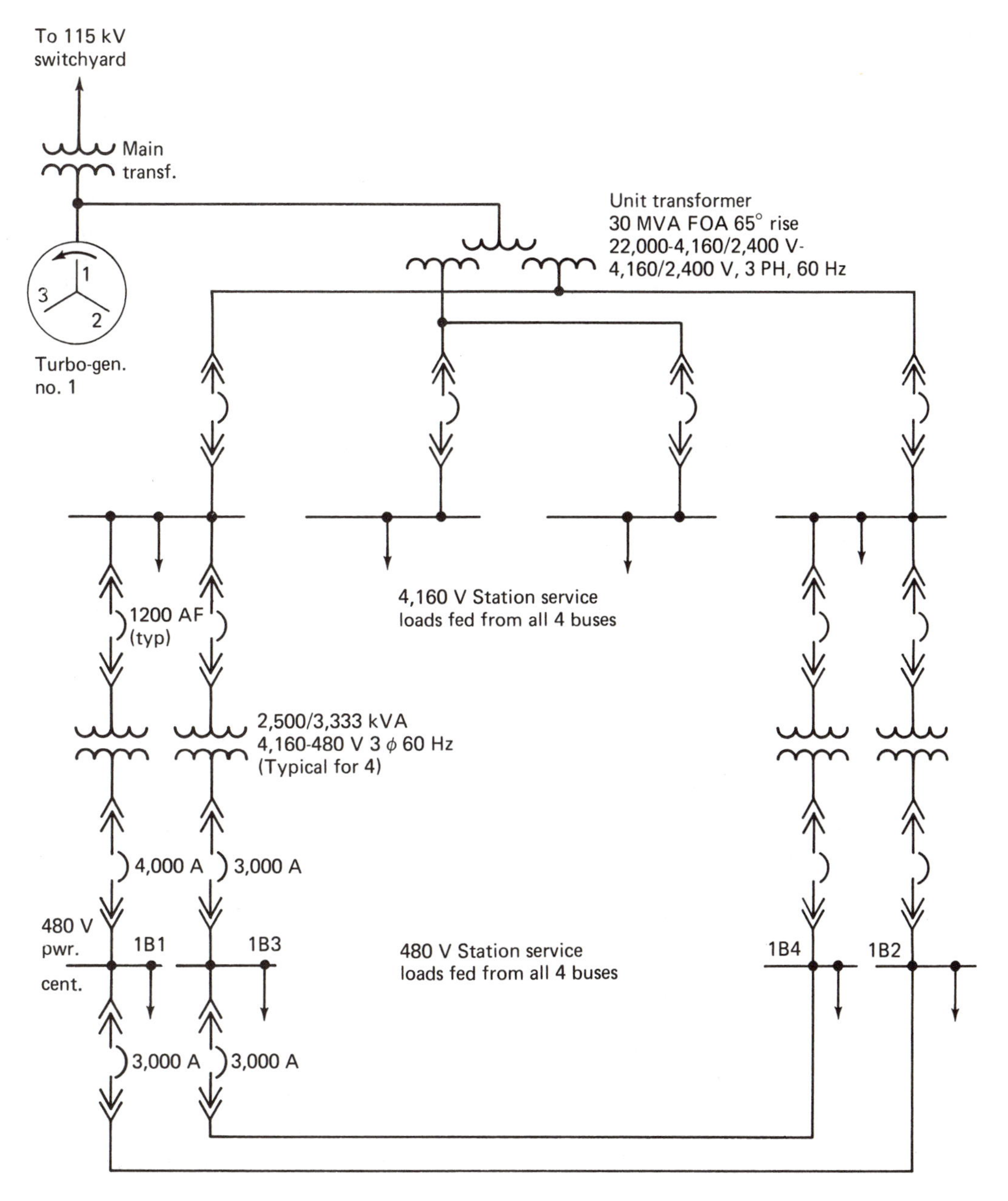

Figure 4-12 Station Unit Power Supply

A connection is made onto the isolated phase bus, on the low-voltage side of the main transformer, to feed the primary winding of the unit transformer. Feeders from the secondary side of the transformer supply power to all four 4,160-V station services busses. These are parallel feeders to those provided from the start-up transformer. The station service power supply from this point on is exactly the same as for the start-up system.

4-7 Station Service Loads

Examination of the loads connected to the 4,160-V busses supplying station power in Fig. 4-1, indicates that some very large motors are required. These motors could have been supplied at either 2,300 or 4,160 V. In this case, the 4,160-V rating was used. Duplicate units are provided in many cases. For instance, there are two forced draft fans, boiler circulating water pumps, condensate pumps, induced draft fans, and three circulating water pumps. These are fed from separate bus sections. This redundancy in equipment is to provide for continuous operation and maintenance of the units. Most of these motors run continuously. They are started by using the circuit breaker as a motor starter rather than supplying separate motor starting equipment. If the motors had frequent starting duty, the circuit breaker might not be used because its contacts might not stand up to the arcing that often accompanies starting.

Smaller station service motors are fed from the 480-V power centers. Another one-line diagram would be developed to show these motors and other loads, including station lighting.

4-8 The Excitation and Regulator Circuit

Excitation of the main generator must be direct current. On a machine of the size shown in Fig. 4-1, the excitation is by means of a revolving field. The armature on this machine is providing the magnetic flux, to sweep across the conductors, mounted in the fixed (stator) part of the machine. The generator needed to supply the excitation power is actually an alternating machine that is connected directly to the shaft of the main generator. The steam turbine, therefore, drives not only the main generator armature but the exciter generator armature as well. Alternating current, from the exciter, is converted to direct current by using a solid state rectifier.

The main generator must maintain proper terminal voltage at all times. As load is put onto the machine, the voltage will tend to change. More or less excitation will be required. A regulator is installed to maintain machine voltage automatically.

The excitation and regulation circuits are shown in Fig. 4-13.

Excitation current is supplied to the main generator field winding from the alternator via the solid state rectifier. Current from the main machine is sensed by the bushing-type transformer and its current signal is sent to the regulator. Voltage in the main line is picked up by the potential transformer and wired to the regulator. Circuits, in the regulator *black box* (too complicated to explain), compare the incoming current and voltage signals to preset requirements and respond to increase or decrease current flow in the alternator field winding. As the alternator field flux is changed, the output of the alternator also changes, which in turn changes flux in the main field winding. This feedback circuit arrangement provides instant response, keeping the main machine voltage properly regulated. Note that the exciter alternator is connected in delta (Δ).

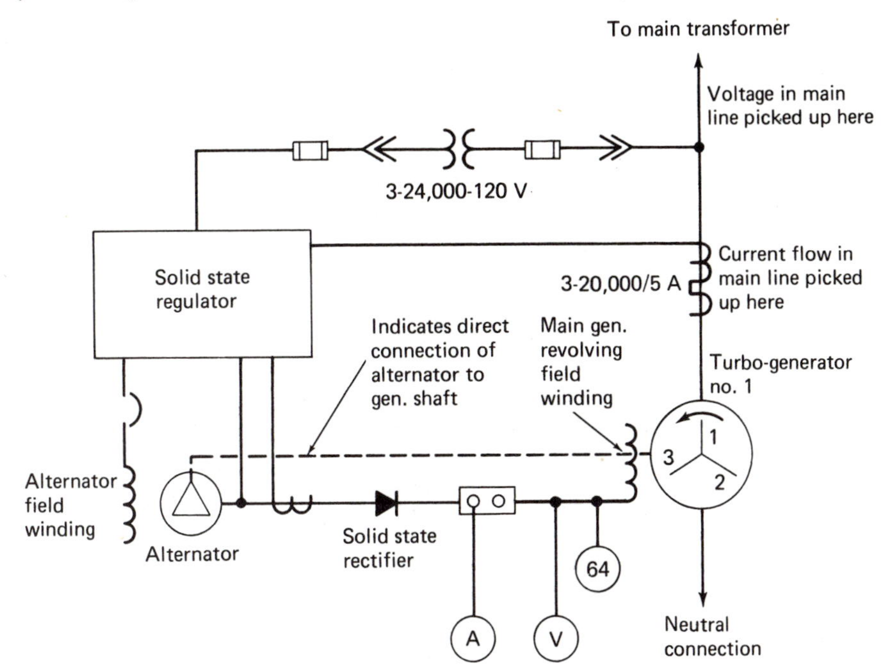

Figure 4-13 The Main Generator Excitation and Regulation Circuits

4-9 The Synchronizing Circuit

Two or more alternating-current generators cannot be put onto the same line and thereby made to operate in parallel unless they satisfy three conditions: (1) their voltages, as registered by a voltmeter, are the same; (2) their frequencies are the same; and (3) the voltage waves are in phase. The last condition is very important because even though frequency and voltage are the same, if the machines are not in phase, an interchange of current will take place between the machines. Once they satisfy the conditions above and are operating together, the machines cannot shift voltage, frequency, or phase.

The elementary principle involved in determing synchronism is indicated in Fig. 4-14.

If the voltage and frequency of generators No. 1 and No. 2 are the same and the machines are in phase, point A will be at the same instantaneous potential as point C and the lamp connected across these points will not light; there is no potential difference to light it. This satisfies the rules of synchronism and the circuit breakers can be closed to operate the machines in parallel. This explanation describes the *dark light* method of synchronism. If the connection at C were made into the same line as the connection at D, then the light would be bright when the machines were in synchronism. The lamp *bright* method requires judgment as to when the lamp is brightest because only at their brightest level are the machines in step. The lamp *dark* method is preferred by many because it is more positive. A broken filament is one drawback to the lamp *dark* method

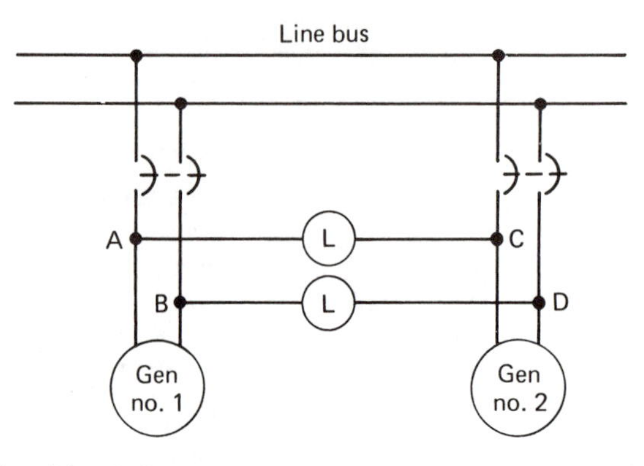

Figure 4-14 Synchronizing Using Lamps

and two lamps in multiple are often used (connections B and D in the diagram of Fig. 4-14 in addition to A and C).

Voltmeters can be substituted for the lamps and when the voltage is at zero between points A to C and B to D, then the machines are in synchronism.

The synchroscope is an instrument that indicates the difference in phase and frequency between the two machines. It shows whether the machine to be synchronized is running fast or slow. Figure 4-15 pictures a typical switchboard-type synchroscope.

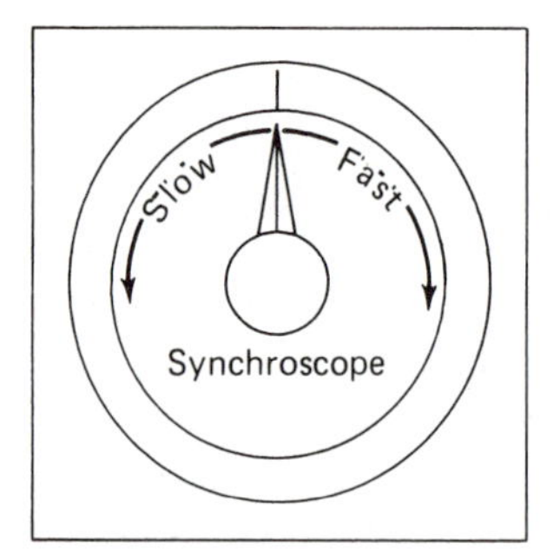

Clockwise rotation of pointer indicates that the incoming machine is operating faster than the running machine; counterclockwise rotation indicates the incoming machine is running slower than the running machine

Figure 4-15 Synchroscope

Synchronizing circuits are turned on by synchronizing selector switches that perform several functions:

1. One contact connects bus voltage to the *running* conductor of the synchronizing bus.
2. One contact connects incoming machine voltage to the *incoming* conductor of the synchronizing bus.
3. One contact completes an *interlock* circuit to the incoming circuit breaker, closing control.

These switches have removable handles so only one synchronizing operation can be made at any one time.

Synchronizing circuits generally operate at approximately 115 V. Where higher line voltages are involved, such as in Fig. 4-1, potential transformers must be used to provide the low-voltage signal for the synchronizing circuit. Figure 4-16 shows the synchronizing circuit portions of the one-line and relay diagram shown in Fig. 4-1.

The synchronizing circuit of Fig. 4-16 includes automatic synchronizing equipment in addition to the voltmeters and synchroscope that are provided for manual operation. The automatic equipment may include a synchronizing relay to give a closing indication to the synchronizing breaker at the correct phase angle, in advance of synchronism, to effect closure of the main breaker contacts at approximately the instant of zero phase difference between the two sources of power, regardless of the frequency difference (within limits). A cutoff relay may be included to prevent breaker closing, unless the phase angle remains at less than a set degree for a selected time. Speed-matching relays are used to adjust the incoming machine speed automatically so the frequency matches the running frequency. Further details on these systems are available from manufacturers of this equipment.

To follow through the manual synchronizing procedures for the diagram of Fig. 4-16, assume that the 115-kV switchyard is in service and power is connected through it from other generators elsewhere in the system. Breaker 13 is closed; breakers 11 and 12 are open. The main generator is up to speed and ready to be synchronized into the system and the disconnect switch in the line

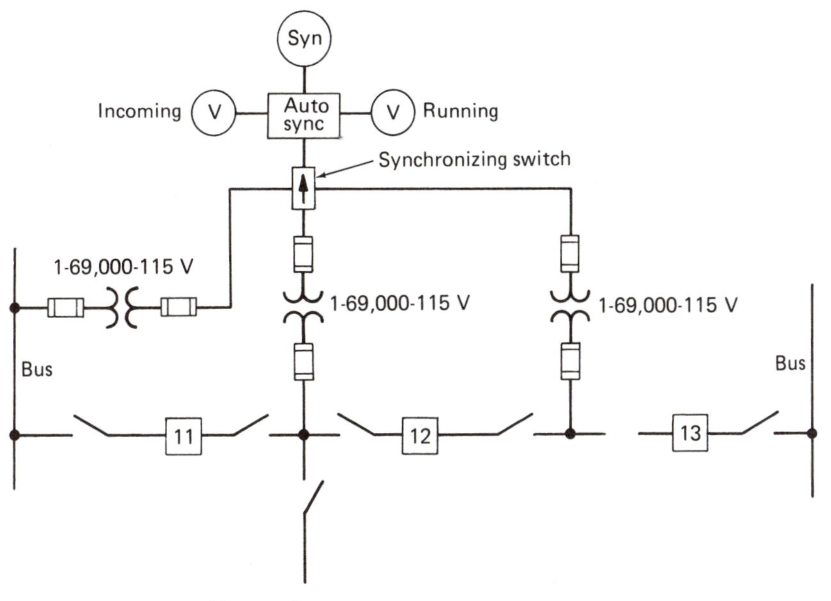

Figure 4-16 Synchronizing Circuit

from the machine is closed. The synchronizing switch is turned on. The synchroscope will immediately indicate whether the machine is running fast or slow as compared to the frequency and difference in phase relation to the 115-kV busses. Adjustment of speed of the incoming machine will cause the pointer on the synchroscope to hold steady at "12 o'clock." The voltmeters will both read zero and breakers 11 and 12 can be closed to connect the machine into the system.

Automatic synchronizing is accomplished by turning on the synchronizing switch if the automatic system has been set up to operate.

4-10 Protective Relays—Definition and Basic Types

A protective relay is a device that may be energized by a voltage signal, a current signal, or both. When energized, it operates to indicate or isolate an abnormal operating condition. Basically the protective relay consists of an operating element and a set of contacts. The operating element takes the signal from sensing devices in the system, such as current or potential transformers (or both), performs a measuring operation, and responds to the results by causing the contacts to open or close. When the relay operates, it may either actuate a signal or complete a circuit to trip a circuit breaker, which in turn isolates the section of the system that is in trouble.

Operating elements of protective relays are generally constructed in four basic types. These are

1. Plunger or solenoid type
2. Hinged armature
3. Induction disk
4. Induction cup

The plunger and the hinged armature type work by magnetic attraction. In these the moving part, called the *armature*, is attracted into a coil or to the pole face of an electromagnet. This type of relay-operating element may be applied on either alternating- or direct-current circuits.

The induction-type relays are magnetic-induction types wherein torque is developed in a moving rotor in the same way that it is produced in an electric motor. These relays can only be used on alternating-current circuits. Figure 4-17 shows the four basic types.

The plunger and hinged armature relays have no inherent time delay and so are used in applications where instantaneous operation is required. The air gap on the hinged armature type changes as the relay operates, which means that the force required to close it changes and the force to allow the armature to drop out again is reduced. For this reason, its measuring quality is impaired. The same is true for the plunger type but to a lesser degree.

The induction disk type is used when time delay in relay operation is required. The time delay is caused by addition of the permanent magnet. The disk rotates between this magnet causing an induction drag.

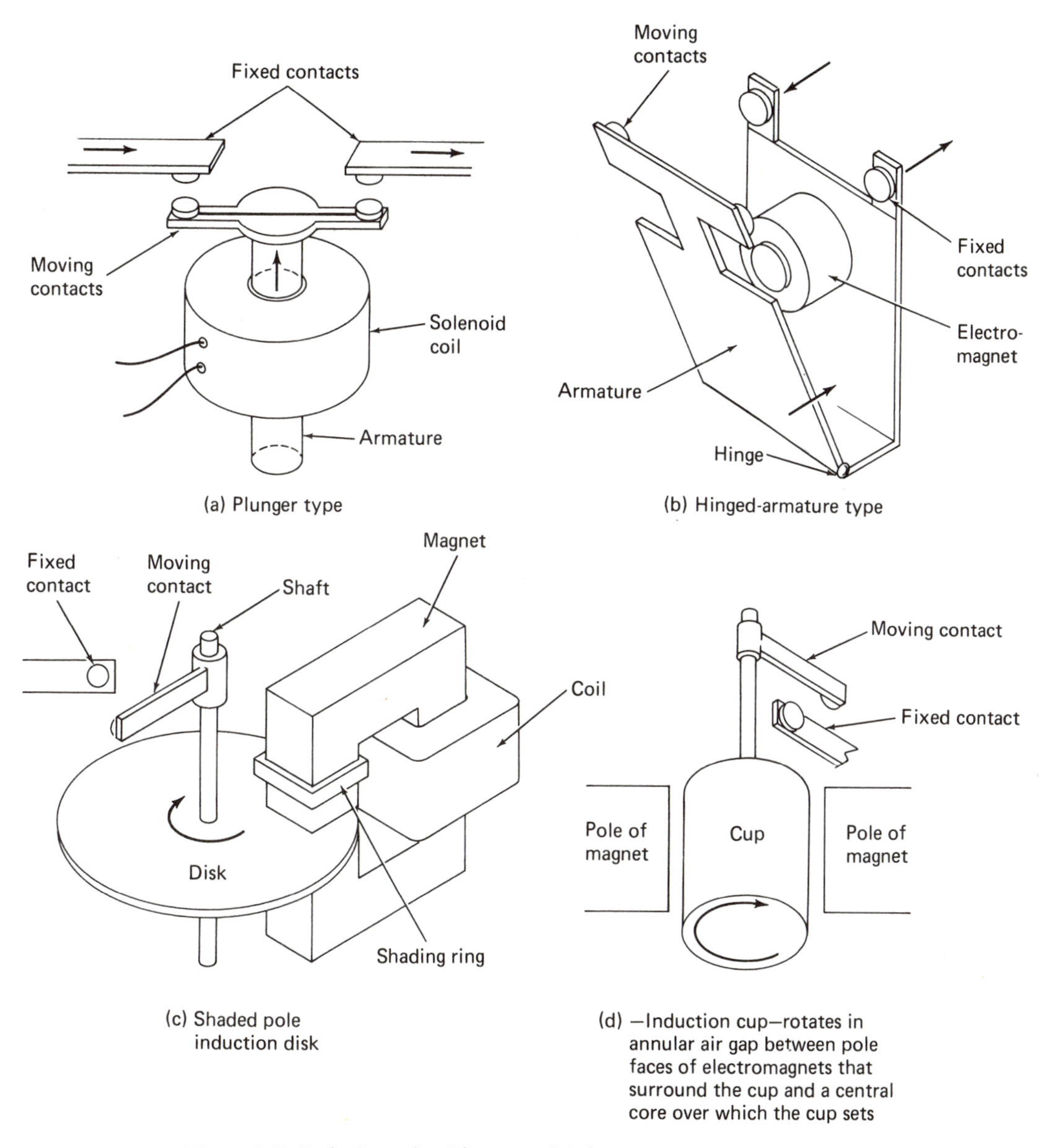

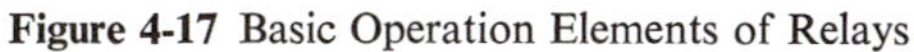
Figure 4-17 Basic Operation Elements of Relays

The induction cup rotating parts are of low inertia so this relay is capable of high-speed operation. It is used for functions requiring instantaneous response.

Elements of the types described above and other ingenious operators are incorporated into the many types of protective relays available. All protective relays have a separate device number assigned to them in the standard adopted for automatic switchgear by the American National Standards Institute (ANSI).

4-11 Main Turbine Generator Differential Relaying

Generators can develop a variety of electrical problems both in the alternating-current stator windings and in the direct-current revolving field windings. These problems can be in the form of phase-to-phase short circuits, phase-to-ground short circuits, overcurrents, overheating, motoring of the generator (when conditions are such that the machine runs as a motor from power supplied to it), and loss of excitation current. Any of the conditions above can cause some damage to the machine but short circuits are the worst! It is particularily important to provide high-speed relay protection for generator short circuits because they can do excessive, costly damage if the short circuit is not promptly cut off from current flowing into it.

The most effective protective relay for generator stator short circuits is a percentage differential relay. This relay has the device No. 87 and is defined as a fault-detecting relay that functions on a differential current of a given percentage or amount.

The main turbine generator in Fig. 4-1 has differential relaying protection, as shown in Fig. 4-18. The differential circuit connections for one phase are also indicated.

Differential relaying is the most selective relaying method because it protects

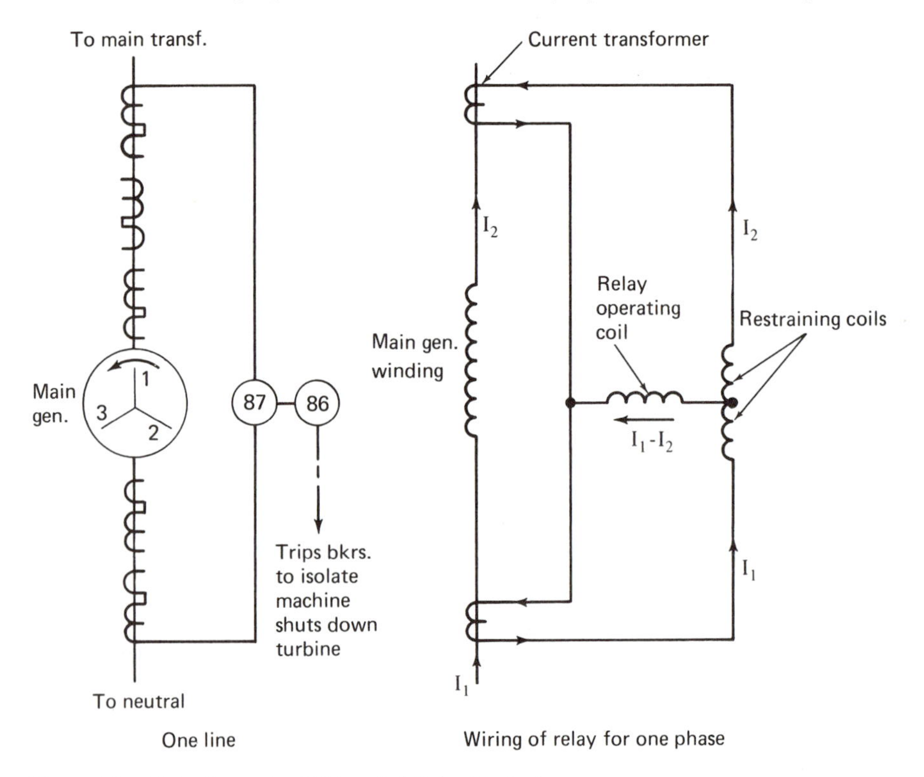

Figure 4-18 Main Generator Differential Relay Protection

the zone enclosed by the current transformers. In this instance, it is the main generator alternating-current windings. This is achieved by connecting the current transformers as shown. As long as current flows normally through the main conductor, the current transformer circuit merely circulates currents around and around and no current flows through the operating coil. If a fault occurred in the main winding, however, a difference current will flow into the operating coil and cause the relay to function. The differential relay is generally of the induction disk type. There is usually only one operating contact; therefore, a lockout relay is generally used with the differential relay. The lockout relay is assigned No. 86 and is defined as an electrically operated, hand or electrically reset, relay or device that functions to shut down and *hold an equipment out of service* on the occurrence of abnormal conditions. This device is a multicontact switch with a spring-loaded mechanism set up to operate when the switch is in the *reset* position. The device, which releases the stored spring action, is an electromagnet. The *reset* position is set by hand (or through an electrical operator) and tripping is done by energizing the electromagnet. When the relay operates, the contacts close (or open). Because there are several contacts available on this device and the contacts are rugged in construction, several functions are initiated simultaneously. In the differential relay circuit of Fig. 4-18, the lockout relay (86) trips all the 115-kV circuit breakers, the generator field breaker, and all the station service secondary transformer breakers and shuts off the steam valves. This action fully isolates the machine. Restoration of service can only be accomplished by resetting this relay; therefore, the system is *locked out* until reset.

4-12 Main Transformer Differential Relaying

The main power transformer must be protected from possible faults because loss of this piece of equipment, through damage from a sustained short circuit, would be as crippling to the system capability as loss of the generator. Transformers are static devices (no moving parts) and have only alternating-current windings so the protection problem is less than for rotating equipment. There are two principle causes for insulation failures in transformers. They are (1) *overvoltage* resulting from lightning, switching surges, etc., and (2) repeated *overheating* due to overloading or failure of the cooling system.

Differential relaying is often used to protect large power transformers. The differential relay circuit for the one line of Fig. 4-1 is shown in Fig. 4-19.

In this particular case, the primary feeder to the primary winding of the unit station service transformer is included in the zone of protection. The current transformers are all connected in a loop arrangement and normal currents will simply circulate. A fault anywhere within the zone of protection will cause a difference current to flow, energizing the relay. The lockout relay will be wired to operate when energized by the differential relay and to shut down the same circuit elements as those shut down for the generator.

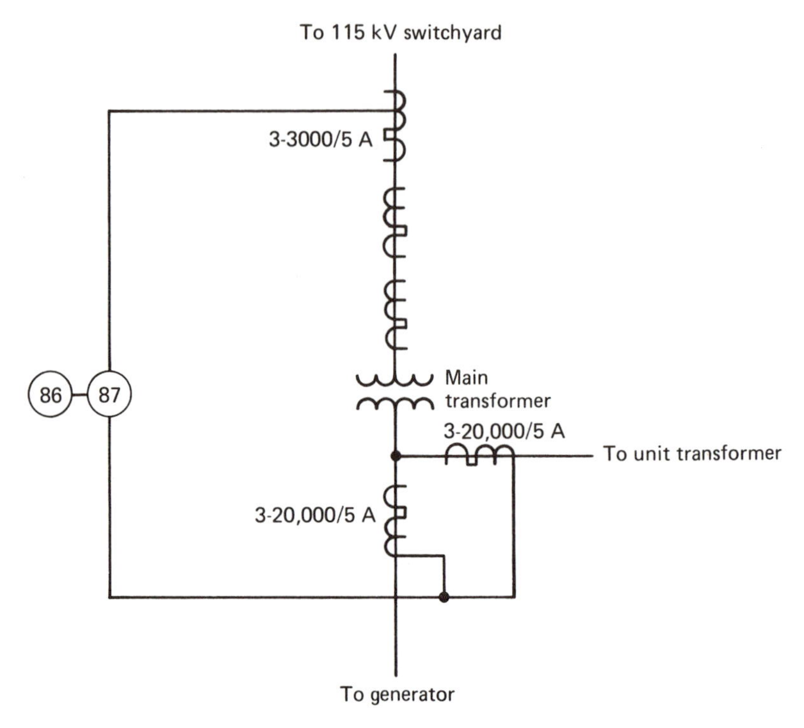

Figure 4-19 Main Transformer Differential Relaying

4-13 Overall Main Generation Circuit Differential Relaying

In addition to the differential relaying protection for each major element, an overall protection is provided, in the diagram of Fig. 4-1, for the total main generation circuit. This circuit is shown in Fig. 4-20. The zone of protection overlaps all the other zones and extends out to cover the feeder to the unit transformer, right up to the transformer terminals. Lockout relay operation will isolate the entire circuit within the zone and shut down the generator. This added protection is installed as backup protection to the other differential circuits as well as to extend protection to the unit transformer. This protection also includes the section of bus between the generator and the main transformer, which is not included in either the machine or main transformer zones. This section of the bus includes the connection to the potential transformers.

4-14 Differential Relaying of Transmission Line

Transmission lines are high-voltage circuits interconnecting two system elements separated by some distance. In the one line of Fig. 4-1, the 115-kV transmission lines run from the terminal of the main transformer to the 115-kV switchyard, which is approximately a mile away. The 115-kV line is run overhead as open bare wire construction on transmission towers. Where the circuit ele-

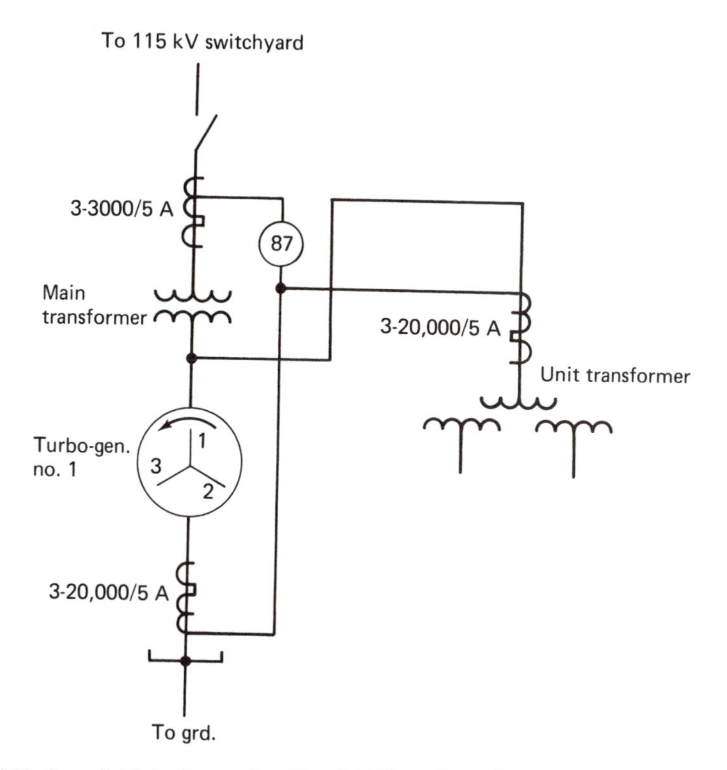

Figure 4-20 Overall Main Generation Circuit Differential Relaying

ments to be protected are close to each other (a few hundred feet apart) circulating current differential relaying can be applied, such as for the generator and the main transformer. At greater distances, however, this method is impractical so it becomes necessary to employ separate relays at each end of this protected zone and to modify the circuits. Pilot relaying either through a pilot wire or by carrier current pilot channel is most commonly used.

With the wire pilot scheme, two pilot wires are connected between the differential relays. These wires can be telephone company wires, which are leased for this purpose. The carrier current system imposes a higher frequency signal onto the transmission line conductors themselves through the application of special equipment. In both cases, the pilot channel, whether wire or carrier current, is used to compare the current flowing at the ends. Figure 4-21 shows the pilot wire scheme for that portion of Fig. 4-1 that is protected by this method.

This pilot wire scheme operates on the opposed-voltage principle. The pilot wires carry current only during a fault. The current transformers are connected to a mixing transformer, which converts the secondary currents from the transformers to a single-phase voltage to be applied to the pilot wires. On a normal through current, opposing voltages are produced at the two ends of the pilot wires and no operating current flows. On an internal line fault (between the current transformers in the zone of protection) the voltages at the two ends of the pilot wires will be additive and a relatively large operating current will flow

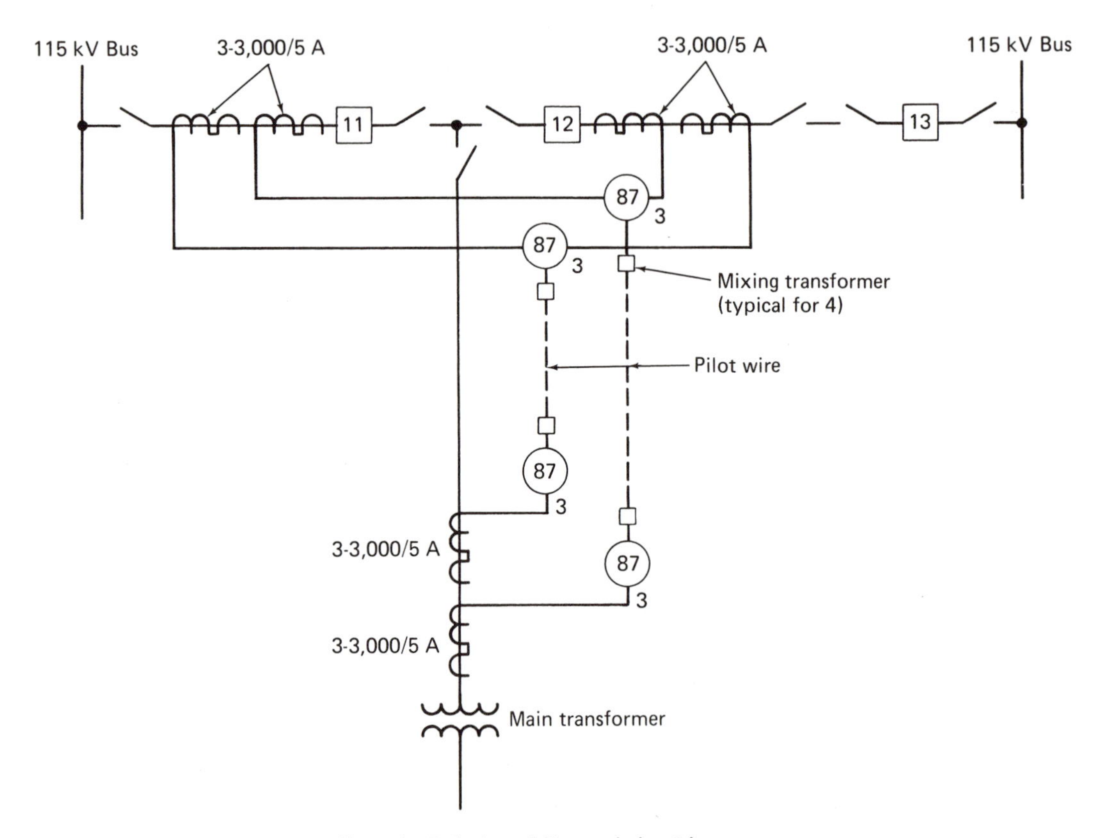

Figure 4-21 Pilot Wire Relaying of Transmission Line

causing the relays to operate. Lockout relays (not shown) working with the differential relays will isolate the transmission line from the 115-kV bus by tripping breakers 11 and 12. The machine and other circuits relating to the machine will be tripped.

4-15 Differential Relaying of the Unit Station Service Transformer

In much the same way as the other differential relay circuits have been described, the unit station service transformer is also protected. The zone of protection extends from the current transformers on the primary bushings to the current transformers located between the feeder secondary circuit breakers and the bus for each section of 4,160-V station service switchgear. This circuit can be readily traced from the one-line diagram of Fig. 4-1. The start-up transformer protection is similar.

4-16 Overlapping of Zones of Protective Relaying

If the separate zones of protection described in the preceding explanations are combined, it can be seen that the zones overlap as shown in Fig. 4-22.

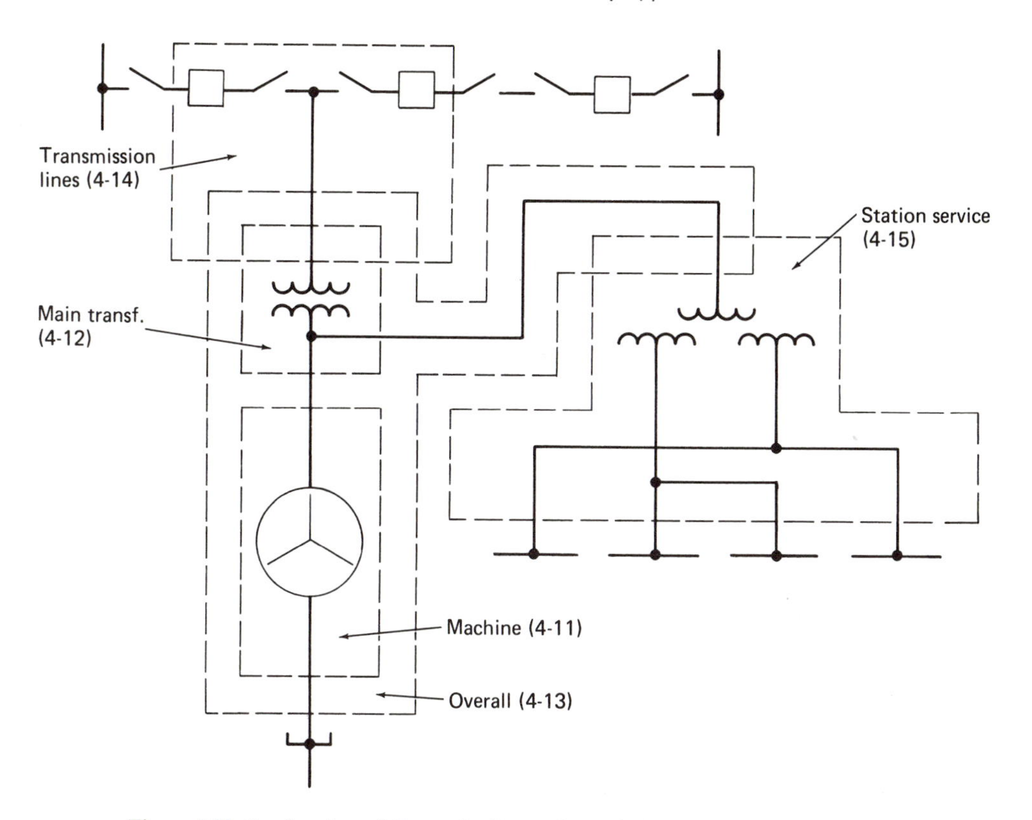

Figure 4-22 Overlapping of Zones for Protective Relaying (Numbers in parentheses refer to paragraphs describing protection in that zone)

4-17 Miscellaneous Main Generation Circuit Protective Relay Applications

There are several other relays shown in Fig. 4-1 other than the differential relay applications for the main generation circuit described above. These relays and their location in the system are as follows:

Ground Protective Relay

Ground protective relay (64) in neutral ground connection is shown in Fig. 4-23.

The 64 relay will detect an insulation breakdown in the machine because current will flow in the transformer. The relay is wired either to an alarm or to shut down the machine. The relay is often made in the plunger-type construction so the action is essentially instantaneous.

Instantaneous Overcurrent

Instantaneous overcurrent, or rate of rise relay (50) and dc overcurrent relay (76) in the generator regulator circuit: the 50 relay functions instantane-

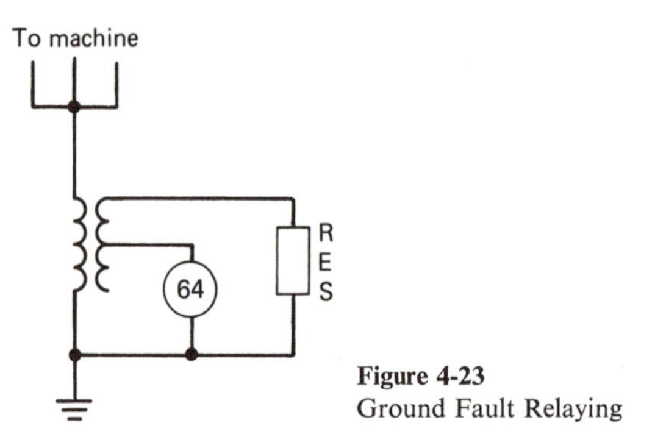

Figure 4-23
Ground Fault Relaying

ously on either an excessive value of current or an excessive rate of rise; i.e., the current in the circuit is climbing to a peak too rapidly indicating a possible fault. The 76 relay operates when the current, in a direct-current circuit, exceeds a given value. These relays connect into the black box regulator circuit to protect the regulator system.

Field Relay (40) and Reverse Phase or Phase-Balance Current Relay (46)

Field relay (40) and reverse phase or phase-balance current relay (46) in the generator neutral circuit: Loss of excitation in a generator may cause severe voltage disturbance to the rest of the power system. The field relay functions on a given or abnormally low value or failure of machine field current. Loss of field results from loss of field to the main exciter (the alternator in Fig. 4-1), accidental tripping of the field breaker, short circuits in the field circuit, poor brush contact in the exciter, or operating errors. This relay may be of the hinged armature type.

Open-circuit or single-phase operation of the generator may be detected by the application of a current-balance relay of the induction disk-type construction. This relay balances each phase current against the other two phases.

Ground Protective Relay (64)

Ground protective relay (64) in the main generator field circuit: Alternating-current machine field circuits are usually operated ungrounded. A single breakdown of insulation resulting in a ground will not result in any damage to the machine. If a second breakdown occurs, however, fault currents will circulate that may cause damage. The application of a ground protective relay will detect the insulation breakdown. The relay can be wired to sound an alarm to warn the operator.

Alternating-Current Time Overcurrent Relay (51)

Alternating-current time overcurrent relay (51) in the grounding conductor to the midpoint connection of the high-voltage winding of the main transformer is shown in Fig. 4-24.

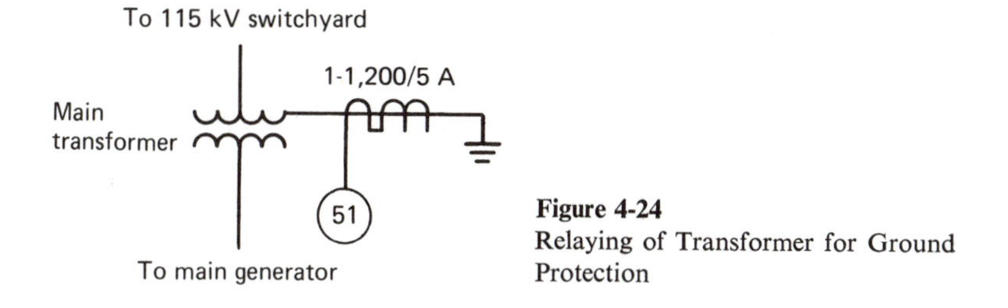

Figure 4-24
Relaying of Transformer for Ground Protection

The time overcurrent relay has a built-in time delay because it is of induction disk construction. The relay functions when the current exceeds a predetermined value. In the application of Fig. 4-24, current flow to ground is an abnormal condition that the relay will detect. This relay acts as a backup to other relays and the time delay characteristic with a long time setting will preclude unnecessary relay action.

4-18 Protective Relaying of Station Service Switchgear

The differential relays protecting the secondary feeders to the 4,160-V switchgear include protection for the main circuit breakers but exclude the switchgear bus and connections to feeder circuit breakers. Time overcurrent relays (51 device) are installed for this protection.

These relays, with their time delay characteristics, can be set to protect the switchgear if a fault occurs within the switchgear but not to trip if the fault occurs on a motor feeder that is protected by its own time overcurrent relay. A typical feeder with its relay is shown in Fig. 4-25.

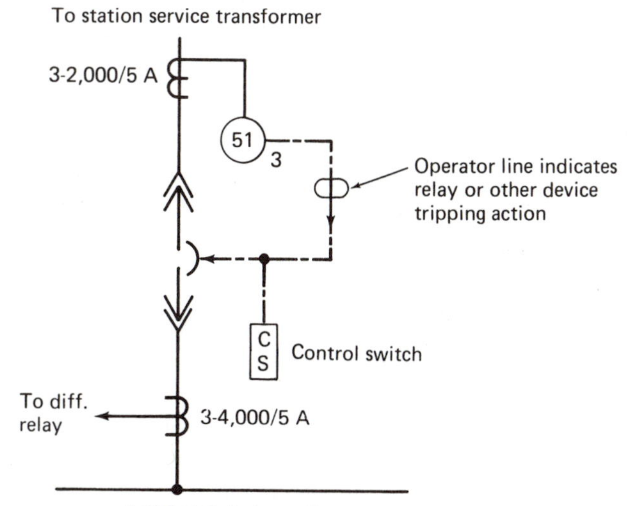

Figure 4-25 Overcurrent Relay Protection of Switchgear

4-19 Protective Relaying of Large Motor Circuits

The station service motors in Fig. 4-1 are protected through use of time overcurrent relays (51) with instantaneous trip devices (50). This combination relay provides protection against overload with the time delay element and protection against short circuits with the instantaneous element. A typical motor circuit with its relay is shown in Fig. 4-26.

Motors of this size and importance in the system are also at times protected with differential relays. The motor must be furnished with both ends of all three windings brought out of the housing so the relays can compare current flow in and out of the motor.

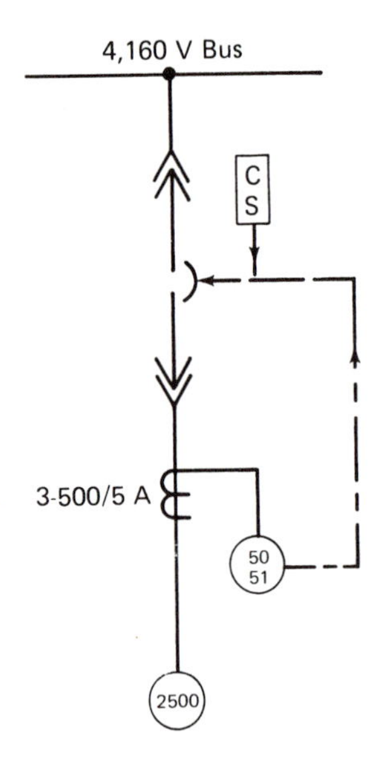

Figure 4-26
Time Overcurrent and Instantaneous Overcurrent Protection of Motor

4-20 Metering and Instrumentation

A certain amount of metering and instrumentation is essential to satisfactory electrical system operation. Beyond the essential needs, additional devices may be desirable and the decision to include them in the design depends on economics. It is a question of whether the information obtainable from the additional meters and instruments will be of sufficient value to justify their extra cost or whether they will just be luxury items. Certainly, in the area of power generation, close monitoring of the system is essential. The meters are read regularly and their information is recorded on daily logs. Some instruments have recording charts for automatic record keeping. In low-voltage systems (600 V or less), metering and instrumentation needs are less and because there is such a multiplicity of circuits at these voltages, extensive metering would be very expensive.

Some of the more commonly used meters and instruments are shown in Figs. 4-27 through 4-33. These are representative diagrams only. Exact wiring for specific meters must be obtained from the meter manufacturer. Several of these devices are included in the one-line diagram of Fig. 4-1.

Meters for monitoring of the main generator in a central station are generally mounted on special panels or consoles that include generator controls as well. Meters relating to switchgear are mounted on the switchgear itself and therefore are supplied with the equipment. The following is a list of commonly used meters and instruments:

A. High Voltage (2,300 Volts and above)

Generator

1. Ammeter—either three meters or one meter and transfer switch
2. Voltmeter—either three meters or one meter and transfer switch
3. Wattmeter—may be calibrated in kilowatts or megawatts
4. Watthour meter—optional, usually used
5. Varmeter
6. Generator field ammeter
7. Exciter voltmeter—optional
8. Temperature meter—optional

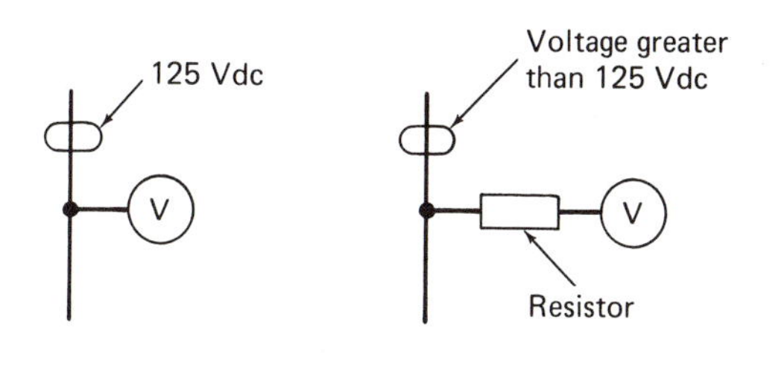

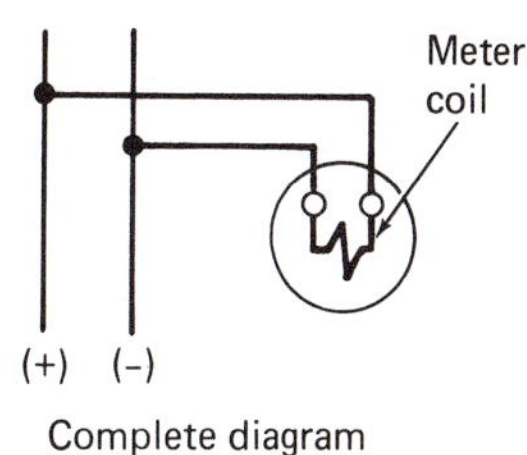

Figure 4-27 Direct-Current Voltmeters

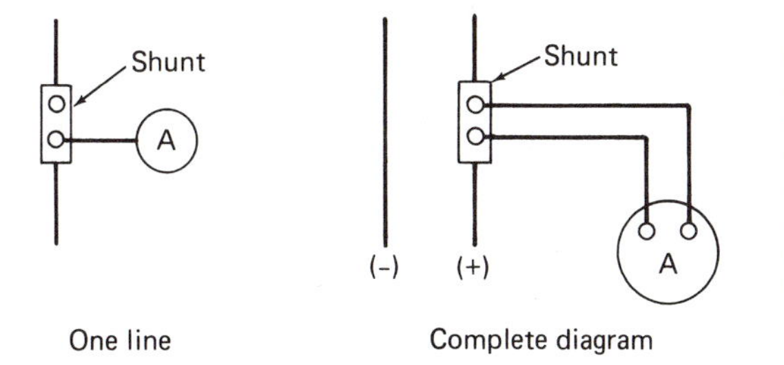

DC ammeters are usually provided with a shunt whose function is to provide adc millivolt signal (to the ammeter) that is proportional to the dc current in the line

Figure 4-28 Direct-Current Ammeters

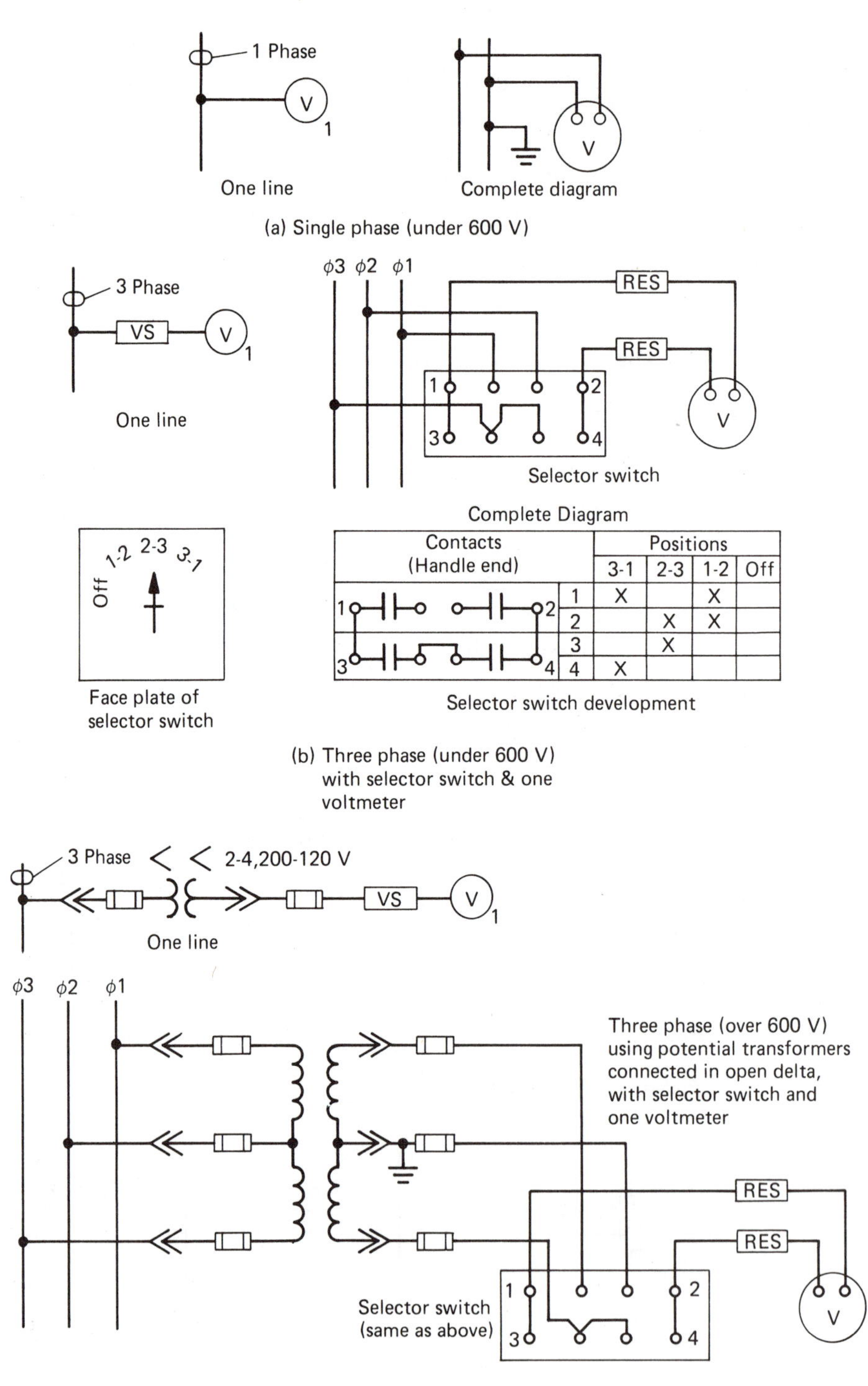

Figure 4-29 Alternating-Current Voltmeters

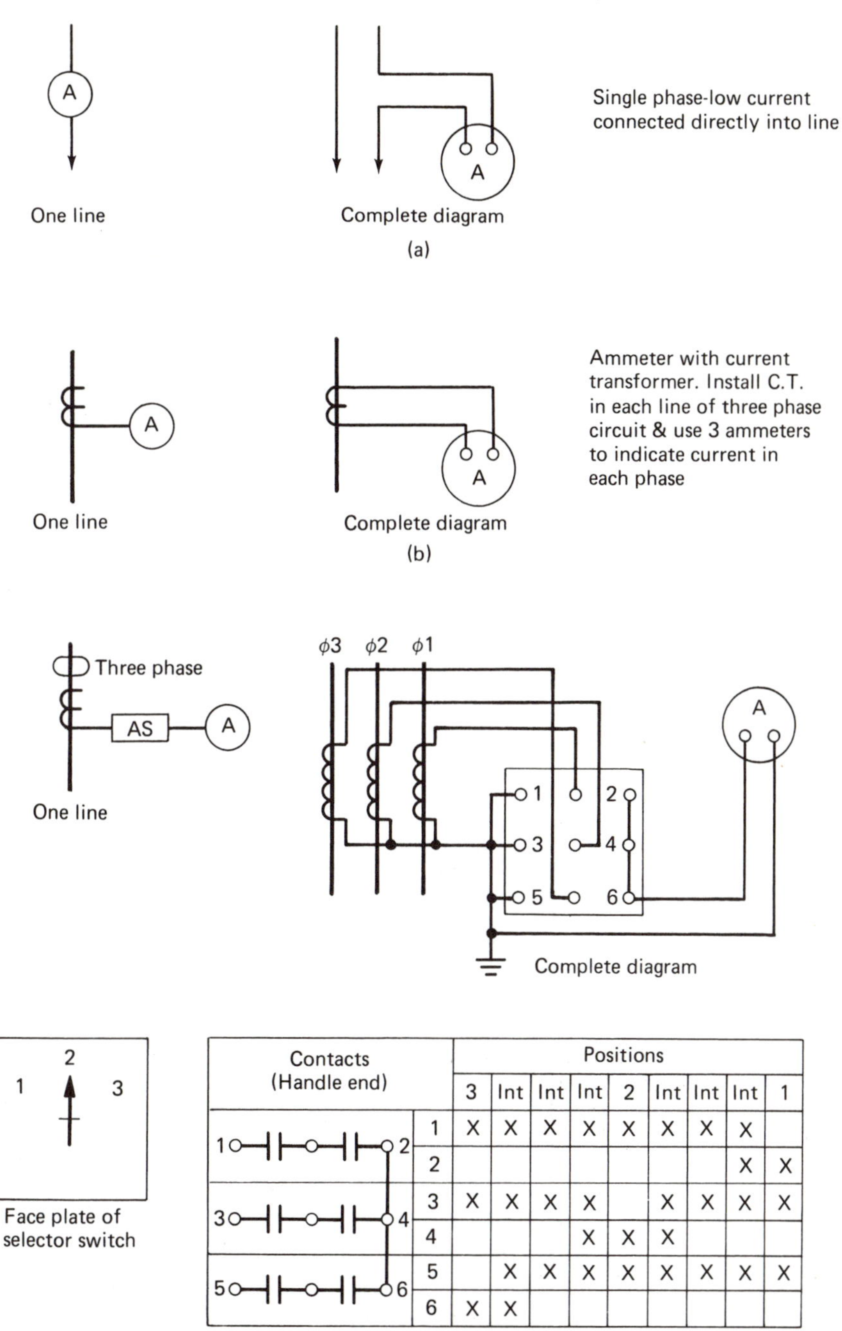

Contacts (Handle end)		Positions								
		3	Int	Int	Int	2	Int	Int	Int	1
1—2	1	X	X	X	X	X	X	X	X	
	2								X	X
3—4	3	X	X	X	X		X	X	X	X
	4				X	X	X			
5—6	5		X	X	X	X	X	X	X	X
	6	X	X							

Figure 4-30 Alternating-Current Ammeters (Important: Any current transformer that is not connected in a closed circuit to the coil in a meter or other device must be short-circuited. Note that in selector switch operation the intermediate (int) and other switch positions keep the current transformer coil not in use short-circuited.)

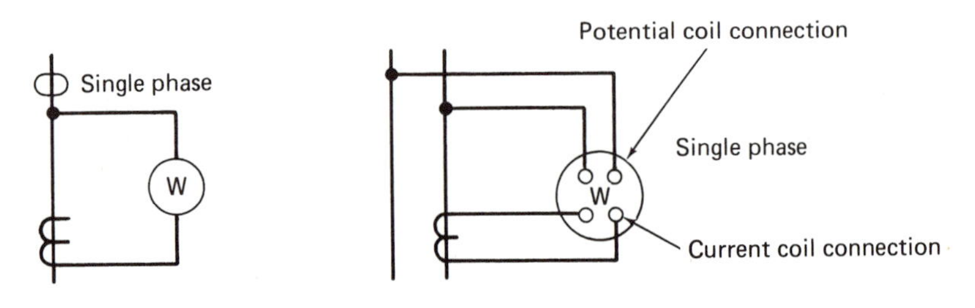

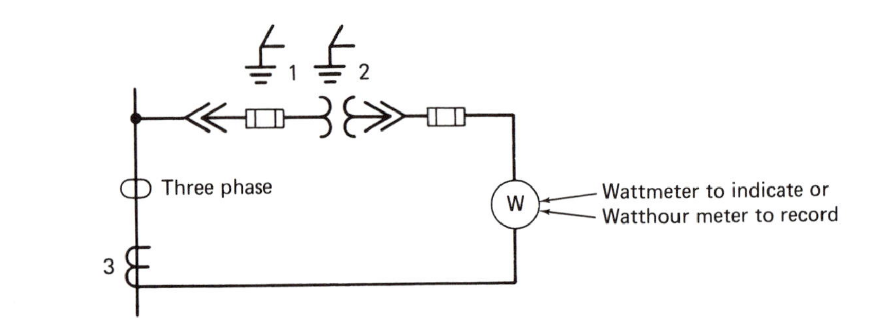

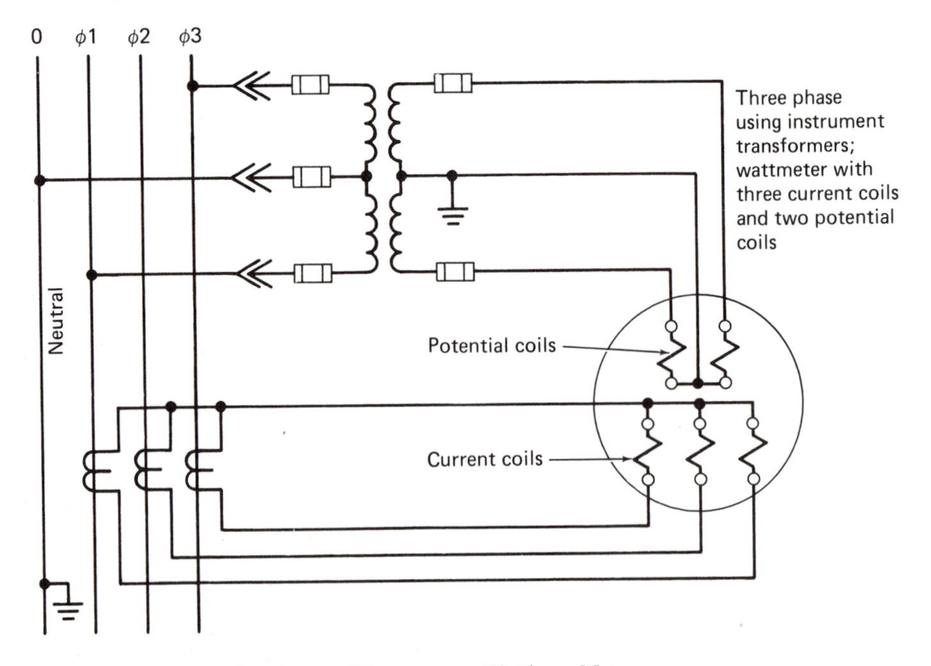

Figure 4-31 Alternating-Current Wattmeter or Watthour Meters

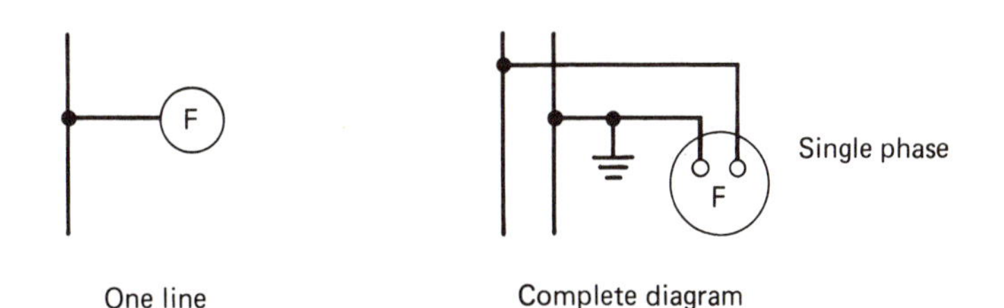

Figure 4-32 Frequency Meter (For three phase connect across any two phase wires. Use potential transformers at higher voltages)

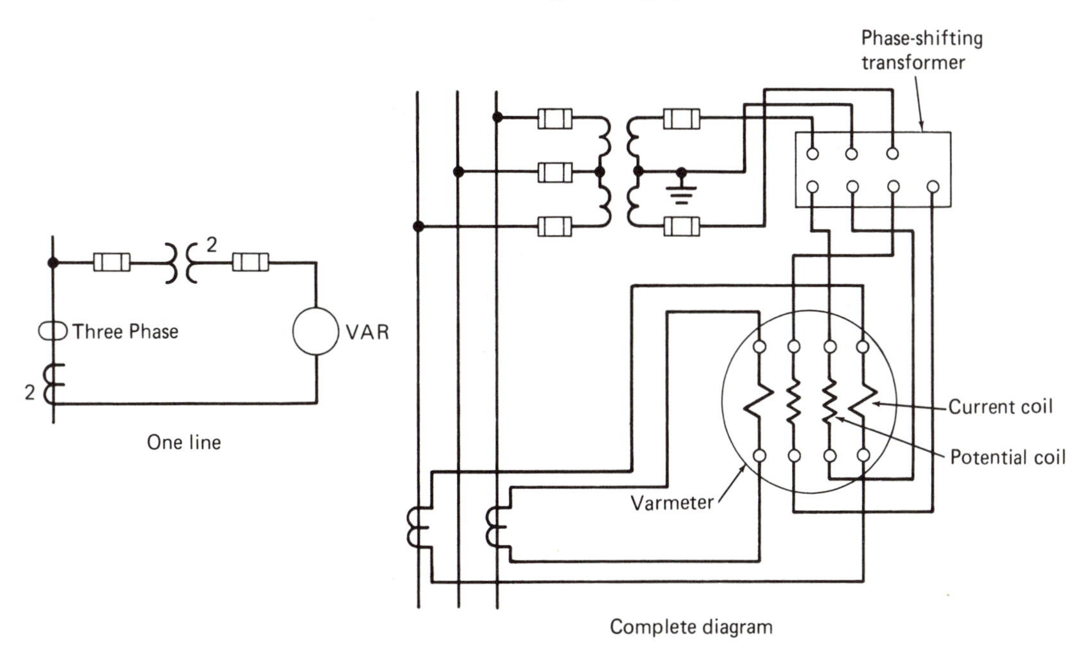

Figure 4-33 Varmeter

9. Synchronizing panel including two voltmeters—one for bus voltage and one for incoming generator, frequency meter, and synchroscope

Feeders

1. Ammeter and transfer switch
2. Watthour meter—optional
3. Varmeter—optional

Incoming Lines

1. Ammeter and transfer switch
2. Voltmeter and transfer switch—generally on transformer secondary if

transformer is involved that allows for checking tap settings on transformer

3. Wattmeter with a 0 center scale for indication of power flow in both directions if reverse power flow is included in design
4. Varmeter—optional

Motors

1. Ammeter and transfer switch

B. Low Voltage (600 V and less)

Generator

Same as for high voltage except temperature meter is generally not required because machines are much smaller

Feeder

Ammeter and transfer switch if cost can be justified, ammeters usually require current transformers, which means cost and additional space allowance in switchgear for meters and transformers, therefore, requirement is optional; an alternate means of acquiring load data on a feeder is through the use of portable instruments

Incoming Lines

Same as for high voltage

Motors

Ammeters on large motors; otherwise, no metering

Chapter Review Topics and Exercises

4-1 How are the coils of the turbo-generator shown in Fig. 4-2 connected?

4-2 Why are both ends of the generator coil wired to the outside of the machine?

4-3 How does the bushing transformer work?

4-4 What is the phase-to-phase voltage on the generator side of the main power transformer? on the switchyard side?

4-5 What is the principal advantage of the breaker-and-a-half scheme?

4-6 Why is start-up power needed?

4-7 What voltages are provided for start-up power? How many busses are provided?

4-8 What kind of current is used for the main generator excitation?

4-9 What conditions are vital to synchronizing systems?

4-10 What does a synchroscope show?

4-11 Describe a *relay*.

4-12 What types of relay construction are used for instantaneous response? which types for time delayed response?

4-13 Why are protective relays installed?

4-14 What type relay is generally used to protect a large generator? What relay is used with it? why?

4-15 Why are overlapping zones of protection provided when applying protective relays?

4-16 What is a 64 relay? How is it applied in the one-line diagram of Fig. 4-1?

4-17 Why are time delay characteristics desirable for some relay applications?

4-18 One voltmeter and a selector switch can be used to check voltage on all three phases. Explain how this is done.

4-19 Where is a shunt used?

4-20 What auxiliary device is used with a varmeter?

4-21 Draft the one line of Fig. 4-1 on a sheet measuring approximately 17″ × 22″. Use $\frac{1}{8}$-in. grid paper under the drawing if available. Develop each system in steps as explained in the chapter. Plan ahead to allow room because the diagram will fit onto the sheet.

5

ELEMENTARY DIAGRAMS

5-1 Definitions

An *elementary diagram* (also called a *schematic* diagram) is defined as a diagram that shows in straight line form the detail wiring of the circuit and device elements without regard to physical relationships. These diagrams are a necessary development of some portions of the one-line diagram data as well as supplemental wiring information that may not show on one-line diagrams. By disregarding physical relationships, the preparation of the diagram is unencumbered and the straight line approach to connection of the circuit elements makes reading the diagram much easier. A comparison of the elementary diagram with the one-line and complete wiring diagram for a very simple circuit is shown in Fig. 5-1.

Figure 5-1 (a) shows the one line for control of a 500-W light by a single-pole switch from a 120-V single-phase source.

Figure 5-1 (b) indicates the complete circuit wiring in a simple straight line form called an *elementary* diagram. It is easy to see that the light is connected through the switch to the ungrounded conductor and directly to the light from the grounded conductor.

Figure 5-1 (c) develops the complete wiring and notes that both the grounded and ungrounded conductors pass through the switch enclosure. Note that this physical relationship is disregarded in the elementary diagram.

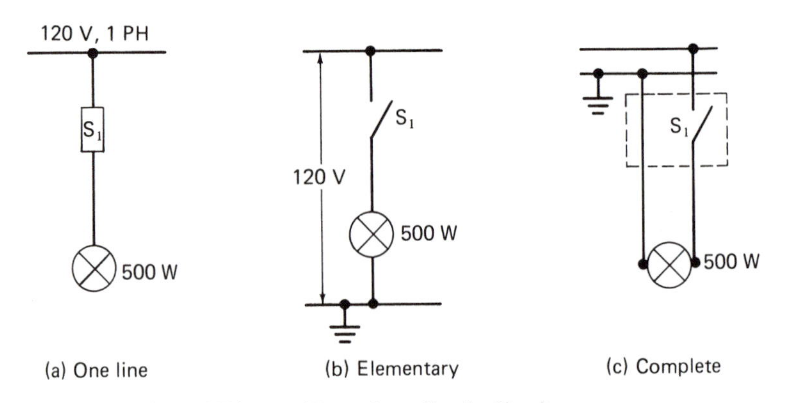

Figure 5-1 Comparison of Diagram Types for a Simple Circuit

5-2 Purpose and Use

Elementary diagrams are initially prepared in order to design the electrical circuit. All electrical circuits are intended to operate in a specific way. To *build* these electrical circuits it is to the designer's advantage to develop his concepts by recording them in as simple a way as possible. This makes design and understanding as easy and uncomplicated as it can be. The diagrams are used to relate understanding of system operations, to develop wiring data, and to make a reference for circuit operation. Such diagrams are often invaluable for *trouble-shooting* because they are much less complicated than the complete wiring diagram.

5-3 Symbols

Graphic symbols are used on one-line and elementary diagrams as a shorthand method of showing the devices and other elements of the circuit. Standard symbols have been developed for this use and are published by engineering societies and others for use in the engineering profession. Wherever possible the standard symbols should be used. Those that are most common in elementary diagrams for the electrical power field are shown below.

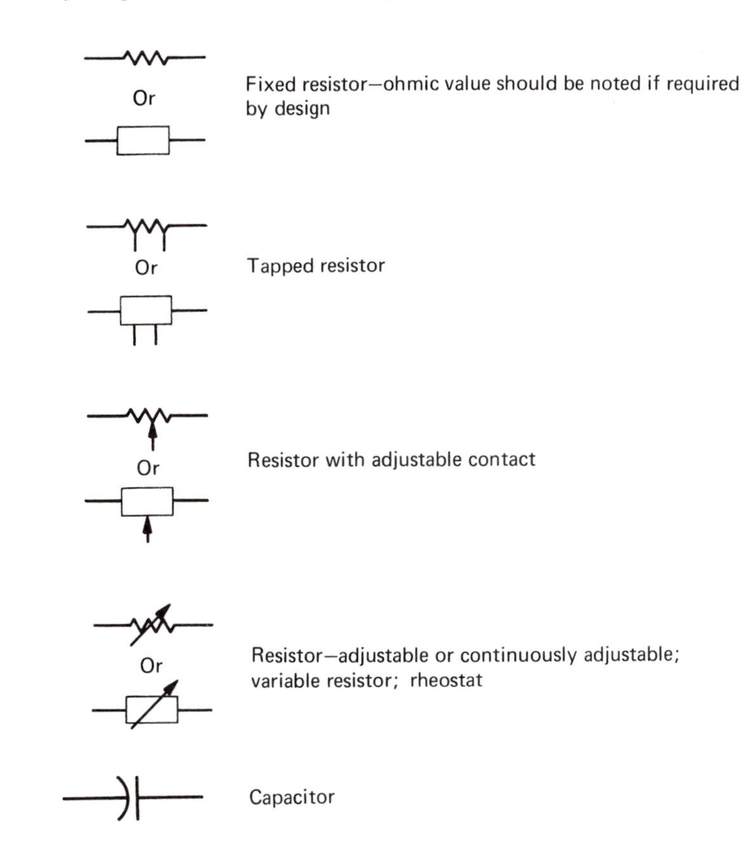

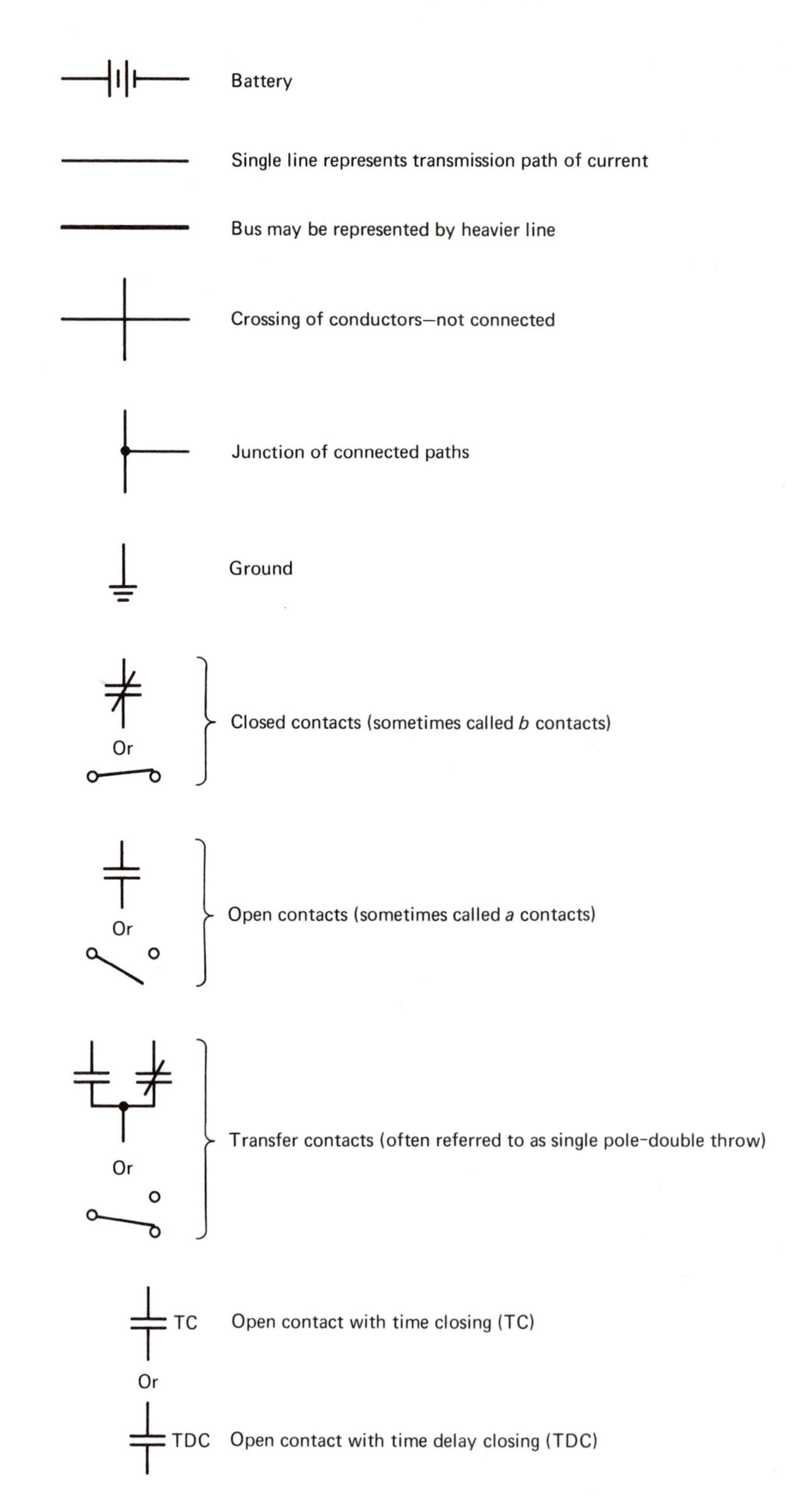
Battery
Single line represents transmission path of current
Bus may be represented by heavier line
Crossing of conductors—not connected
Junction of connected paths
Ground
Or
Closed contacts (sometimes called *b* contacts)
Or
Open contacts (sometimes called *a* contacts)
Or
Transfer contacts (often referred to as single pole–double throw)
TC
Open contact with time closing (TC)
Or
TDC
Open contact with time delay closing (TDC)

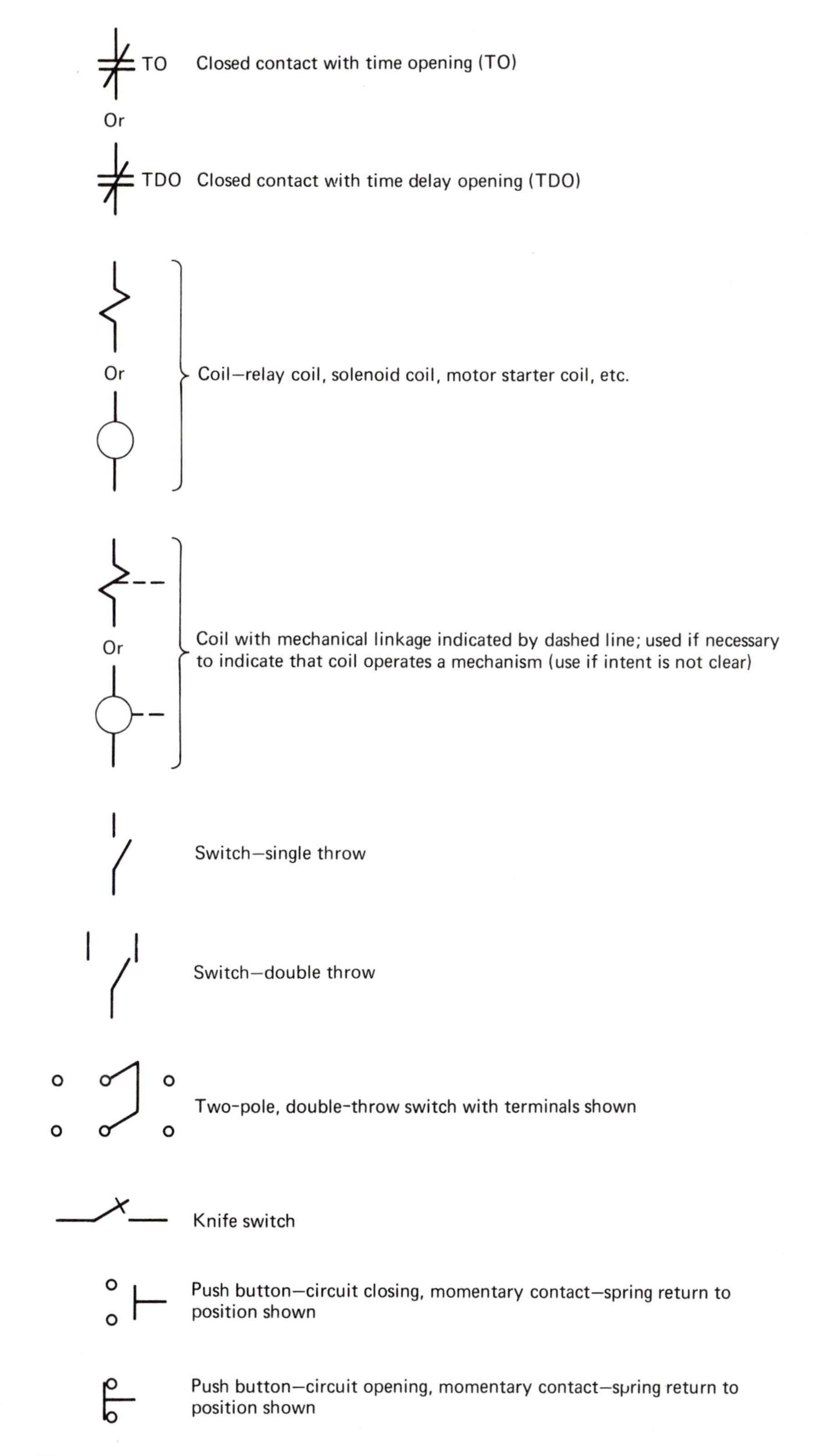
TO
Closed contact with time opening (TO)
Or
TDO
Closed contact with time delay opening (TDO)
Or
Coil—relay coil, solenoid coil, motor starter coil, etc.
Or
Coil with mechanical linkage indicated by dashed line; used if necessary to indicate that coil operates a mechanism (use if intent is not clear)
Switch—single throw
Switch—double throw
Two-pole, double-throw switch with terminals shown
Knife switch
Push button—circuit closing, momentary contact—spring return to position shown
Push button—circuit opening, momentary contact—spring return to position shown

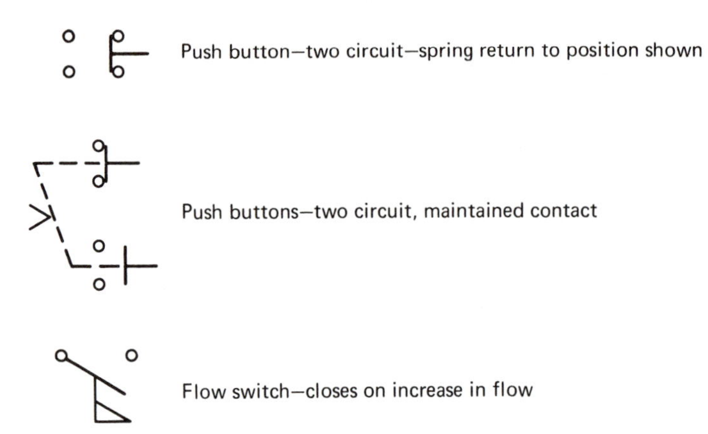

Push button—two circuit—spring return to position shown

Push buttons—two circuit, maintained contact

Flow switch—closes on increase in flow

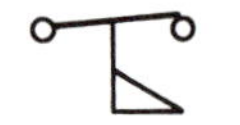

Flow switch—opens on increase in flow

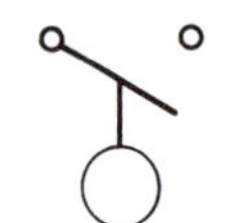

Liquid level switch—closes on rising level

Liquid level switch—opens on rising level

Limit switch—normally open

Limit switch—normally closed

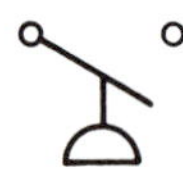

Pressure switch—closes on rising pressure (also used for vacuum)

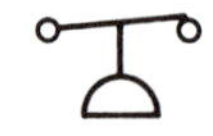

Pressure switch—opens on rising pressure (also used for vacuum)

Temperature switch—closes on rising temperature

Temperature switch—opens on rising temperature

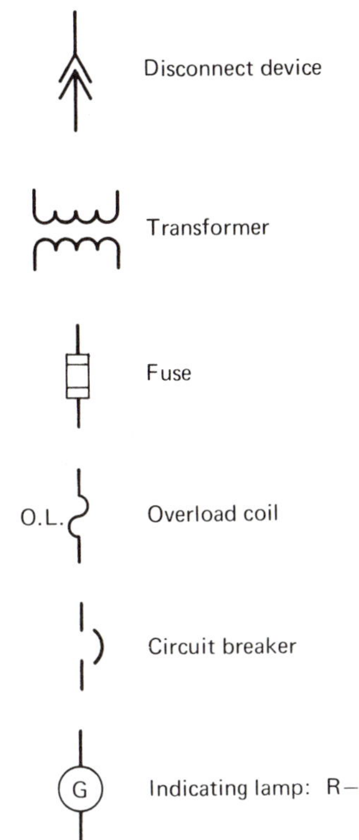

Many symbols require further clarification; this should be noted near the symbol. As an example a pressure switch might require a note saying *closes at 15 psi—opens at 30 psi.* A float switch might require the note *closes at extreme high level—opens at low level.* The inclusion of these notes is important to reading the diagrams and should not be overlooked.

5-4 Relays

A relay is an electrical–mechanical device that operates by being either energized or de-energized. The electrical action causes a mechanical action to take place that *relays* further electrical action. These devices are made in simple form and also in complicated and very precise form. For purposes of discussion in this chapter, the simple form of relay will be described.

The relay's mechanical action can be arranged to function in many different ways. The results are basically the same. One of the simplest schemes is the plunger-type relay shown in Fig. 5-2. An electromagnetic coil surrounds a steel shaft called the *armature*. Mounted on this armature, or plunger, are the moving contacts. Mounted on a fixed portion of the relay are the fixed contacts. The armature is pulled up through the coil by magnetic action when the coil is energized. It falls back by gravity when the coil is de-energized. Normally,

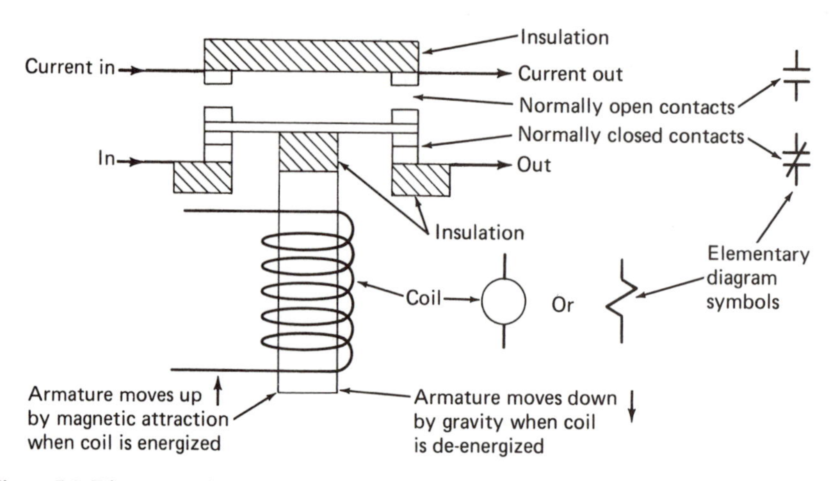

Figure 5-2 Diagrammatic Operation of a Relay

open contacts are closed when the coil is energized; normally, closed contacts are opened.

5-5 Solenoids

A solenoid is an electrical–mechanical device that operates to perform a mechanical function when energized or de-energized. These devices are used extensively to facilitate operation of mechanical devices such as valves, brakes, and clutches. They are similar in basic construction to the relay except that the armature action drives the mechanical action directly. As an example, a solenoid-operated valve is shown in Fig. 5-3.

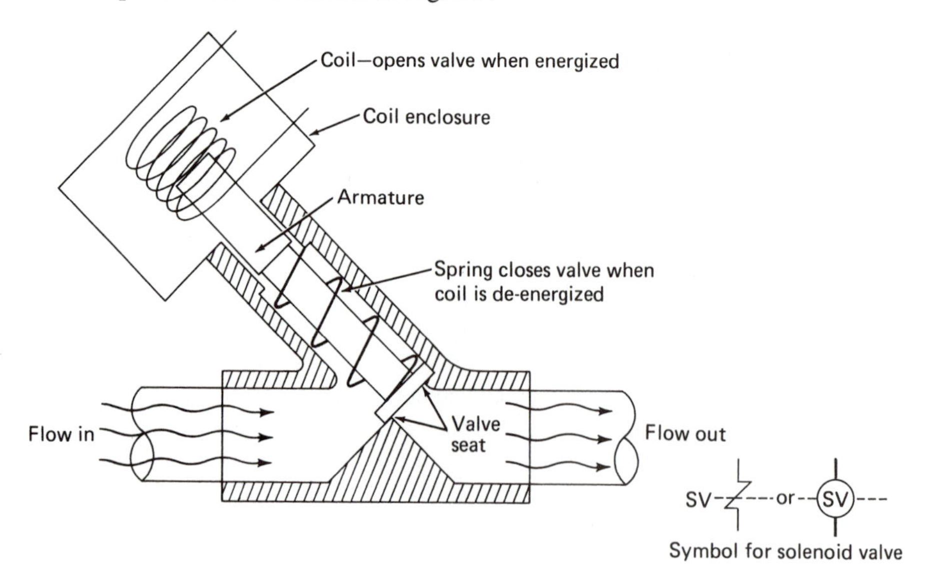

Figure 5-3 Typical Solenoid-Operated Valve

5-6 Mechanically Operated Switches

The symbols listed previously showed pressure switches, float switches, and temperature switches that are made to operate through mechanical action. As an example, Fig. 5-4 shows the general idea behind a float switch.

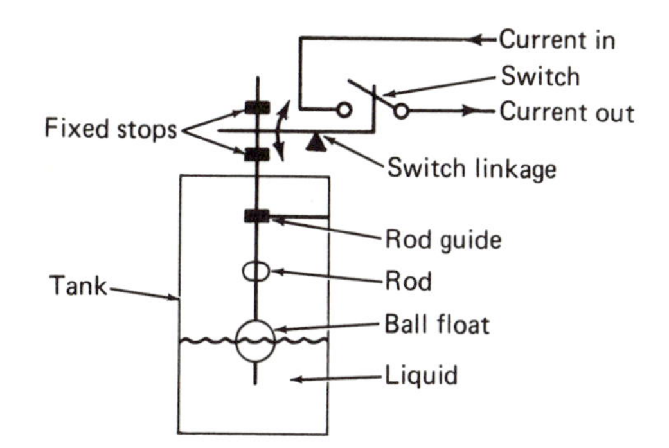

Figure 5-4 Schematic Diagram of Float Switch Operation

As the liquid level rises in the tank, the ball float rises and the fixed stops on the rod force the switch linkage up, which pulls the switch closed. When the switch closes, it starts a pump that reduces the liquid level in the tank. As the level is pumped down, the ball float settles, which causes the fixed stop on the rod to force the switch linkage down and opens the switch, thereby stopping the pump.

Limit switches are mechanically operated and as their name implies, they open or close upon some limit of operation. For example, in the machine tool industry they are used to stop or start particular machine operation once the tool or other device has reached its limit. A door can be monitored by security personnel at a remote location through the application of limit switches. If the door is opened, it triggers a limit switch, which could be wired to an alarm. There are many types manufactured with a variety of operating mechanisms and contact arrangements.

These mechanically operated switches are used extensively and therefore often appear in elementary diagrams. Their specific function is generally noted near the device symbol to clarify its operation.

5-7 Selector Switches

Many control schemes require selective operations. For example, it is common to see motor control that allows for selecting *Hand—Off—Auto* operation. A selector switch is used to provide this feature. In the *Hand* position, control is in the hand of the operator; in the *Auto* position, the control is by means of automatically operating devices such as pressure switches or temperature switches. Typical selector switch operators for use in push-button station-type devices are shown in Fig. 5-5.

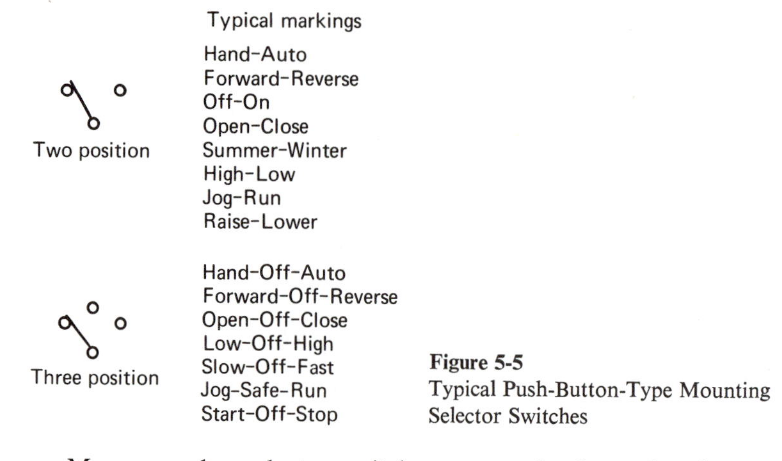

Figure 5-5
Typical Push-Button-Type Mounting Selector Switches

More complex selector switches are used where the elementary circuit requirements demand added contacts or additional positions, etc. An example of a selector switch that might be used to control a large air circuit breaker is shown in Fig. 5-6.

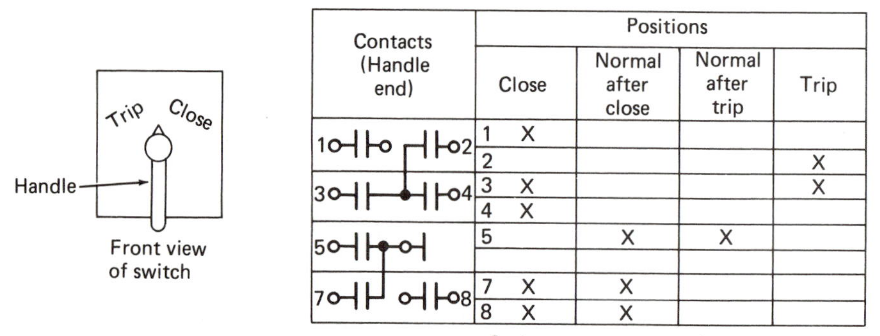

Contacts (Handle end)	Positions: Close	Normal after close	Normal after trip	Trip
1	X			
2				X
3	X			X
4	X			
5		X	X	
7	X	X		
8	X	X		

Figure 5-6 Selector Switch Development for Circuit Breaker Control

Contacts that are closed in any given position are indicated with an X. In the close position, contacts 1, 3, 4, 7, and 8 are closed. This arrangement provides the additional contacts required by the circuit.

5-8 Overload Relays for Motor Starters

Motor starters are almost always provided with a thermally operated device, called an *overload relay*, that protects the motor from the danger of overload currents. These are commonly called the *heaters*. They are constructed in several different designs but their function to shut down the motor on excessive overload currents is the same for each design. Figure 5-7 shows the principle behind the heater design.

Power to the motor flows through the heater coil. Under normal operating

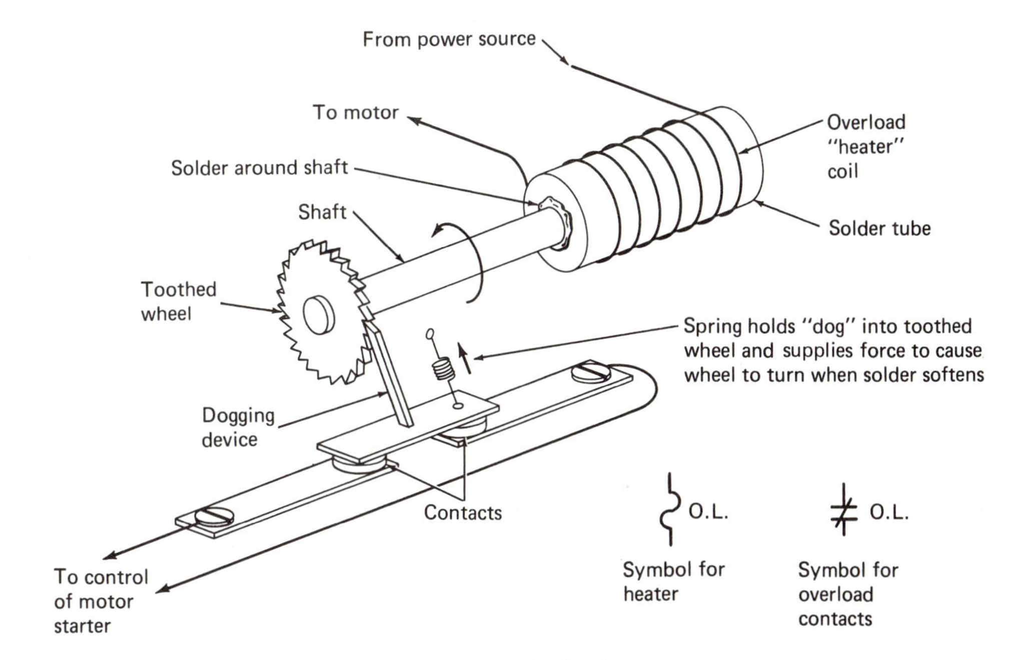

Figure 5-7 Schematic Arrangement of Overload Relay

conditions the current flow has no effect on the solder in the solder tube that holds the shaft from turning from the force of the spring. On overload currents the solder softens and the shaft rotates, which causes the contacts to open. When the contacts open, the motor is shut down. As soon as current flow is stopped, the solder starts to harden and seize the shaft firmly again. The dogging device is pushed back into place on the toothed wheel, which closes the contacts, restoring overload protection. Pushing the dogging device back into place is called *resetting* and is accomplished by pushing a button on the front of the motor starter.

5-9 Sources of Information for Preparing Elementary Diagrams

The principal source of information for preparing elementary diagrams is the one-line diagram. Basic control data are indicated in abbreviated form on those diagrams and can readily be developed into elementaries. For example, Fig. 5-8 shows the one-line diagram for control of a motor. The symbols indicate that the motor is protected by a combination fused disconnect and an across-the-line motor starter that will have built-in overload relays. A start-stop push-button station is to be used complete with a red indicating light (motor running) and a green indicating light (motor shut down). An emergency stop push button has also been provided. Near the starter is the letter T, which indicates that a control power transformer is provided for low-voltage control power supply.

From the data indicated in Fig. 5-8, the experienced designer can easily prepare an elementary diagram.

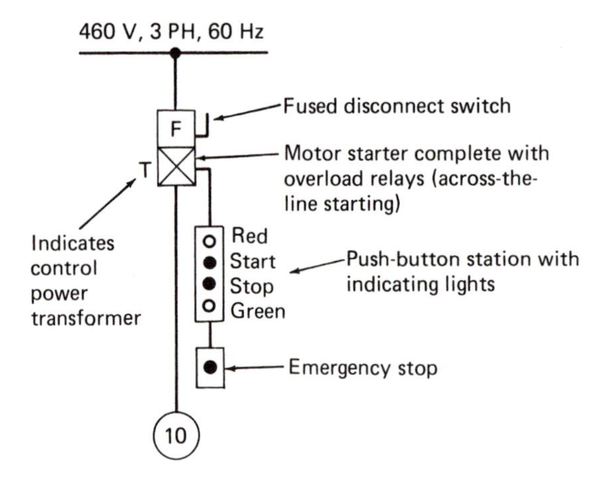

Figure 5-8 One-Line Diagram for Control of Motor

Other sources of information are usually furnished about the mechanical operations required, as well as supplemental wiring requirements that develop during design. For instance, it may develop that the motor of Fig. 5-8 must be interlocked with another motor and prevented from starting until the other motor is running. This condition happens frequently on conveyor-belt systems to make certain that all belts are running before material is loaded onto the first belt. This information becomes available as the mechanical systems are engineered and the electrical designers are advised of how a system must function. Changes are continually taking place and the alert designer must keep informed to be sure that the electrical system will be properly designed.

Some firms use a logic diagram system, which is a diagrammatic tool for outlining the way in which systems are to be designed to operate. It is a convenient method of recording what is permitted to happen, not to happen, or only to happen if other events have occurred, etc. Therefore these logic diagrams are also a source of basic data for preparing electrical elementaries.

5-10 Preparing the Control Power Source

It is a generally accepted practice to start elementary diagrams by first laying out the source of power that provides the energy to operate the devices in the circuit. The source of power for this control bus may be either alternating or direct current, depending on the system involved. In power plants, substations, and other locations where a reliable source of control power is essential, the source is often a bank of batteries. In less critical facilities, it is usually taken from the power supply for all other loads. Usually this is an alternating-current supply. It is good practice to keep the control power voltage at 120 V ac; direct-current supplies are usually 125 V. When the main alternating-current power is higher than 120 V ac, a small control power transformer can be installed to provide the necessary voltage reduction. The control bus is drawn

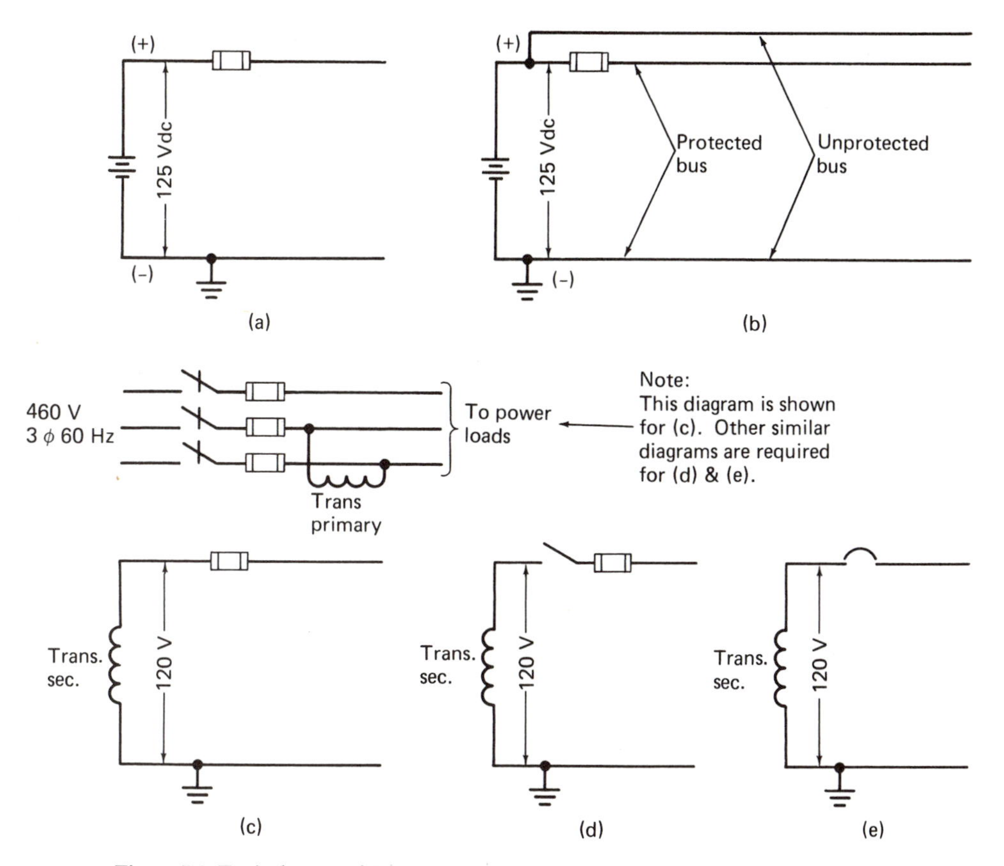

Figure 5-9 Typical Control Bus Power Supply Notations

either as two horizontal lines or as two vertical lines, depending on personal preference. The source of power may be noted, if desired (it is helpful), and the voltage should be noted. Any control or overcurrent protection of the control bus should be indicated. Figure 5-9 shows several control bus arrangements.

Note that all the alternating-current supplies have their source through a control power transformer. The freedom allowed elementary diagrams, of separating parts of devices in order to simplify the diagram, is indicated here. The transformer primary coil is shown near its supply and the secondary coil is shown as the source for the control power. Actually the two coils are physically wound on the same frame.

5-11 Developing an Elementary Diagram

With the foregoing data in mind, it is possible to develop a simple elementary diagram. The control diagram for the one-line diagram of Fig. 5-8 can be used as an example.

First it is important to recognize that there are two seperate circuits involved in the operation of the motor. There is the *power circuit*, which supplies 460-V, three-phase, 60 Hz power to the motor, and there is the *control circuit*, which provides the means for controlling the motor. A complete elementary diagram will show both circuits so it can be readily seen how the wiring of one relates to the wiring of the other. Figure 5-10 shows the power circuit, including the connections to the primary winding of the control power transformer.

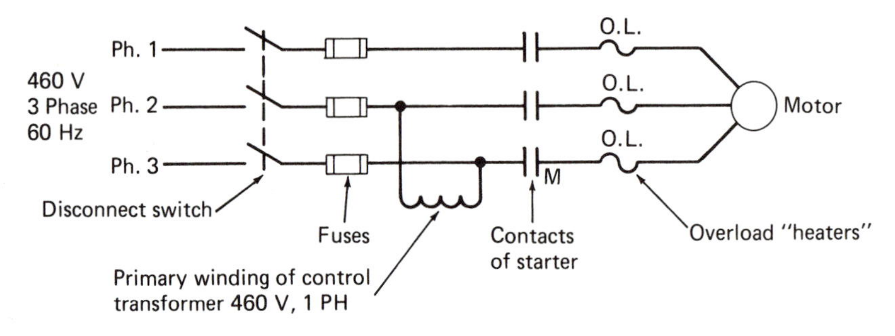

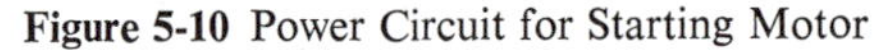

Figure 5-10 Power Circuit for Starting Motor

The control circuit is drawn in steps in the following figure in order to illustrate the thought process that the designer might follow.

Step no. 1 draw control bus

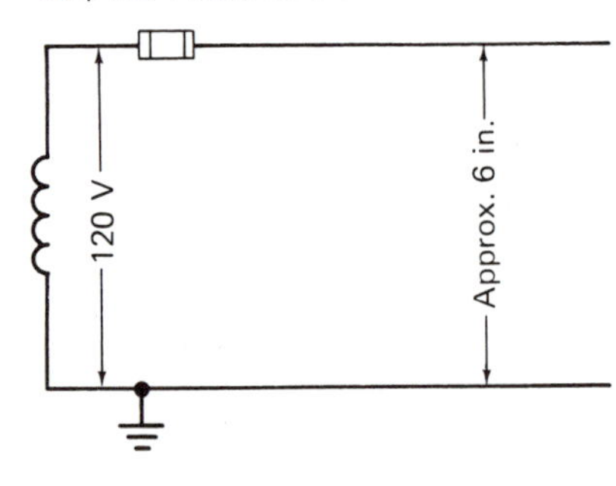

On a full-size drawing a good spacing between the horizontal (or vertical) lines is 6 in.

Step no. 2 draw in stop button

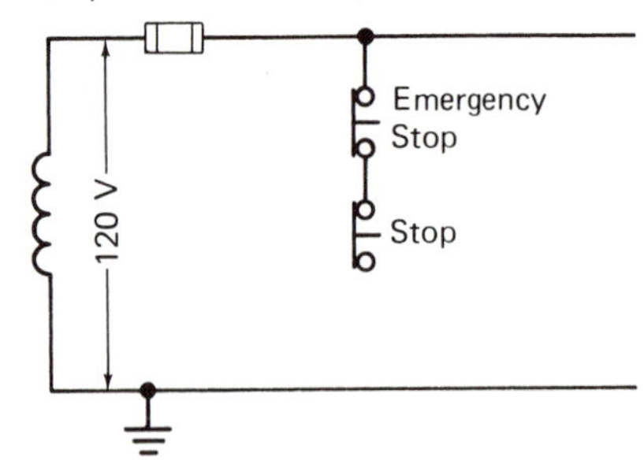

Stop buttons are always normally *closed.* The symbol shows a bar across the terminals and a push rod symbol connected to the bar. Pushing on the end of the rod will open the circuit. These buttons are spring returned to their normal position. Stop buttons are usually the first devices connected to the control bus and stop buttons are always in series with each other. Pushing *any* stop button will open the circuit.

Step no. 3 draw in start button

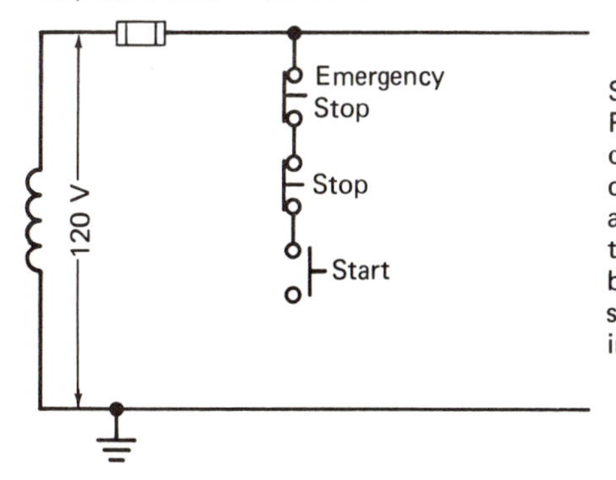

Start buttons are always normally open. Pushing on the button will cause the bar to connect across the terminals and close the circuit. Note that the stop & start buttons are connected next to each other because they are in the same location. If the stop button at the motor is placed next to the start button extra wiring would be required in the actual installation.

Step no. 4 draw in motor starter coil

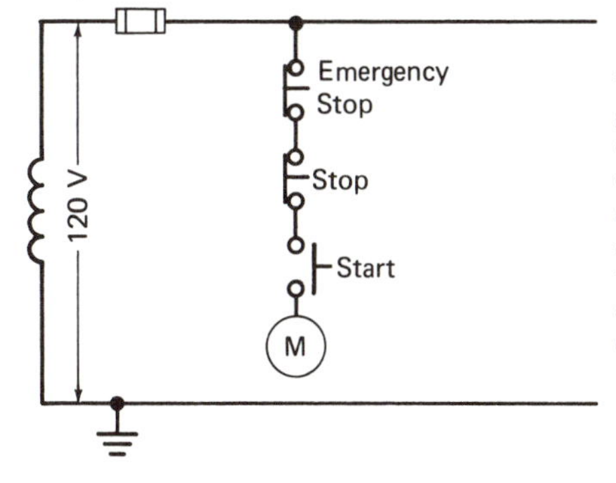

The coil of the motor starter is actually a *relay* (see fig. 5-2). When the coil is energized the contacts that are caused to open or close will do so. The coil therefore is the principal element in the circuit in terms of causing an action to take place. This step in developing the circuit completes connection to *one* side of the coil. The other side must be connected, of course for the coil to be energized.

Step no. 5 connect coil to control bus through overload contacts

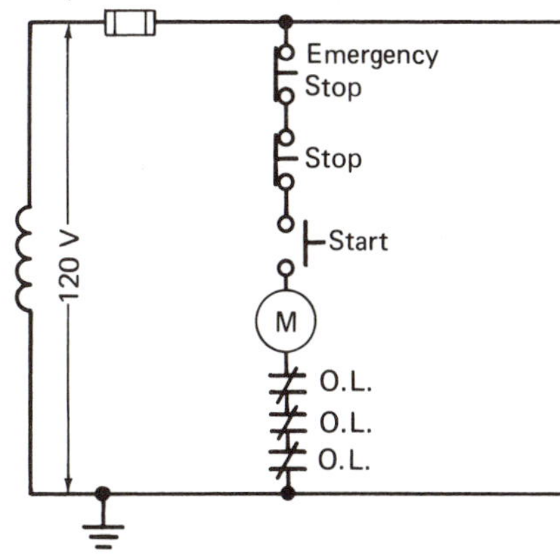

Contacts of the overload heaters (see fig. 5-7) are connected in series between the coil and the line of the control bus. If anyone of these contacts opens the coil will be de-energized and the motor will stop. This completes the circuit to the coil and now pressing the start button will energize the coil. But because the start button is spring return to open when the button is released, the coil will be de-energized and the motor will stop. *This is no good* so step 6 must be made to overcome the problem.

Step no. 6 connect seal-in contact around start button

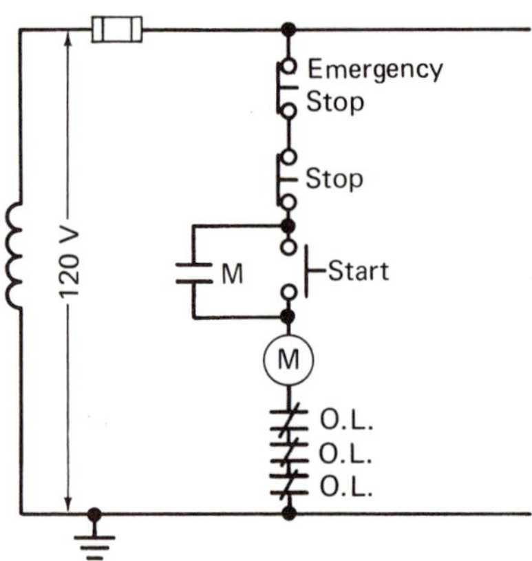

A "seal-in" contact "M" is connected in parallel with the start button so that the start button can be released after it is used to start the motor by energizing coil M. Once coil M is energized it will cause the M contact to close. This will seal-in the circuit to the coil and keep it energized until a stop button is pressed or an overload contact opens. When coil M is de-energized the M seal-in contact will open.

Step no. 7 add indicating lights

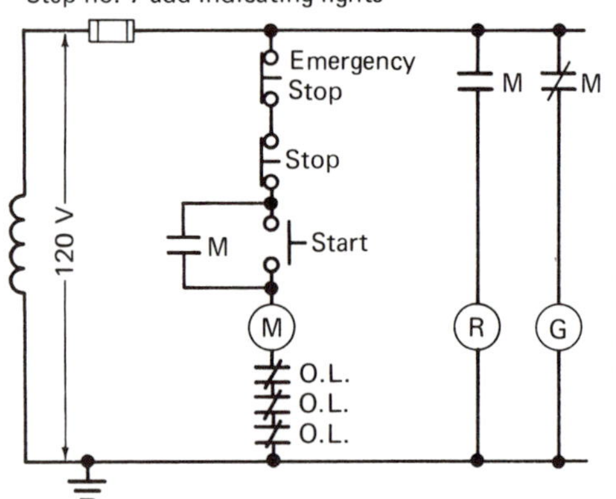

Additional contacts of the M coil relay are used to control the indicating lights. When the coil is energized, the normally open contact wired to the red light closes and the light is energized. At the same time the green light, which was on when the motor was shut down, will go out as the normally closed contact opens.

Step no. 8 add power circuit diagram to complete the diagram

460 V
3 PH
60 Hz

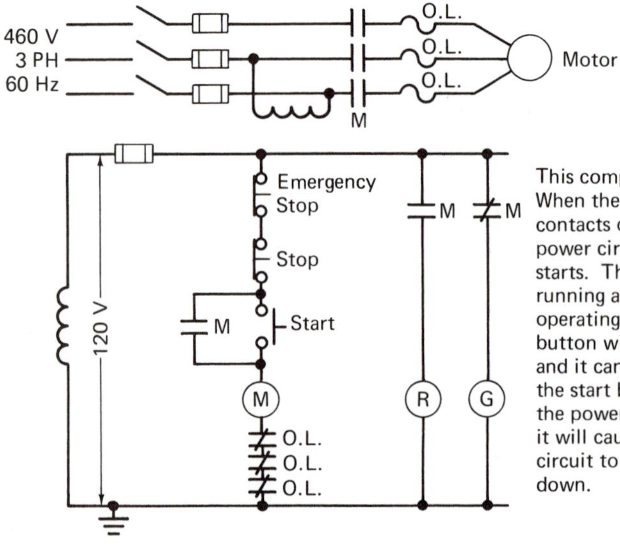

This completes this elementary. When the coil is energized, the M contacts operate. Those in the power circuit close and the motor starts. The seal-in contact keeps it running and the lights show the operating condition. Pressing a stop button will shut the motor down and it can only be restarted with the start button. If a heater ∿ in the power circuit senses an overload, it will cause its contact in the coil circuit to open and shut the motor down.

5-12 Elementary Diagram Incorporating Automatic Control

The motor starter, in the previous example for developing the elementary, used a simple start–stop control. This is a very common control, but automatic operation of motors is required in many cases. The one-line diagram and the elementary diagram for automatic control of a pump through a float switch is shown in Fig. 5-11.

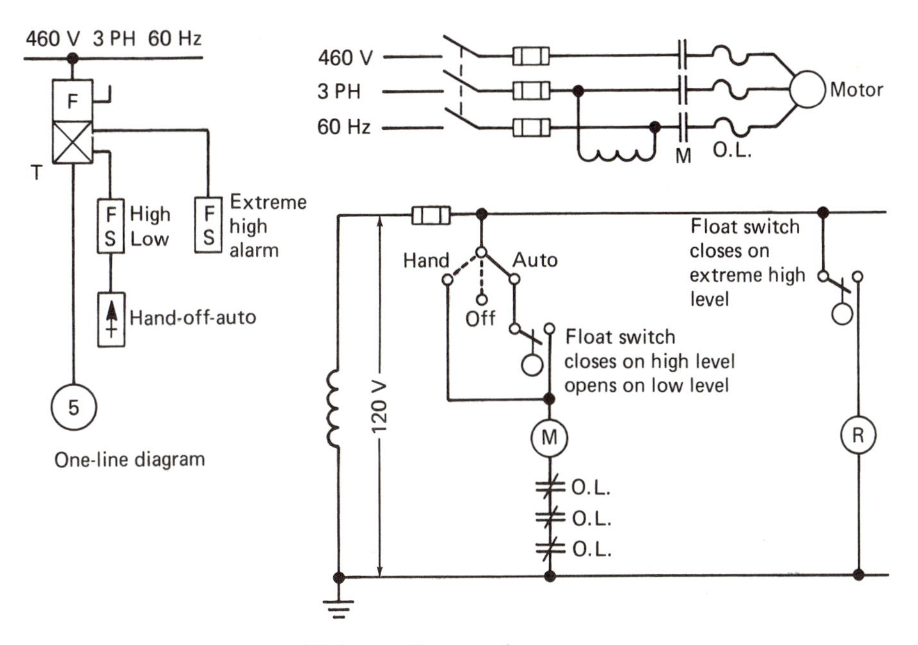

Figure 5-11 Motor Starter with Automatic Control

Control of the motor in the automatic position is through the float switch, which starts and stops the motor as required to pump down the liquid level. Hand operation bypasses the float switch and thereby provides for emergency use or for testing the motor operation. A red high-level alarm light has been incorporated into the design so if the pump fails, a high-level condition is indicated.

5-13 Manual Motor Control

Manual starting switches are designed for starting and protecting small alternating-current and direct-current motors rated 1 hp or less. These switches have the outward appearance of the common toggle switch used for local switching of lights. The switch is an on–off snap switch combined with a thermal overload device operating on the soldered ratchet principle. Typical elementary diagrams for these switches are shown in Fig. 5-12.

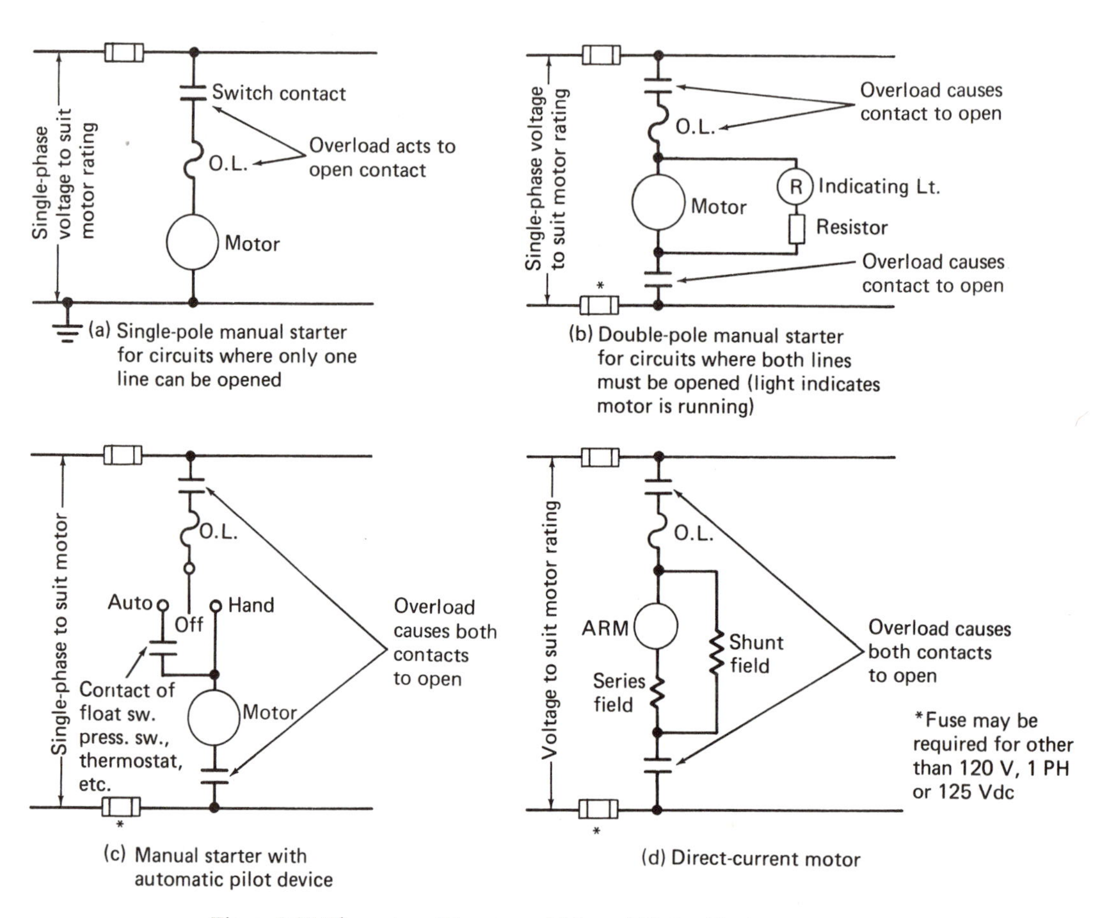

Figure 5-12 Elementary Diagrams of Manual Motor Starters

5-14 Variations of Motor Starting with Start–Stop Push-Button Stations

As examples of different control schemes that are possible through elementary diagram study, the following diagrams show some of the variations possible.

In Fig. 5-13 all motors are started with a single push button. When M1 coil is energized, it starts the M1 motor and also closes the M1 contact to the M2 coil, starting the M2 motor, which in turn closes the M2 contact to the M3 coil, starting the M3 motor. An M3 contact seals in the start button to keep all motors running. The stop button will shut down all motors, as will the opening of any overload contact. For instance, if an overload in the M2 coil circuit opened, the M2 coil would be de-energized, which in turn would open the M2

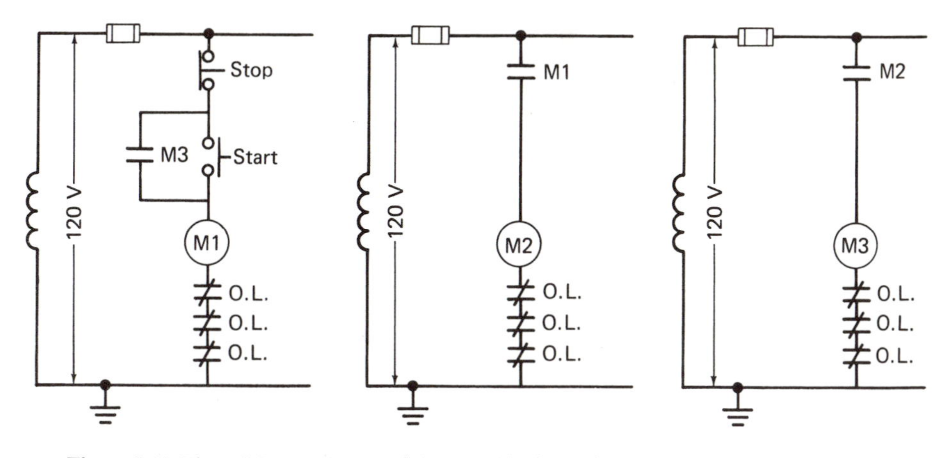

Figure 5-13 Three Motors Operated from a Single Push-Button Station (Power circuit for each motor is similar to Fig. 5-10)

contact to de-energize the M3 coil, and this would open the seal-in around the start button to de-energize the M1 coil. As each coil is de-energized, its motor would stop.

The push-button control for each motor in Fig. 5-14 is standard control. The overload contacts are wired in series so an overload on any motor will shut down all motors. The master stop button is an optional feature that provides added convenience for shutdown. This circuit might be used on a conveyor system, where loss of power to any part of the system would require complete shutdown so that material on the conveyor would not pile up.

There are systems that do not have sufficient capacity to start several motors simultaneously. If the several motors are to be started by a single push button, under these conditions, a time delay relay can be used to meet this requirement. Figure 5-15 shows the elementary diagram for such a scheme. The M1 and TR (time relay) coils are energized simultaneously by the start button. The M1 motor starts, but the timing relay contacts hold the TR contact to the M2 coil open until the preset time has elapsed. When the TR contact closes, the second motor starts. An overload on M2 will shut down only the M2 motor; whereas an overload on M1 will shut down both motors. The stop button also shuts down both motors.

A very common control scheme finds application when a single motor is to be operated from different remote locations. Figure 5-16 shows three push-button stations. Note that all stop buttons are in series so any button will shut down the motor; start buttons are wired in parallel; pushing any button provides a path to the coil. Indicating lights at each station are energized through contacts of the coil and are wired in parallel. Operation of any push button will indicate whether the motor is running (red light) or shut down (green light) at all the stations. This is a desirable feature that aids in operation control.

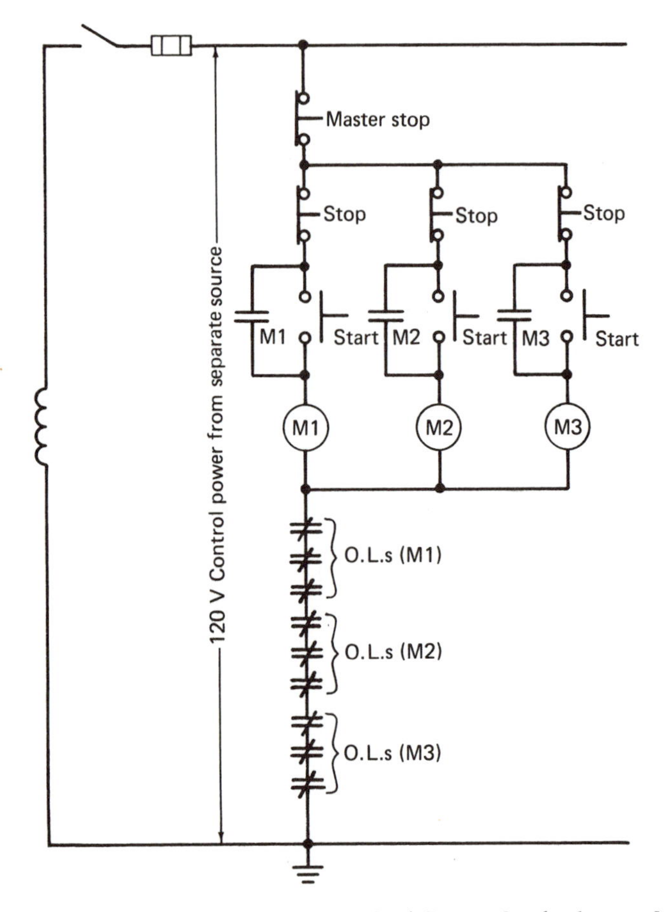

Figure 5-14 Three Motors Each Operated by Own Push-Button-Overload on any Motor Shuts Down all Motors (Power circuits similar to Fig. 5-10)

5-15 Two-Wire Control Circuits

The elementary diagram described in Sec.5-12 and shown in Fig. 5-11 is called a *two-wire control circuit* because control of the pump motor is through a two-wire device—the float switch. There are many variations of this circuit because of the many types of pilot devices that form the two-wire control. Additional two-wire control elementary diagrams are shown in Fig. 5-17 and 5-18.

The two-wire control device can be any device that operates automatically to close and open its contacts. The high-pressure cutout switch is a safety device to shut down the circuit if an unsafe condition develops. (Note that the high-pressure cutout operates for both hand and auto positions.) The disadvantage to the circuit is that a no-voltage condition will cause the motor to stop, but then it will start again immediately when voltage is restored. There is a hazard in this possibility because a maintenance worker might be investigating the

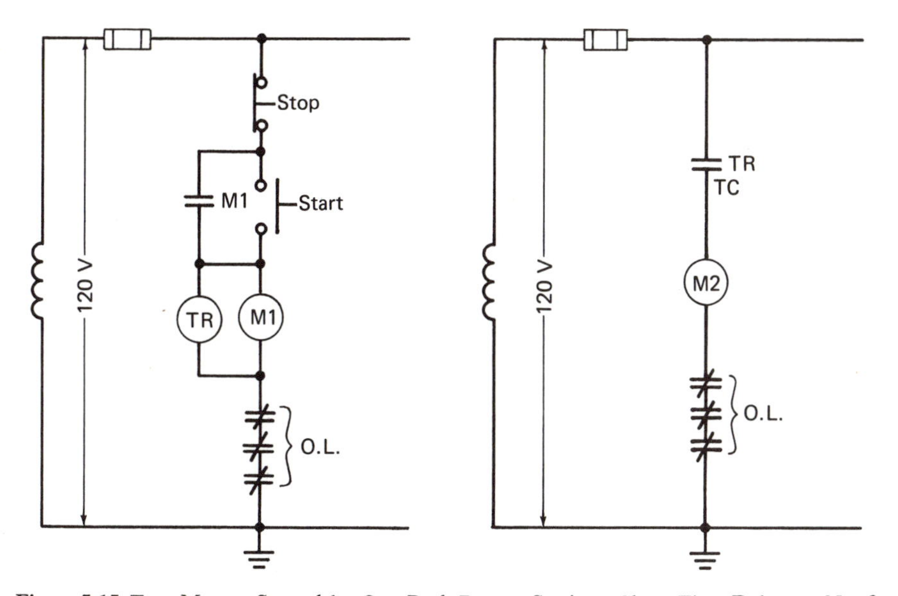

Figure 5-15 Two Motors Started by One Push-Button Station—Short Time Delay on No. 2 Motor Starting (Power circuit similar to Fig. 5-10)

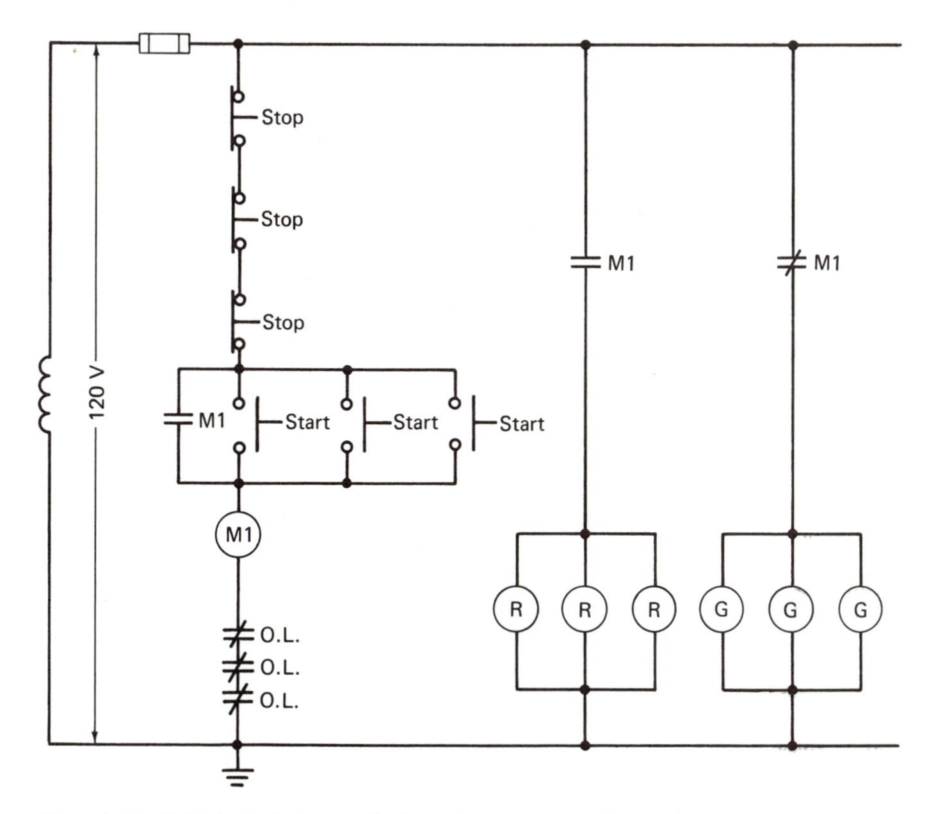

Figure 5-16 Multiple Push-Button Stations Operating One Starter (Power circuit similar to Fig. 5-10)

shutdown and not expect the system to start up again without warning. To overcome this problem, some two-wire circuits are designed with a control relay and reset push button as shown in Fig. 5-18.

To set up the automatic two-wire control for the circuit in Fig. 5-18, the reset push button must be pressed, energizing the CR relay coil, which in turn

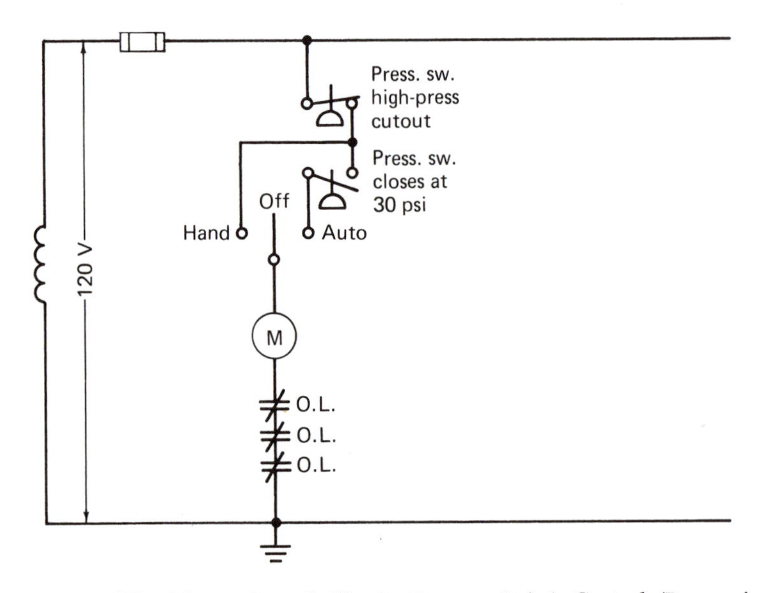

Figure 5-17 Two-Wire Motor Control Circuit—Pressure Switch Control (Power circuit is similar to Fig. 5-10)

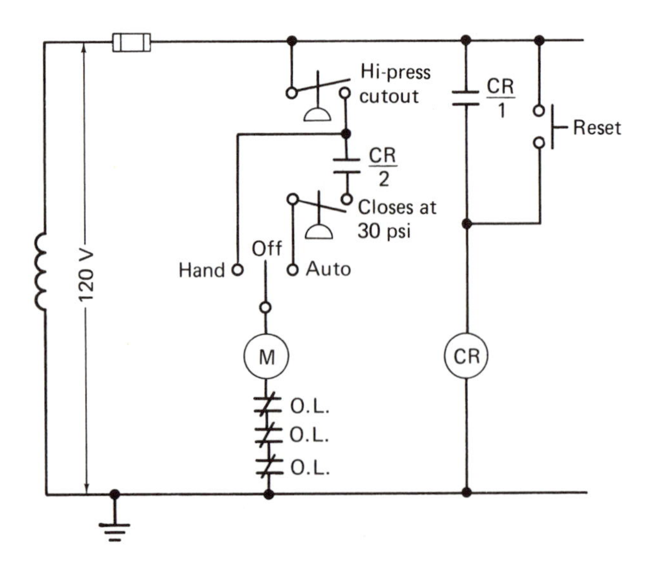

Figure 5-18 Two-Wire Motor Control Using Control Relay for No-Voltage Protection (Power circuit is similar to Fig. 5-10)

closes CR/1 to hold the coil energized and CR/2 to allow for automatic control through the pressure switch. Loss of voltage will de-energize the CR coil and force use of the reset.

5-16 Sequence Control

The need for sequence control of motors and devices is common throughout industry. Conveyor systems are an example of such a requirement. Another is when it is desired to have a second motor start automatically when the first motor has stopped. This might occur when the second motor is needed to run a cooling fan or a pump. Figure 5-19 shows the elementary diagram for a typical scheme.

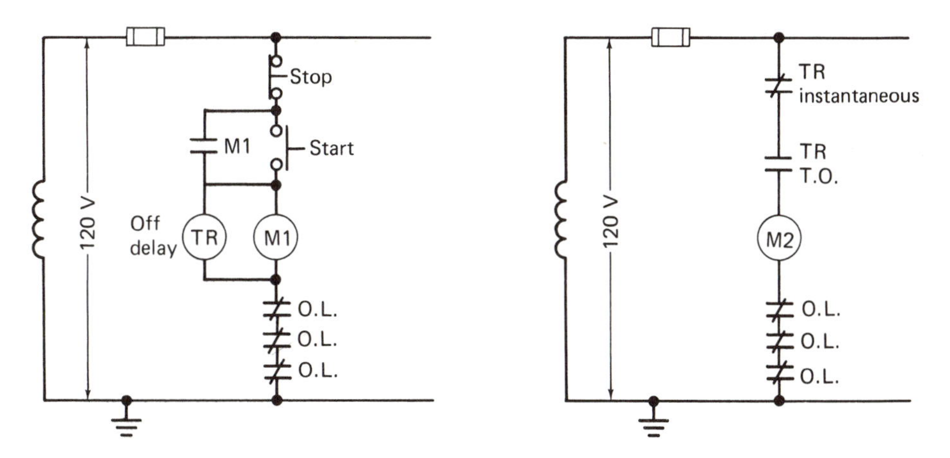

Figure 5-19 Sequence Control of Two Motors—One to Start and Run for a Short Time after the Other Stops (See Fig. 5-10 for power circuits)

When the start button is pressed in Fig. 5-19, both coils M1 and TR are energized. The M1 motor starts and the M1 contact seals around the start button. The TR/instantaneous contact, in the circuit to the M2 coil, opens and stays open. The TR, time open contact closes, but because the instantaneous contact opened, the M2 coil is not energized and the M2 motor cannot start. As soon as the stop button is pressed, both M1 and TR coils are de-energized. The M1 motor stops. The TR/instantaneous contact returns to its normally closed position and because the time opening contact of the TR relay is closed, the M2 coil is energized and the M2 motor starts. The M2 motor will run until the time opening contact times out and opens.

Sequence control for a conveyor system is shown in Fig. 5-20.

The schematic arrangement of conveyors shown with the elementary diagram illustrates that conveyor No. 1 must be running before No. 2 is started; otherwise materials on No. 2 would pile up at the base of No. 1. The control circuit is simple. Additional motors could be started in the same way. Automatic start-up could be easily arranged through the use of time delay relays to replace

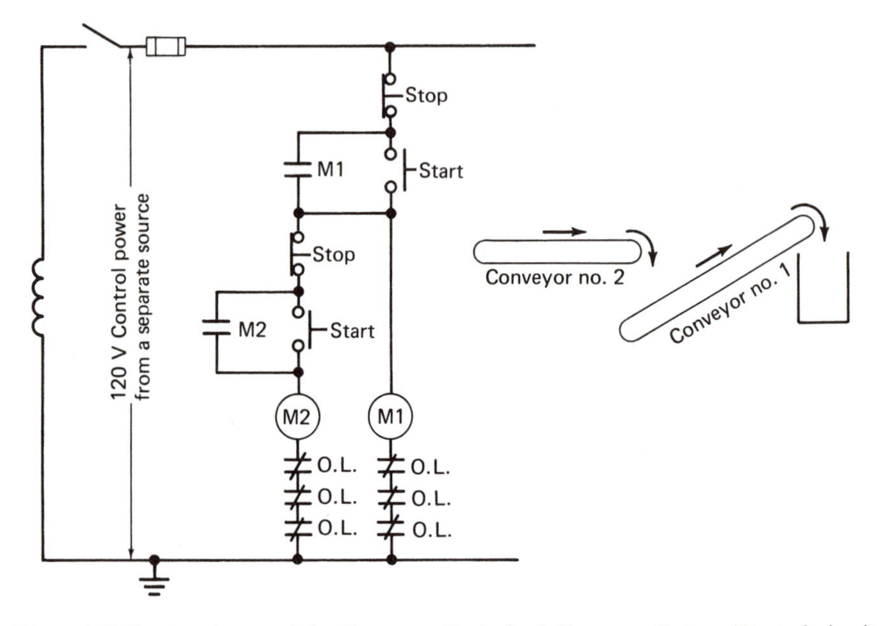

Figure 5-20 Starters Arranged for Sequence Control of Conveyor System (Control circuit similar to Fig. 5-10 except control power is from separate source)

the start buttons, which follow in sequence the starting of the first conveyor by hand.

5-17 Reversing Control of Three-Phase Motors

In order to reverse the direction of rotation of a three-phase motor, any two conductors to the motor must be interchanged. Motor starters can be purchased to perform the interchange operation. Figure 5-21 shows the one-line diagram, the power circuit, and the typical one-line diagram for a typical reversing starter.

The power circuit of Fig. 5-21 shows that when the reversing contacts (REV) are closed, the phase 1 line lead connects to the phase 3 motor lead in forward operation and the phase 3 line lead connects to the phase 1 motor lead. Phase 2 does not interchange. The control diagram provides push buttons for both directions plus a single stop button. Note that a normally closed electrical interlock contact from each coil is placed in series with the opposite coil directly after the start button. If the forward button is pushed, the coil cannot be energized when the motor is running in reverse. The same is true for the reverse operation. This is a safety feature to prevent a short circuit at the power contacts. In addition, most reversing starters are provided with a mechanical interlocking device; this prevents closing of the opposite direction contacts, when one set is closed. This control scheme requires stopping the motor from running in either direction, by use of the stop push button, before direction can be reversed.

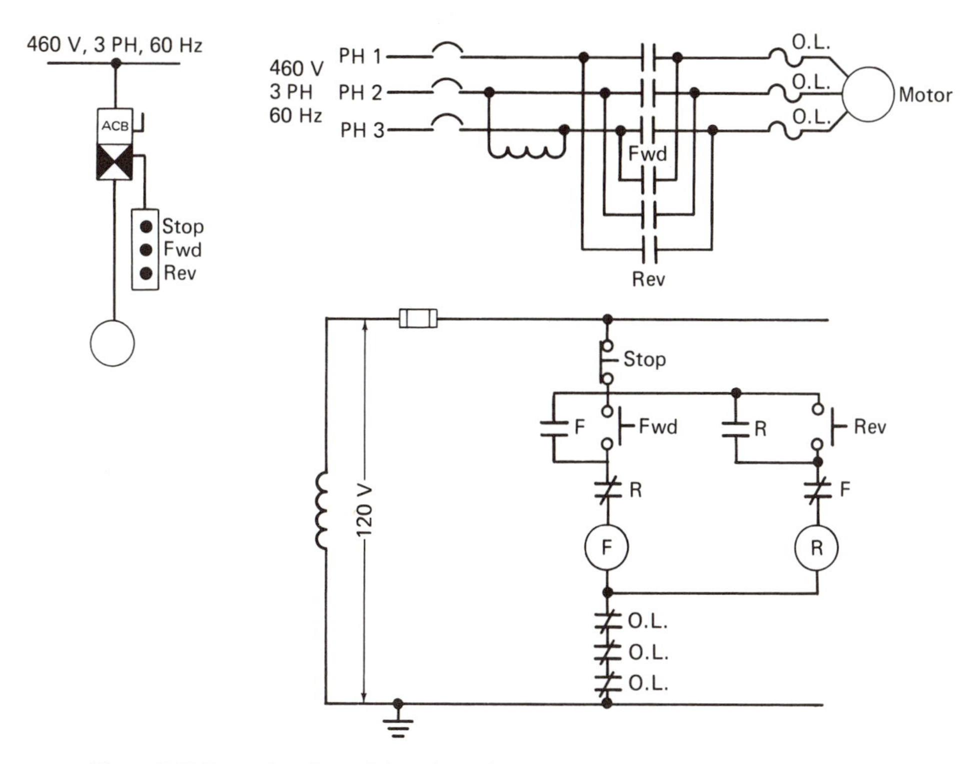

Figure 5-21 Reversing Control for Three-Phase Motor

5-18 Assigning Wire Numbers on Elementary Diagrams

Elementary diagrams are developed and used in order to see the circuits in their simplest form. To utilize the diagram in the practical sense of wiring between devices and in keeping track of the individual wires, an identifying number should be assigned to each wire. It is important that only one number is given to the same wire even though there may be several branches to that wire in the circuit. Wire numbers should change when the circuit passes through a device. Standard wiring of motor starters by manufacturers generally assign wires numbered L1 and L2 to the control bus with wire No. 1 as the line lead after the bus protection and wires No. 2 and 3 to the wires on either side of the seal-in contact. There is no fixed standard for wire numbers beyond those generally used by manufacturers, as described above, so consecutive numbers can be assigned to suit the designer's requirements. A typical elementary with wire numbers is shown in Fig. 5-22.

The legend for devices in Fig. 5-22 is as follows:

T: Clutch coupling temperature cutout

LLS: Lower limit switch—closed except when door is fully closed

RLS: Raise limit switch—closed except when door is fully open

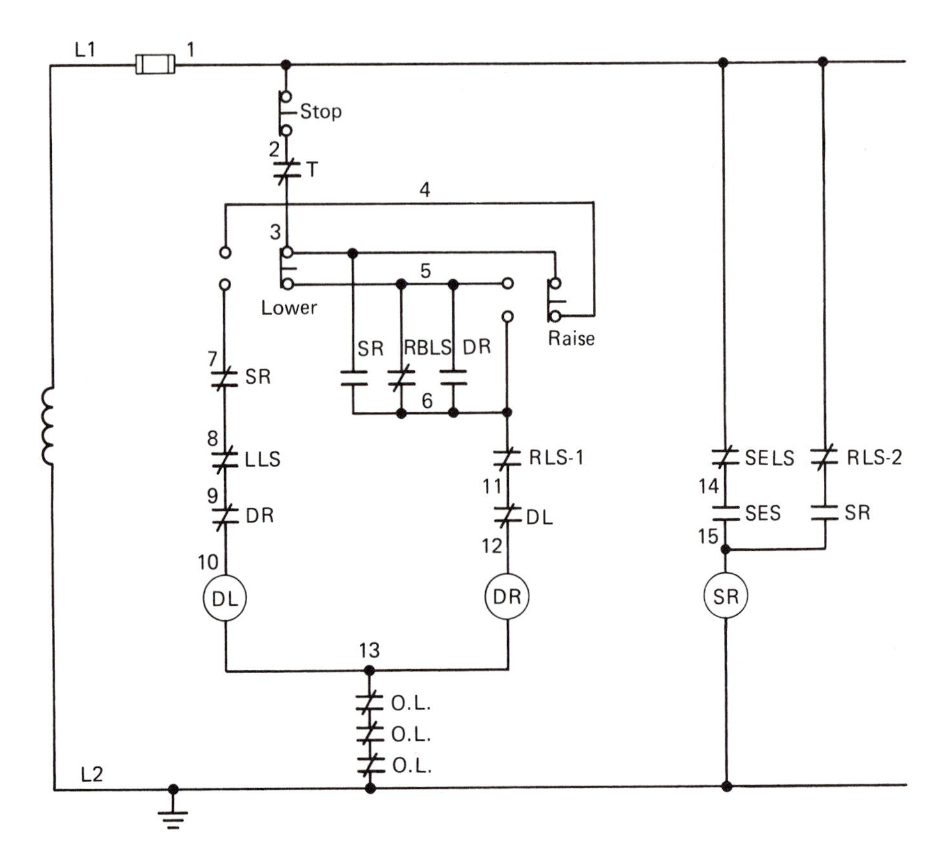

Figure 5-22 Elementary Diagram for Overhead Door Operator Showing Assignment of Wire Numbers

RBLS: Rollback limit switch—closed except when door is within 6 in. of the floor

SELS: Safety edge limit switch—closed except when door is within 6 in. of the floor

SES: Safety edge switch—momentary contact—closes when door hits object

SR: Safety edge relay

Close examination of Fig. 5-22 will show that wire numbers change only if the circuit passes through a device and the number is shown *only once*. Repeating numbers, even though on the same wire, can lead to possible numbering error if a device were cut into the wire due to a design change. Modifications to wire numbers such as postscript A, B, C, etc., identifications can be used if it is felt they serve some useful purpose. As an example, assume that an additional interlock contact were cut into the circuit right after the T interlock; it might be useful to identify the wire between T and the new interlock as 2A rather than using No. 17, which is the next consecutive unused number. This is a matter of judgment in maintaining relationships between devices.

The elementary diagram of Fig. 5-22 is for use with motor-operated truck court doors. The control system is designed so that the door may be manually stopped at any point during the raising or lowering cycle. A safety edge switch is provided on the bottom of the door so that if the bottom of the door strikes any object at a point more than 6 in. above the floor the door will automatically rise to the fully open position.

Examination of the elementary will show that there is no seal-in circuit around the lower push button so this button must be held closed to lower the door. This requirement forces the operator to operate the door manually.

5-19 Determining Wire Requirements by Reading Elementary Diagrams

In addition to making it as simple as possible to understand a control scheme, the elementary diagram is the principal source of information in counting up the wiring requirements. As an example, the safety relay in Fig. 5-22 is to be mounted on the wall adjacent to the door. How many wires go to it? Review of the elementary will provide the answer.

The wires going to the SR relay are

L2 and 15 = 2 (coil leads)
7 and 8 = 2 (b contact in DL coil circuit)
3 and 6 = 2 (a contact seal-in around raise p.b.)
16 = 1 (lead between RLS-2 b contact and SR a contact in SR coil circuit)
Total 7

This counting of wires can be done for all elements in the circuit so that physical wiring layouts can be developed.

Chapter Review Topics and Exercises

5-1 What principal advantage is there in showing a circuit in elementary form?

5-2 Why is it necessary to note the operating condition of such devices as pressure switches and float switches beside the device symbol when shown in an elementary diagram?

5-3 Describe a relay.

5-4 Describe a solenoid.

5-5 Why are selector switches used?

5-6 Describe how an overload relay operates.

5-7 What are the usual sources of information needed to prepare elementary diagrams?

5-8 What function does a control power transformer perform?

5-9 What are the two basic circuits required for motor control?

5-10 Explain the steps to developing an elementary diagram for typical motor control.

5-11 Sketch an elementary diagram using a pressure switch for automatic motor control and a hand–off–auto selector switch for selective control. Show a pressure switch to alarm a high-pressure condition.

5-12 Referring to Fig. 5-13, explain how an overload relay opening in the M3 motor circuit would shut down all three motors.

5-13 Why are stop buttons wired in series?

5-14 Why are start buttons wired in parallel?

5-15 Why is no-voltage protection a good idea?

5-16 How is sequence control set up?

5-17 What does a reversing contactor in a motor starter do? Why?

5-18 What basic rule is followed in assigning wire numbers on an elementary diagram?

5-19 In Fig. 5-22, why must the lower button be held closed by hand when the door is being lowered?

5-20 Explain how the number of wires running from one control device to another can be counted by using an elementary diagram.

5-21 Draft any four elementary diagrams shown using drafting tools and techniques. Try drafting one with the control bus run vertically rather than horizontally. Is it easier to read?

6

RISER DIAGRAMS

6-1 Definition

The word *riser*, tends to define these diagrams because they are drawn to show how an electrical system *rises* up through a multistory building and distributes service. Because they show equipment and the means by which the equipment is interconnected, they tend to become an *interconnection* diagram too. They allow for indicating the characteristics of the interconnections and so may also be used as diagrams to indicate conduit and wire and cable data. The definition evolves into the following: *A riser diagram is a drawing that indicates by means of single lines and simplified symbols the distribution of electrical systems in a multistoried structure showing the major equipment interconnections and the characteristics of the interconnections.*

These diagrams are semiphysical in layout because they locate equipment on a floor-by-floor basis and show the routing of interconnecting systems from floor to floor.

These diagrams are generally required only for the wiring design of multistoried buildings although they can be easily used where distribution is confined to one floor.

Riser diagrams are not a substitute for one-line diagrams because they do not show the same information. Where one-line diagrams show the electrical circuit, riser diagrams show how the wires and cables are distributed in the system in accordance with the one-line diagram.

Riser diagrams are used a great deal to show other systems as well as systems for power and lighting. Such other systems include telephone service, fire alarm protection, smoke detection, public address, time clocks, etc.

6-2 Purpose

Riser diagrams provide a simple pictorial-type layout of system distribution that are easy to read and immediately help all concerned to *see* the physical relationships that the one-line diagrams do not show. Their purpose is to show the interconnecting system in a simplified way that relates in some degree to physical locations where the complexity of several floors of a structure are involved.

6-3 Sources of Information

Sources of information to draw the riser diagrams are primarily the one-line diagram for power and lighting and an understanding of system requirements for other systems such as telephone, fire alarm, etc. Of course a general knowledge of the building structure is necessary. This information is generally available from the architectural drawings.

Whereas the one-line diagram is generally developed by engineers, riser diagrams are almost always developed by the designer or draftsman.

If the characteristic of the interconnections is to be indicated, then sizes of raceway, wires, cables, etc., must be determined (see Chapter 14).

6-4 Layout of the Riser Diagram

Layout of the riser diagram is fairly simple. Floors are designated by a single horizontal line drawn across the area of the drawing judged large enough to show the diagram. If distribution of a system is confined to *cores* that have a definitive location in the building, then vertical line separations can be made to identify, for instance, the east core and the west core or some similar definable area. As an example, the building shown in Fig. 6-1 was constructed so that the cores were external to the main floor areas and therefore did not take away valuable space in the center of the building. These cores provided space

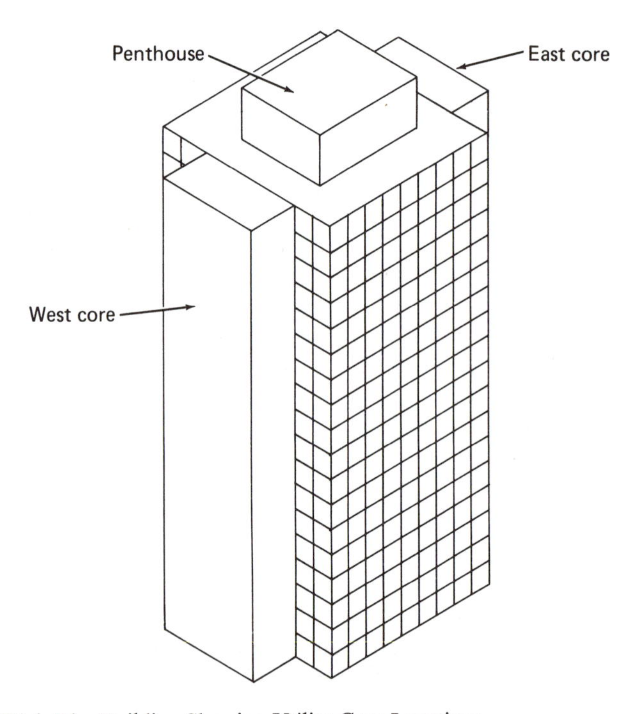

Figure 6-1 High Rise Building Showing Utility Core Locations

for stairways, elevators, toilets, and electrical and communications closets. They therefore could be classified as the utility section of the building and called *utility cores*.

With this background knowledge, the riser diagram could be outlined as shown in Fig. 6-2. The ground floor lobby and an upper mezzanine and concourse level are shown below ground. Vertical lines separate the diagram into an east core and a west core area. Spacing of lines is *not* important. It may be necessary to have more space between the lower mezzanine line and the lobby line because more equipment is located on the lower mezzanine. Some rough layouts will generally be needed to determine the spacing sequence.

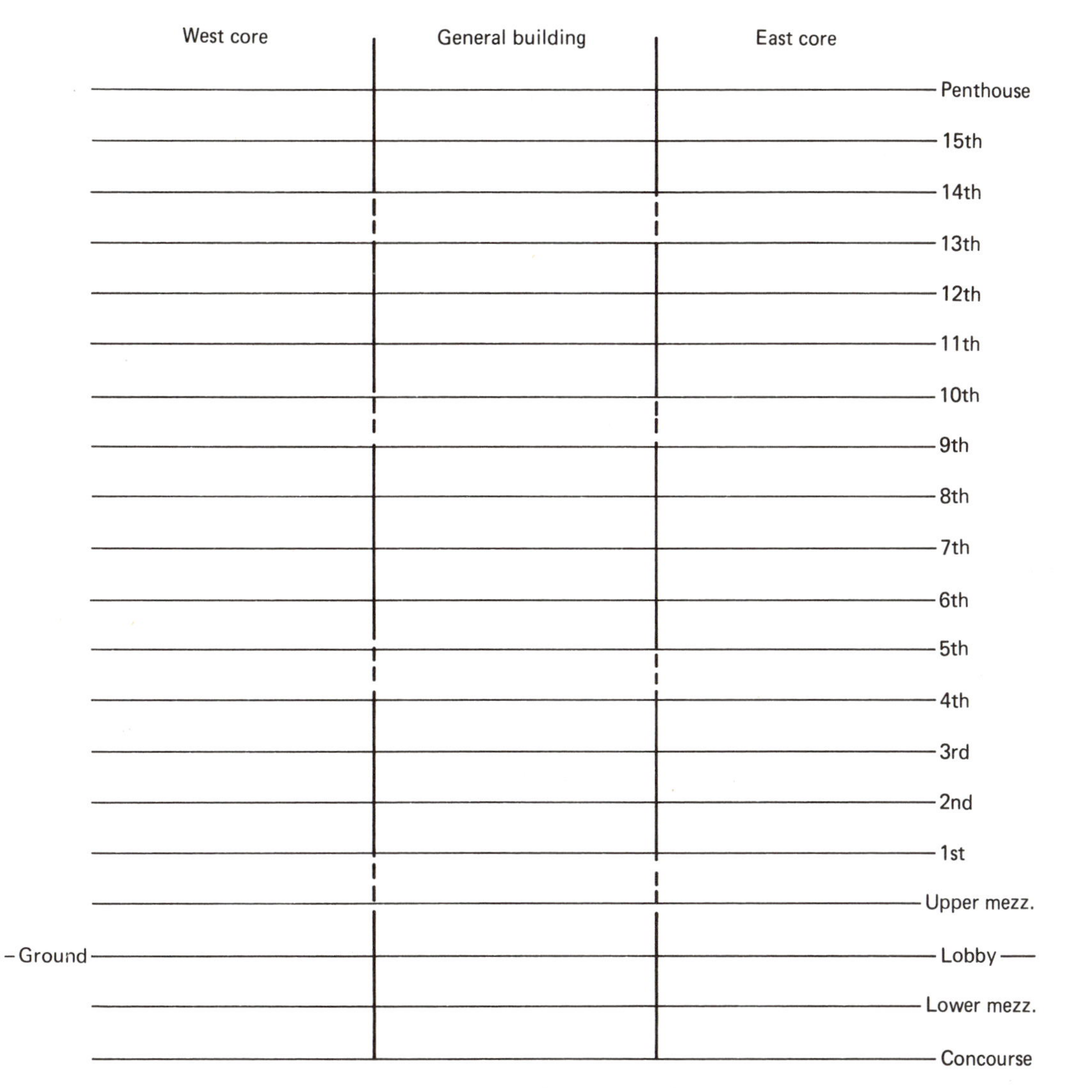

Figure 6-2 Outline Layout for Riser Diagram of Multistory Office Building

6-5 Symbols for Riser Diagrams

Symbols for riser diagrams are generally not standard although one-line diagram symbols can be used if they suit the need. Usually equipment is shown as a rectangle or a square; drafting templates are very handy for this need. Some engineering firms may have developed riser diagram symbols for their use. The types of symbols may therefore vary widely in practical use. A legend as part of the drawing showing the symbols used should overcome any problem in understanding the drawing. Interconnecting lines are always single line with proper identification to indicate their actual characteristic.

6-6 A Suggested Step-by-Step Procedure for Developing a Riser Diagram

1. Layout outline showing floors and vertical separations such as cores similar to Fig. 6-2.

2. Locate major equipment using rectangles or squares or applicable symbols on the proper floors.

3. Following the one-line diagram, connect the major equipment by straight line drafting; i.e., there is no need to show curved lines to depict bends in conduit or cable.

4. Locate panelboards, motor control centers, bus duct, and similar equipment. Use rectangles, squares, or (in the case of bus duct) a heavy line or parallel lines close together to show a reasonable representation of the bus duct.

5. Connect panelboards, motor control centers, etc. into the major equipment at the physical point in the system from which they are fed. For example, a panelboard on the upper mezzanine may receive its feed from a plug-in switch located in the electrical closet on the first floor.

6. Show and connect any miscellaneous items that may relate to other riser diagrams. For example, the fire alarm system will be provided with a power source that shows only on the emergency power riser diagram so picking up and showing the source of supply should be shown on the emergency power riser diagram with a reference made "for continuation" to the fire alarm riser diagram.

7. Label all equipment, interconnections, etc.

6-7 A Typical Power Riser Diagram for the Office Building

The one-line diagram described in Chapter 3 and Fig. 3-38 would be developed as follows:

1. Prepare the basic riser diagram similar to the outline shown in Fig. 6-2. Bear in mind that spacing between horizontal lines is a function of how much needs to be shown on any particular floor.

2. Locate the major equipment as shown in Fig. 6-3. From the one-line diagram, the 15-kV service entrance switchgear is located in the lower mezzanine electrical equipment room. The 480-V unit substation number LM-1 is also in the lower mezzanine electrical equipment room. Unit substations P1 and P2 are located in the electrical equipment room in the penthouse.

3. Because physical relationships are beginning to develop at this time, it becomes important to know if the incoming lines enter the switchgear through the bottom or the top. It is not particularly significant to show which end of the gear they enter because the riser diagram does not orient the equipment in plan view. It may be helpful, however, to plan that the rectangle or square designating the equipment is the front view. In any case, the top and bottom are defined and therefore should be taken into account.

These incoming lines and interconnecting feeders are shown connecting the major equipment in Fig. 6-4.

Because the high-voltage cables feeding up to the unit substation in the penthouse must run vertically in conduit better than 200 ft from the lower mezzanine to the penthouse, these cables must be supported inside the conduit. The riser diagram is an ideal place to call attention to this requirement. The National Electrical Code makes specific rules regarding the spacing of supports and their location. It also defines the types of supports that are acceptable.

In this case it was decided to use an open-mesh sleeve-type support that wraps around the cable and tightens onto the cable as more weight or pull is placed on it. This particular type of grip was originally made up as a basket weave and used as a Chinese toy or puzzle. Pulling on the sleeve made it tighten on a finger and if a finger of each hand was inserted in opposite ends of the sleeve, it would tighten in both directions and make an amusing puzzle as the hands were pulled apart when trying to remove a finger. The harder one pulled, the tighter the sleeve grasped the finger. Completely harmless as a toy, its modern application as a cable grip is widely used. Figure 6-5 shows how this grip was used in supporting the cables in the vertical conduits passing up through the floors in the office building. The support box at the floor level allows for installation and inspection of the support. The coil springs were installed in order to allow for expansion and contraction of the cables under varying temperatures and electrical loading. Grounding-type bushings were installed on the conduits to ensure conduit ground bonding across the box and grounding of the cable grips themselves. Note that a ground conductor runs in the same conduit with the cable (see Chapter 8, p. 199).

4. Locate power centers, motor control centers, the plug-in bus ducts, and any other distribution equipment that are supplied power by feeders from the major equipment. This additional equipment and the feeders supplying them are shown in Fig. 6-6. There is a great deal more equipment in the building; however, this explanation, because of space limitations, is confined to information that has been shown in the one-line diagram Fig. 3-38.

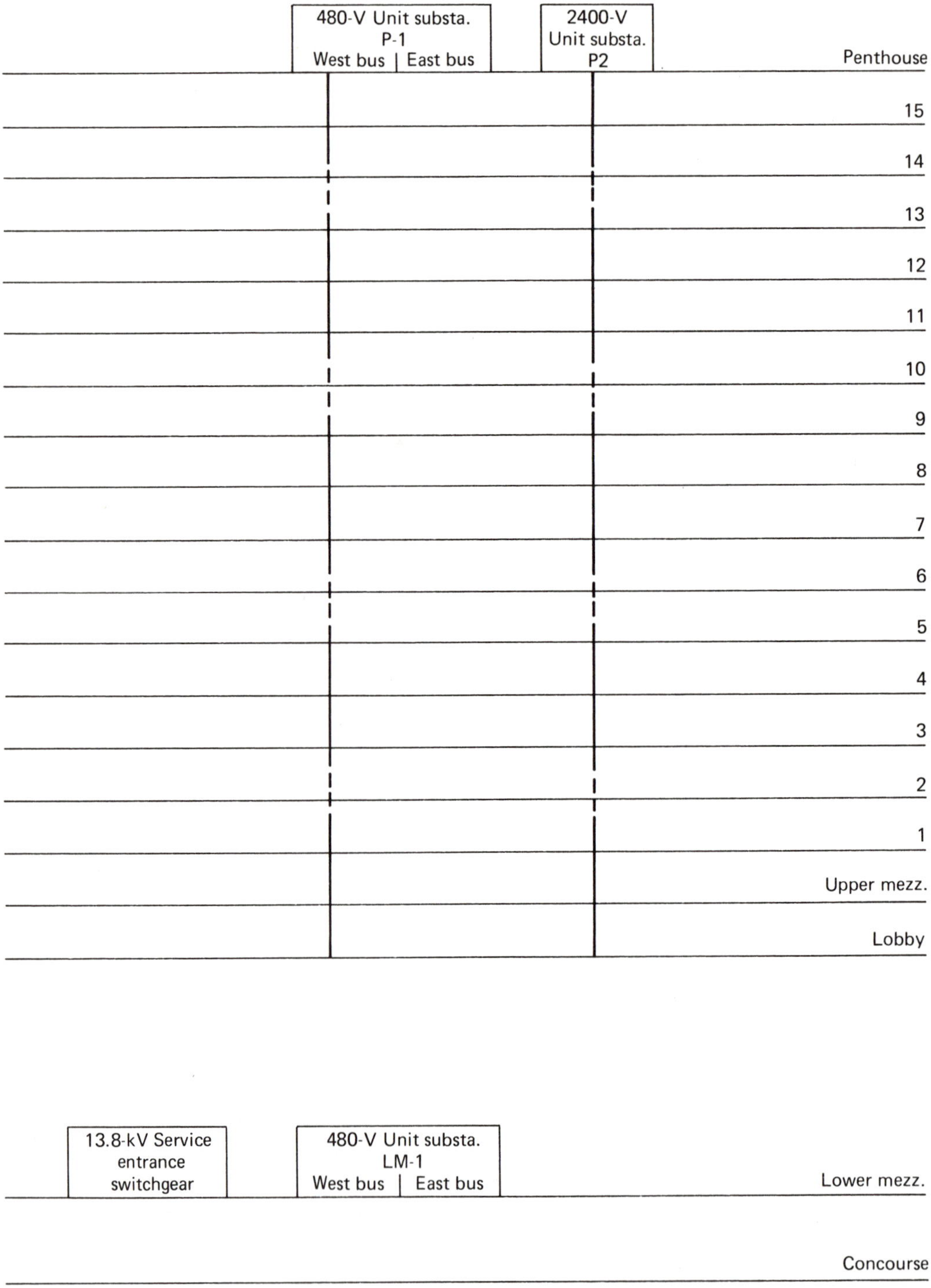

480-V Unit substa.
P-1
West bus | East bus

2400-V
Unit substa.
P2

Penthouse

15

14

13

12

11

10

9

8

7

6

5

4

3

2

1

Upper mezz.

Lobby

13.8-kV Service
entrance
switchgear

480-V Unit substa.
LM-1
West bus | East bus

Lower mezz.

Concourse

Figure 6-3 Locating Major Equipment on Riser Diagram Outline

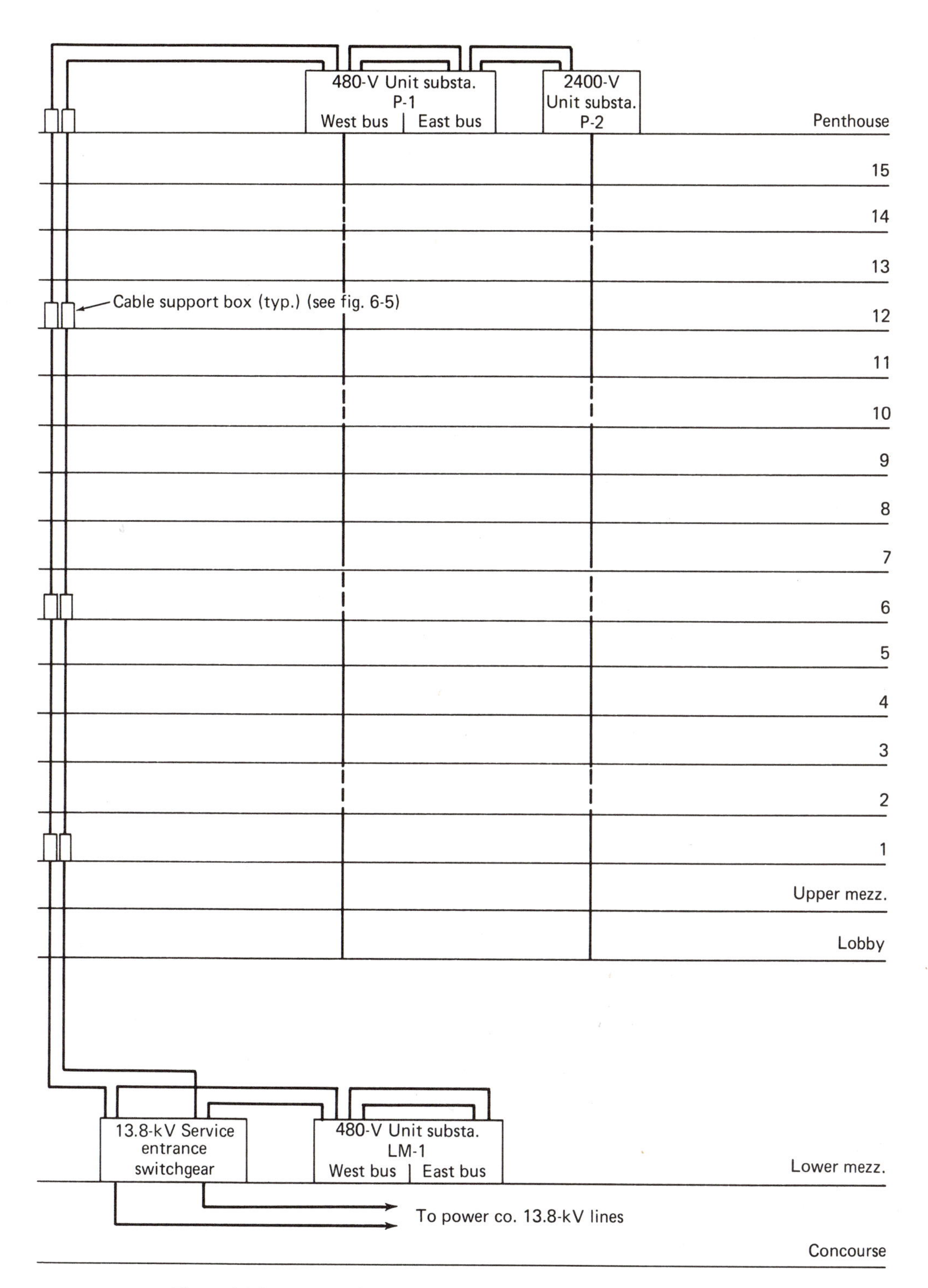

Figure 6-4 Interconnection of Major Equipment on Riser Diagram

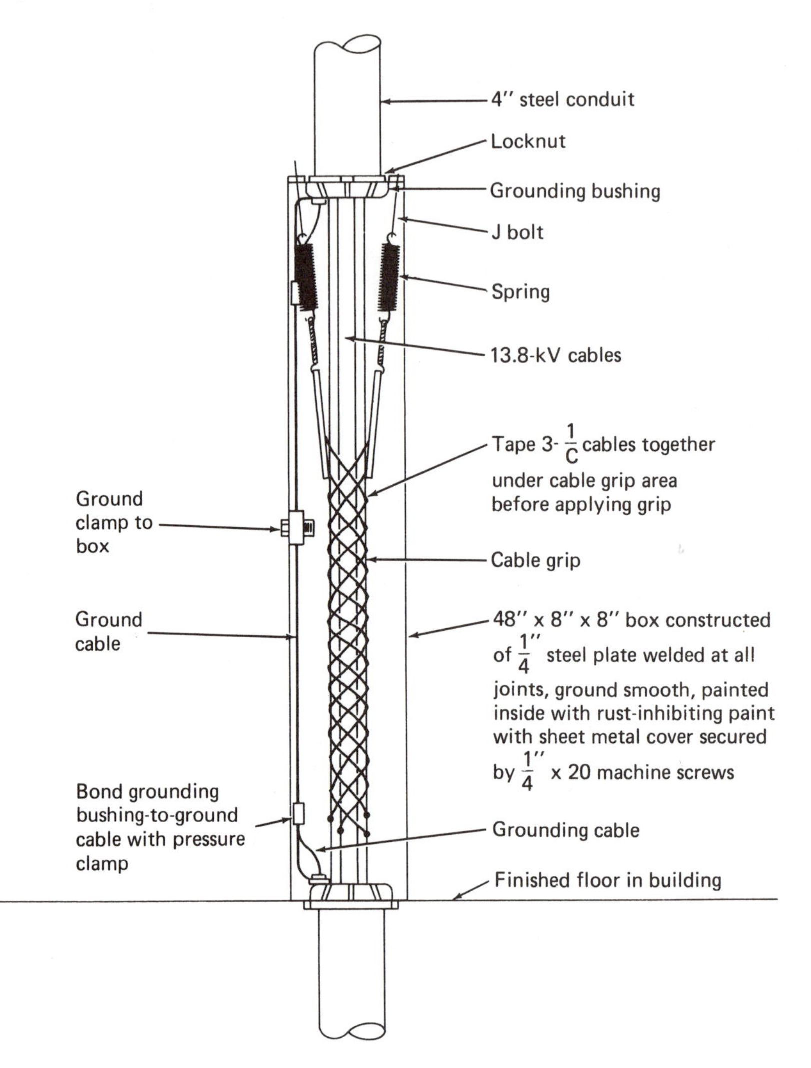

Figure 6-5 Method of Supporting Vertical Cable Runs

5. Locate and show connections into the main feeder distribution equipment, the panelboards, motor starters, transformers, etc., that are the final point of distribution to individual loads by way of branch circuits. Generally this is the extent of distribution shown on riser diagrams. Branch circuit distribution becomes too extensive and complicated to carry this information on riser diagrams. An example of the type of data shown on a particular floor is given in Fig. 6-7 for the office building first floor, west utility core.

This section of the riser diagrams shows distribution from the plug-in bus duct as follows, starting with the top switch in the diagram and working down.

(a) A 60-A fused plug-in switch feeds a magnetic contactor that controls

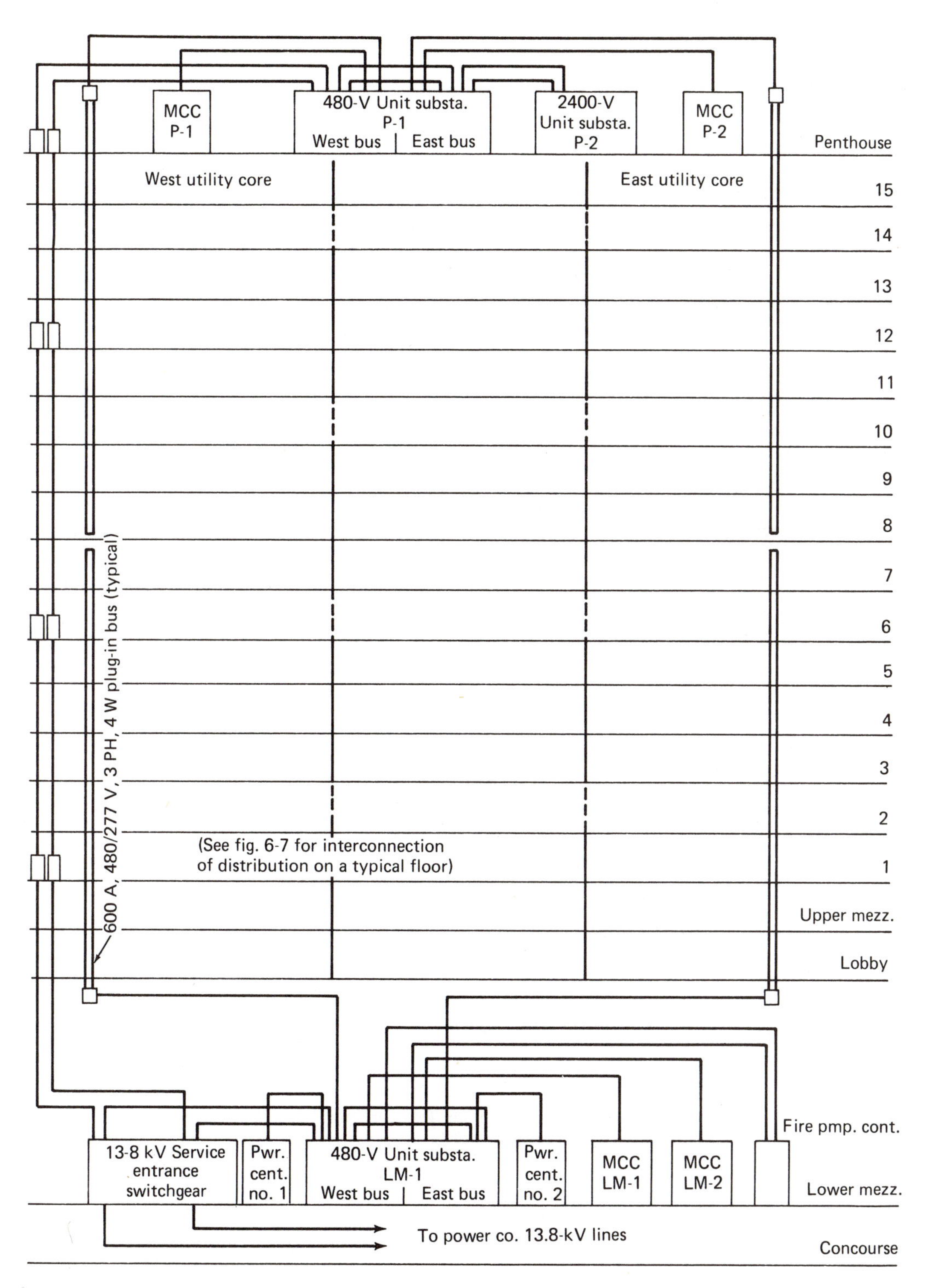

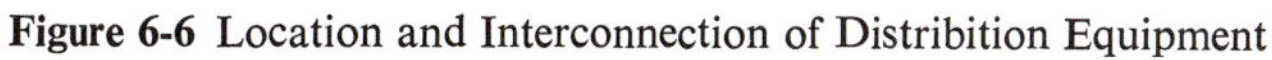
Figure 6-6 Location and Interconnection of Distribition Equipment

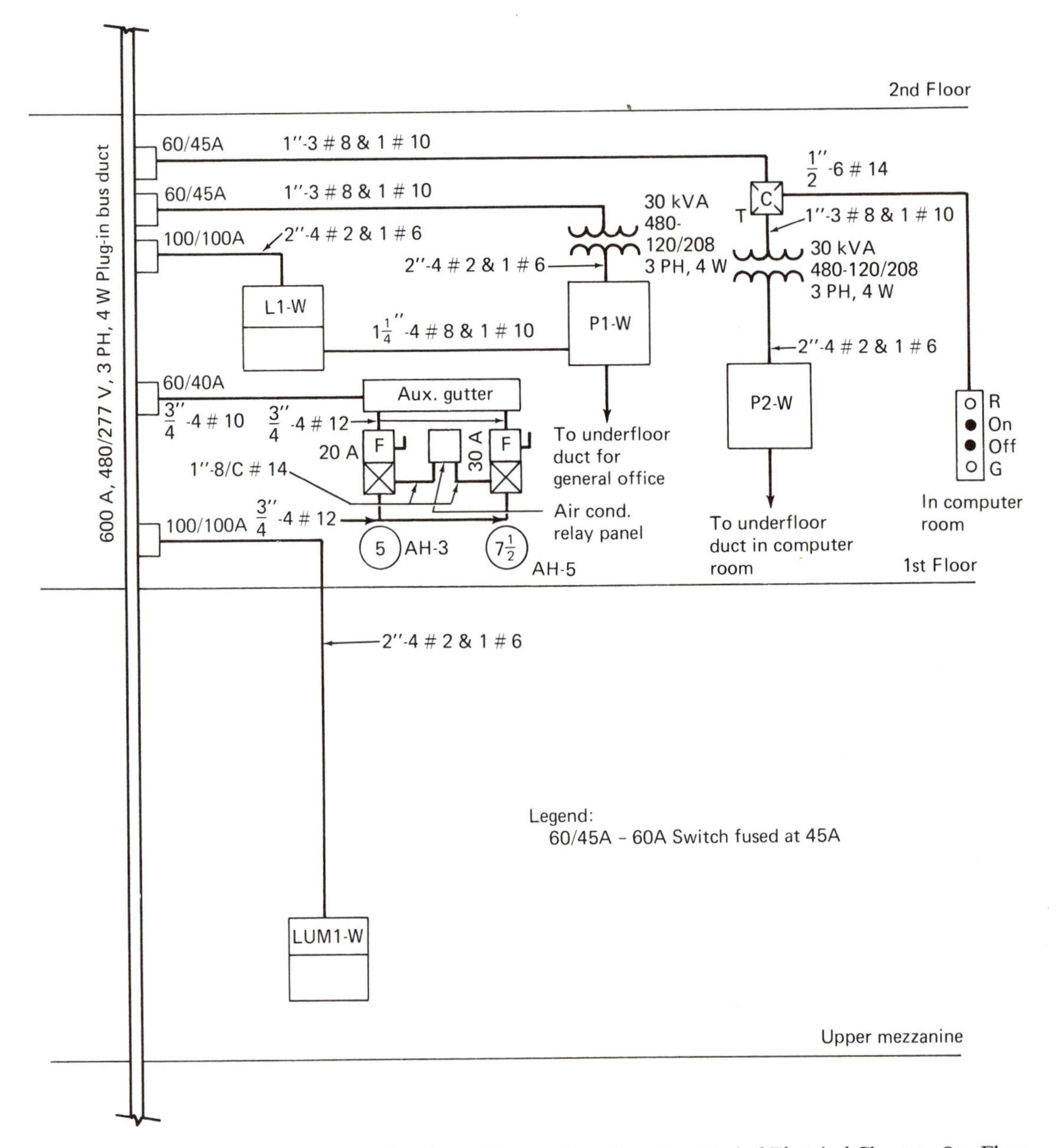

Figure 6-7 Example of Riser Diagram Type Data for a Typical Electrical Closet on One Floor in Office Building

power to a 30-kVA transformer rated 480–120/108 V, three phase, four wires. The transformer feeds into power panel number P2-W. The contactor is controlled by an on–off push-button station having a red light to indicate that the circuit is energized and a green light to indicate a de-energized condition. Panel P2-W feeds branch circuits into the under-floor duct in the computer room. The under-floor duct system is an integral part of the floor that allows for distribution of power and other services through cells running through the structural floor itself. There are many types used. Figure 6-8 shows a section through the type used in the office building.

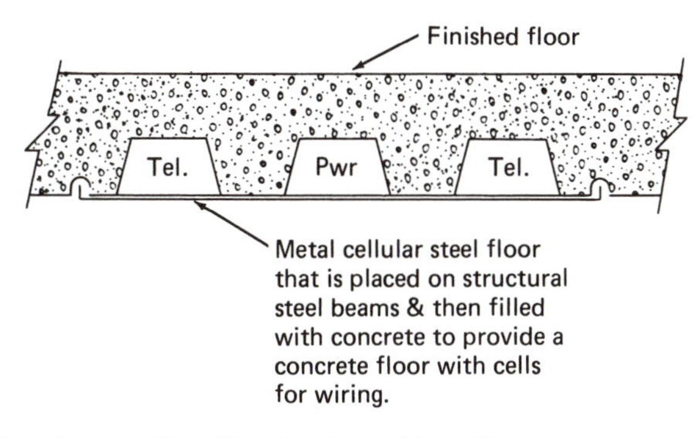

Figure 6-8 Section through Floor Showing Under-Floor Duct

The cells in the floor system can only run in one direction so *header ducts* are installed at a right angle to the cells in order to feed into the cell system and allow for cross wiring from cell to cell. This arrangement is shown in Fig. 6-9.

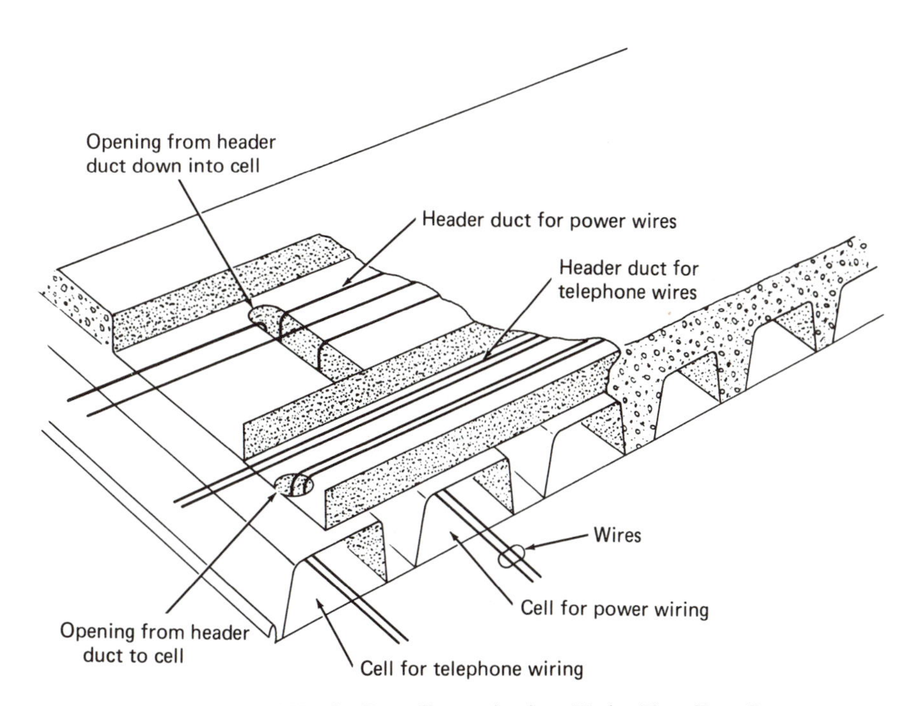

Figure 6-9 Arrangement of Header Duct Connecting into Under-Floor Duct System

The header duct itself is often carried directly up to the panelboard feeding the under-floor system through the use of fittings that are available from the manufacturer. Fittings for tapping into the cells in order to bring wiring to the surface of the floor are the means by which the floor system is used to provide low-voltage outlets and telephone service to almost any location on the floor.

The push-button station controls the contactor that acts electrically to open or close the power supply to the transformer. This provides for emergency shutdown of the power supply to the computer; i.e., if the contactor is de-energized, it will open the feed to the transformer and thereby de-energize the panelboard and all the branch circuits feeding the computer machines.

(b) A 60-A fused plug-in switch feeds directly to a 30-kVA transformer rated 480–120/208 V, three phase, four wire that in turn feeds into panel P1-W. Panel P1-W feeds the under-floor system for the general office area and also feeds the lower section of lighting panel L1-W.

(c) A 100-A fused plug-in switch feeds 480/277 V, three-phase, four-wire power to panel L1-W. This panel has an upper section for the higher voltage and a lower section for 208/120 V, three-phase, four-wire power that is fed from panel P1-W. This is the lighting panel and provides branch circuits at 277-V single phase for general office lighting. The lower section provides 120 V single phase to lighting fixtures that are rated at 120 V such as incandescent lights. It also serves time clocks and miscellaneous 120-V loads such as water coolers.

(d) A 40-A fused plug-in switch feeds an auxiliary gutter that houses the cables from the conduit run and provides the space for making taps so that feeds can be made to the two motor starters. These starters control the motors for air handling units No. 3 and 5, which are part of the air conditioning system Control of the starters is from the air conditioning relay panel.

(e) A 100-A fused plug-in switch feeds lighting panel LUM1-W, which is located on the upper mezzanine Note that the switch, however, is on the first floor.

If equipment for other floors is similar to that shown in detail for the first floor, it is timesaving merely to show the equipment to be installed on each floor and to note the similarity that exists. This is shown

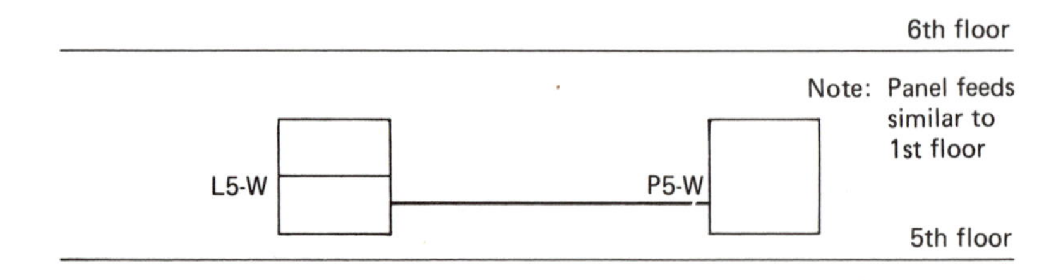

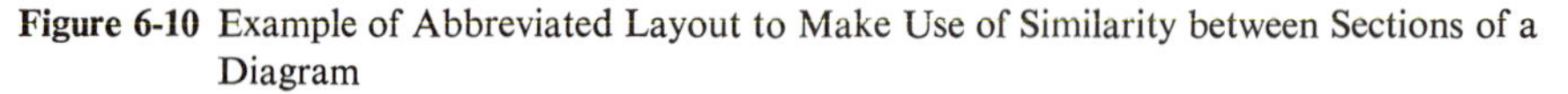
Figure 6-10 Example of Abbreviated Layout to Make Use of Similarity between Sections of a Diagram

in Fig 6-10. The panels and distinguishing numbers are shown and the note refers to the first floor for additional data that does not need to be repeated.

6. To complete the riser diagram, all equipment and characteristics of interconnections are given. Usually it is better to hold the job of identifications until all the line work is complete because invariably labels get in the way of the drafting and have to be erased and done over again.

7. As mentioned earlier, a legend may be necessary in order to facilitate reading of the drawing. Symbols used in the example given are as follows:

T C — Magnetic contactor with 480-120-V control transformer

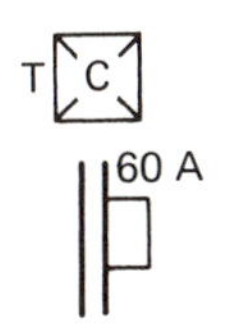

Plug-in bus duct with 60-A, 3-pole, 4-W plug-in fused disconnect switch, unless otherwise noted

1" - 3 # 8 & 1 # 10 — Exposed run of 1" rigid steel conduit with three # 8 & one # 10 wires

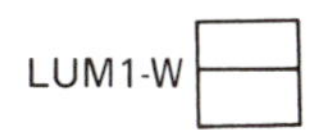

Lighting panelboard or power panelboard split bus—upper section 480/277 V, 3 ϕ 4 W; lower section 208/120 V, 3 ϕ 4 W unless otherwise noted; LUM1-W = lighting, upper mezzanine, panel 1, west core

Dry-type transformer—size as indicated

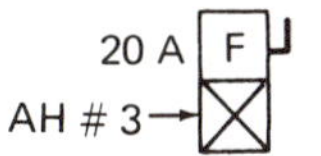

Combination fused disconnect & magnetic motor starter for air-handling unit no. 3

Motor control center

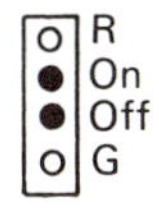

Push-button station—on-off with red & green indicating lights

(5) — Squirrel cage induction motor—5 hp

When one-line diagram symbols are used on riser diagrams, because it is a natural and convenient thing to do, it is always debatable as to whether rating data of the device should be repeated if given on the one-line diagram. The danger lies in the possibility that the information on the one line may change for some reason and not be changed in other places where it appears. A good rule to follow is *not* to show the same information in any other place if at all possible. This can become difficult because the rating information may be the only data available to identify the difference between similar circuits clearly. In the example shown in Fig. 6-7, the ratings are shown only for those devices that do not appear on the one-line diagram.

6-8 Riser Diagrams Are Not a Substitute for Layout Drawings.

The layout drawings show actual physical locations of equipment and routing of raceway systems. The electrical contractor might very well be able to install the equipment by choosing his own locations. He could also run raceways by use of the riser diagram but without knowledge of the possible interference from equipment, piping, ductwork, other trades, etc., mass confusion could develop on a large job. The engineers' and designers' work includes coordination of all services and therefore physical drawings as well as one-line and riser diagrams are vital.

6-9 Telephone System Riser Diagrams

Telephone system riser diagrams are prepared in much the same way as the power diagram except that any one-line diagram showing system requirements must be prepared by the telephone company. From such a diagram and through consultation with the telephone company engineers, the requirements for telephone distribution are determined. This is implemented into the riser diagram and physical drawings that provide for the installation of empty raceways and terminal boxes. The electrical contractor is required to install only the raceway system and boxes and the telephone company generally installs all terminal cabinets, wiring and equipment, telephone sets, etc.

Similar to the electrical power closets on each floor, the design also includes a communications closet which is completely separate from the power closet. The vertical communications shaft up through the core in the building is similar in purpose to the power bus duct except that it is composed of multiconductor cables only that run between terminal boards on each floor. Connections for the telephone sets on each floor are provided at the terminal boards. In some ways, this is similar to tapping the plug-in bus except that it is not so simple because a great many more conductors are used.

6-10 A Typical Telephone Riser Diagram

A typical telephone riser diagram for the office building being used as an example is shown in Fig. 6-11.

This diagram shows the incoming cables entering the building from the telephone company manhole through three 4-in. conduits at the concourse level. These conduits stub up through the lower mezzanine floor into the west core communications closet. Telephone cables, installed by the telephone company, are run through conduit sleeves in the floor up through successive closets to the telephone equipment room on the eighth floor. Conduits are required to be installed from the room into the core area. The riser diagram shows them in diagrammatic form; the physical conduit layout drawing locates the actual routing of these runs. Cables are run from the telephone equipment room to sleeves in

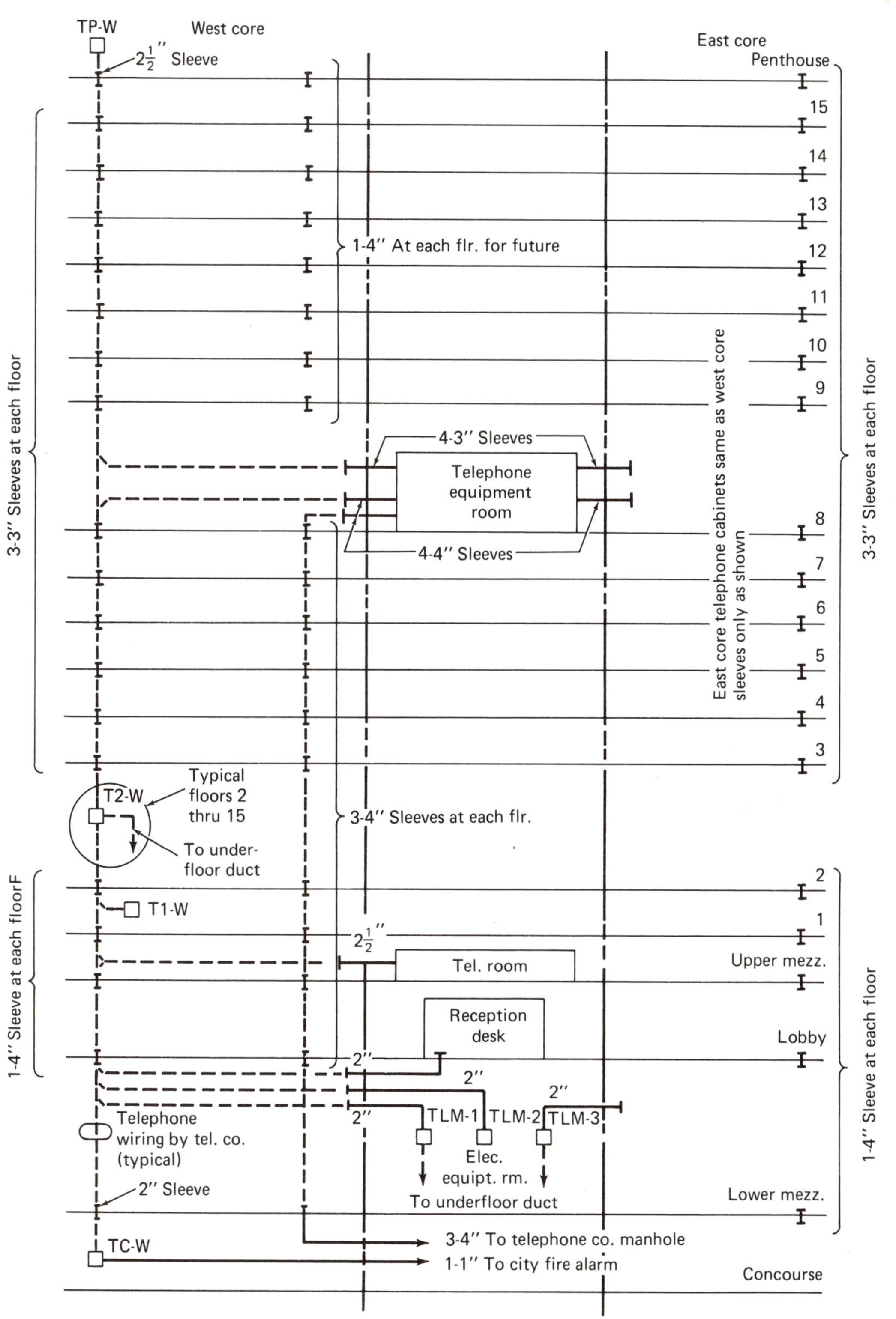

Figure 6-11 Telephone Riser Diagram for Office Building

the floor that provide for distribution up and down to terminal cabinets at each floor. Distribution from these terminal cabinets to the individual telephone sets throughout each office floor area is generally via the underfloor duct system (see Fig. 6-9). Other telephones are served from the terminal cabinets by running wires or cables through conduits, in ceiling spaces, etc. as may be necessary to reach the telephone set location.

The symbols applying to the telephone riser diagram are as follows:

□ T1-W	Telephone cabinet furnished & installed by telephone co. T1-W = cabinet on 1st flr., west core
—I—	Conduit sleeve through floor-bushings on both ends-size & number as noted
¦	Telephone wires or cables-furnished & installed by telephone company

Note that the east core wiring is covered by a note rather than through repetition of the west core drafting. This saves drafting time and expense. Also a typical terminal cabinet detail is shown for the second floor and an appropriate note is used to cover other floors.

This diagram provides a simple, easily understood display of the telephone distribution that would be difficult to understand without the riser diagram.

6-11 Fire Alarm Systems

Fire alarm systems have been developed by several manufacturers who are generally pleased to provide engineering assistance to help design a system best suited to the building's needs. Data on the various types of systems available are generally available from the manufacturers. Compliance with codes and insurance rating requirements must be closely observed. After the engineer has determined the system to be used, the raceway wiring requirements are furnished by the manufacturer in sufficient general detail so that the design team can prepare the riser diagram. The diagram is prepared in much the same way as any other riser diagram.

6-12 A Typical Fire Alarm System Riser Diagram

A typical fire alarm system riser diagram for the multistoried office building is shown in Fig. 6-12.

This diagram outlines the system so that the general arrangement and operation can be readily seen.

Power supply for the alarm system is provided from an emergency distribution panel. Standby battery power is very often required to maintain the system in operation for several hours (60 hr is often specified).

The fire alarm control panel and annunciator is located in the concourse

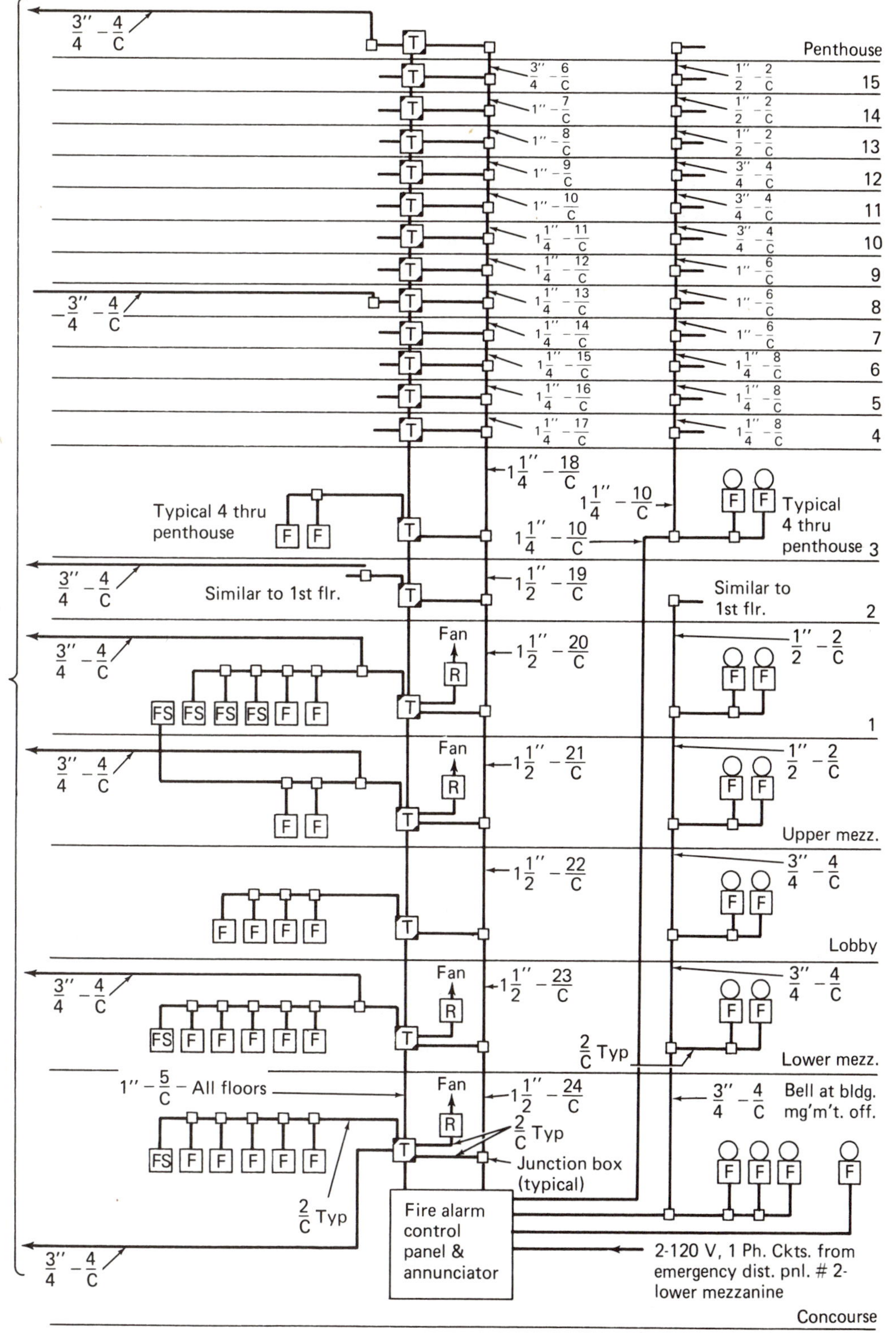

Figure 6-12 Fire Alarm System Riser Diagram for Office Building

area. Automatic coded transmitters are located on each floor. Fire alarm stations, provided at several locations on each floor, are wired to the transmitter.

Actuating any station will cause the transmitter to send a signal via the signal cable run down to the control panel. The location of the fire alarm station that has been pulled will be annunciated (shown and audibly alarmed) at the control panel and fire alarm gongs will be actuated as required.

Heat-sensitive devices called *firestats* are located in the discharge ductwork of supply and return air fans so that a fire in the ductwork will be alarmed and located. The fans are generally wired to shut down if the firestat is actuated. This is accomplished by using an auxiliary relay wired to the automatic coded transmitter.

The fire alarm system works directly with the smoke detection system and therefore the two riser diagrams are usually shown on the same drawing.

Note that the diagram is abbreviated wherever duplication of data is required. The use of notes saying *typical* is common but should be used with care because the atypical must be properly identified.

The symbols applying to the fire alarm riser diagram are as follows:

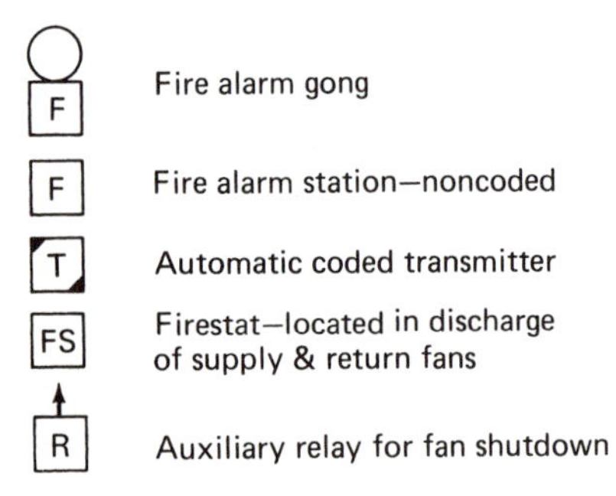

6-13 Emergency Power and Lighting

Emergency power and lighting supplies must always be considered in the design of any facility. The systems can be very extensive with duplicate sources of utility power and the installation of an on-site emergency generator. Hospitals in particular require careful analysis of emergency power requirements. The emergency power and lighting supplies must be designed to provide power almost immediately upon loss of the normal supply. Careful review of all code requirements is mandatory for preparing the design of any emergency supply system.

The riser diagram is ideal for displaying the emergency system because it shows the extent of distribution to all levels throughout the building.

6-14 The Emergency Supply System for the Office Building

The emergency supply system for the office building being used as an example is shown in Fig. 6-13. The symbols are as follows:

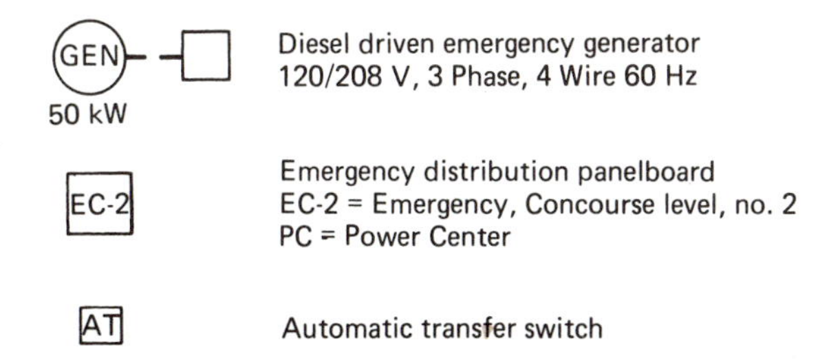

As previously stated, the emergency system must operate immediately and must of course be automatic. There are two primary elements that are critical in meeting these needs. The emergency source must of course be reliable. Several types of emergency generator sets are available as manufactured units for this type of service. The prime mover can be a gasoline engine, a diesel engine, a propane gas engine, a natural gas engine, or some other power source that will start automatically in a very short time. The other primary element is the automatic transfer switch that has built-in controls to cause transfer of connections from the normal source of power over to the emergency source and then back to the normal source when the normal power supply is restored. This device is simple in theory. It is shown in one-line form in Fig. 6-14.

Several makes of transfer switches are available from manufacturers. Most employ some method by which the contacts are held closed by mechanical means rather than through the use of electrical solenoids. Descriptions and application data are readily available from the manufacturers. These switches use controls to tell that the normal power has failed and a transfer must be made. When normal power fails, of course, its control is de-energized and the emergency source control is energized. This causes the controls to function and to move the switch from one position to the other. The reverse of this procedure takes place when normal power is restored. Usually some time delay feature is incorporated in the switch control so that momentary dips in voltage do not cause transfer.

In the office building system, a 50-kW diesel-driven emergency generator is installed on the concourse level. This unit will supply 120/208-V, three-phase power to the main emergency distribution panel, which is also located on the concourse level. This panel has several circuit breakers that protect the feeders going up to different floors in the building. The emergency generator is equipped with an automatic starting control that is wired to all the transfer switches through the two No. 12 wires (2 #12) included in the conduit carrying the emergency cables to the transfer switches. Loss of normal power at any transfer switch will cause the automatic start control to call for the diesel engine to start on its battery and thereby drive the emergency generator.

Power from the generator is distributed to automatic transfer switches on the lower mezzanine level, the fourth floor, and the eleventh floor. Normal power is supplied to the transfer switches from power centers on the lower mezzanine level and from the plug-in bus duct through dry transformers at the fourth and eleventh floors. Main emergency distribution panels are supplied

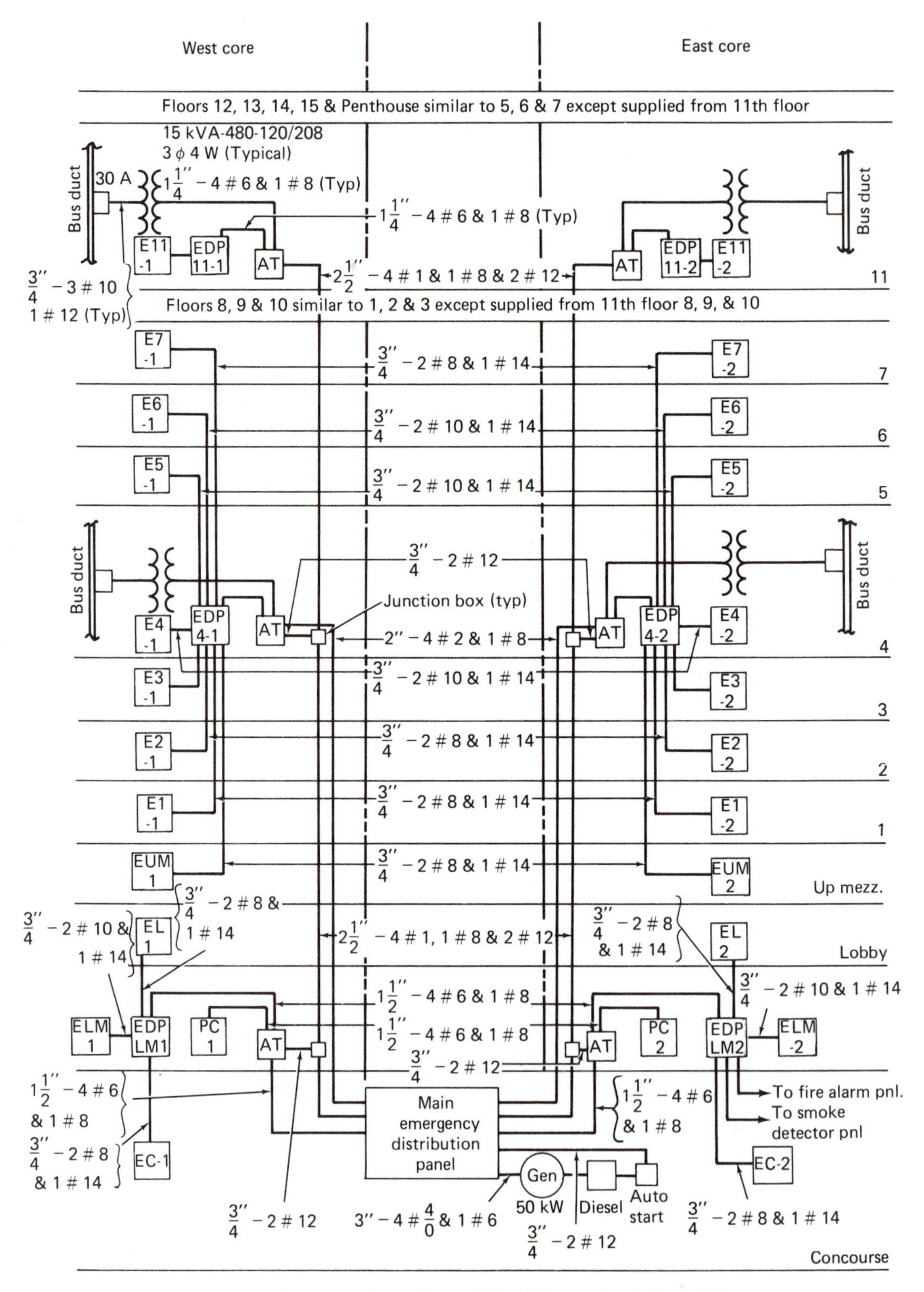

Figure 6-13 Emergency Power System Riser Diagram for Office Building

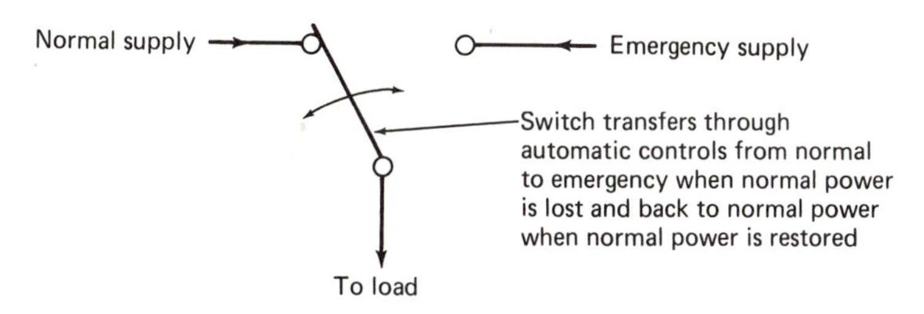

Figure 6-14 Simplified Automatic Transfer Switch

power from either the normal source or the emergency source by way of the automatic transfer switch. Feeders from these panels supply emergency power to emergency panels on each floor.

The emergency power load consists mostly of emergency lighting in corridors and in stairways. Exit signs are also supplied from the emergency lighting circuits. These emergency lights are on at all times and are normally supplied from the normal source of power and automatically kept on by transfer to the emergency source so that people can see to leave the building if an emergency requires that they do so.

6-15 Other Miscellaneous Systems

Other miscellaneous systems that can be readily shown on riser diagrams include smoke detection, time systems, intercommunications systems, closed circuit television, and building security. The suppliers of these systems generally provide the riser diagrams that are required. Their contract may include installation as well as materials. If raceway systems are required, the electrical contractor usually installs them and the building designer will supply the drawings. This is similar to the requirements for the telephone system discussed in Sec. 6-10. Of course, security system diagrams are not widely circulated due to the very nature of their function.

Chapter Review Topics and Exercises

6-1 How does a riser diagram differ from a one-line diagram?

6-2 Explain how riser diagrams are started, i.e., the layout procedures.

6-3 Describe how feeder cables are run up through the building and the special attention given to these runs.

6-4 What is the method for distributing 480/277-V power to each floor in the building?

6-5 How is power made available at each floor?

6-6 Describe the under-floor raceway system used in the office building. How are conductors for telephone sevice run?

6-7 When should identifying *labels* be drafted onto the drawing? Why?

6-8 Why can riser diagrams not be used by themselves without other drawings?

6-9 Who runs the telephone wires and cables?

6-10 Where does a fire alarm system generally get its power supply? Why?

6-11 What other riser diagram system is generally coordinated with the fire alarm system?

6-12 How can repetition of the same information be avoided for each floor? Are there dangers in doing this? Why?

6-13 What does an automatic transfer switch do as used in an emergency power system?

6-14 What are the usual emergency loads in an office building?

6-15 Who generally installs such systems as closed circuit television and building security?

6-16 Draft the power riser diagram using the steps explained and include the detail of an electrical closet on the first floor. Use grid paper under the drawing if available. Drawing size should be at least 17″ × 22″.

7

WIRE AND CABLE SCHEDULES; CONNECTION AND INTERCONNECTION WIRING DIAGRAMS

7-1 General

The subject of wire and cable schedules and connection and interconnection wiring diagrams are combined in this discussion because they interrelate so closely to each other. Some methods combine wire and cable schedule and interconnection on the same drawing.

The schedules provide for installing the majority of conductors on a large project and the wiring diagrams show how the conductors are to be connected to apparatus and devices. One operation logically follows the other.

7-2 Descriptions and Definitions

The following descriptions and definitions are given as a guide to the general character of the schedules and diagrams covered in this chapter. Specific types vary in form, content, and complexity.

Conduit and Wire Schedules

Usually a preprinted form in a size smaller than full-sized drawings but larger than a standard $8\frac{1}{2}'' \times 11''$ page that provides for data on about 30 conduits and their cables on a sheet (a common size measures $15'' \times 21''$). The ruled columns allow space for noting the following data:

Conduit

1. Conduit or cable number
2. Size and material
3. Length

Conductor

1. Total number
2. Number of spares
3. Normal current
4. Alternating current or direct current
5. Working voltage
6. Area in MCM or size AWG
7. Type of insulation
8. Type of finish
9. Outside diameter—decimal of inch
10. Measured total length

Routing

1. From
2. To
3. For

These schedules provide data for installing wires and cables. Interconnection diagrams show the wiring connections to all equipment and devices.

Cable and Interconnection Sheet

A preprinted, $8\frac{1}{2}'' \times 11''$ form made to cover both installing and connection data for an individual cable or circuit. This is an alternate method for cable and wire *and* interconnection data from the conduit and wire schedules and interconnection wiring diagram scheme. This form is handy in size for the electrician to handle (full-size drawings can be cumbersome in the field) and by limiting data to one cable per sheet the chances of connecting errors by the electrician are minimized. A typical cable and interconnection sheet form is shown in Fig. 7-1.

Connection Diagram

The connection diagram confines itself primarily to showing the connection between devices and accessory items such as terminal boards, fuse blocks, and resistors, etc. *on a piece of equipment*. These diagrams show the relative location of the devices on the equipment. Rather than showing the actual routing of conductors, an origin- and destination-type code arrangement is used to direct conductor connections. As an example, Fig. 7-2 shows the rear view of a panel with the connection wiring between several devices and the devices and a terminal block.

Each device on the panel is assigned an arbitrary identification letter or pair of letters. The instruments and device terminals are numbered. Connections are indicated by designations that are composed of the device letter and the

CABLE & INTERCONNECTION SHEET

Cable No. ______________ For: ______________________________

Size conductors ______________________________ AC ☐ DC ☐

No. spares __________

Conduit no. ______________ Size ______________ Tray only ☐

From ____

To ____

Routing via ____

Ref. dwg. ________ ________ Term. blk ________	Wire no.	Color	Wire no.	Ref. dwg. ________ ________ Term. blk ________
This space used to show wiring connections to device terminals or terminal block numbers		Black		This space used to show wiring connections to device terminals or terminal block numbers
		White		
		Red		
		Green		
		Orange		
		Blue		
		White black		
		Red black		
		Green black		
		Orange black		
		Blue black		
		Black white		
		Red white		
		Green white		
		Blue white		
		Black red		
		White red		
		Orange red		
		Blue red		
		Red green		
		Orange green		
		Black white red		
		White black red		
		Red black white		

Remarks:

Figure 7-1 Typical Form Used for Cable and Interconnection Data for a Single Cable

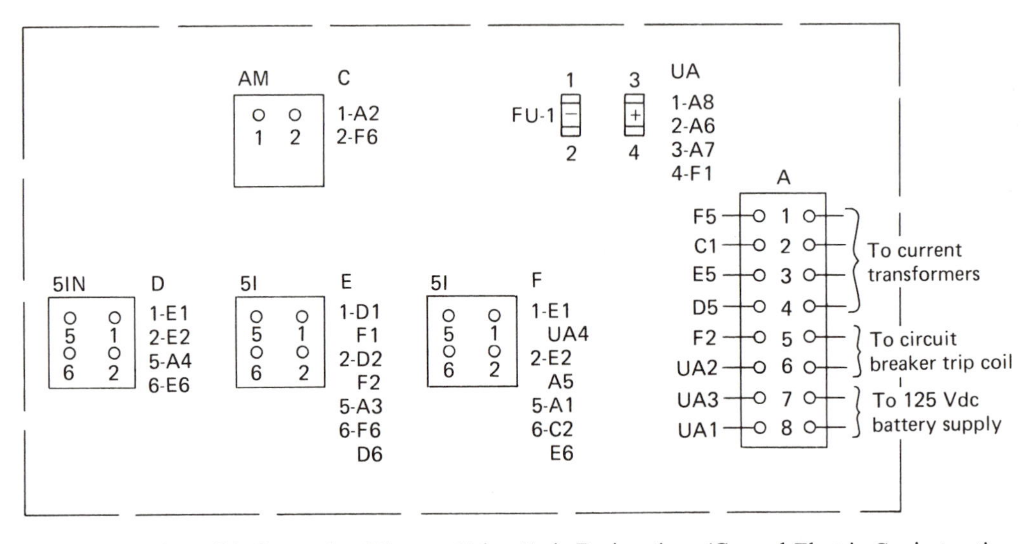

Figure 7-2 Connection Diagram Using Code Designations (General Electric Co. instruction data)

terminal number. For example, device D, terminal 5 is indicated as connecting to A4. At terminal board A, terminal 4, the indication is D5. Both devices designate the connection to be made. This system is sometimes called the *opposite end callout system.*

The connection diagram is used primarily by manufacturers to wire switchboard panels, etc.; however, the method can be used effectively on any wiring diagram problem. This method is not very useful as a means of reading how a circuit functions—the elementary diagrams serve that function.

Interconnection Diagram

An interconnection diagram is one that shows the complete connections between equipment units or unit assemblies and associated apparatus. The internal connections of the units are shown on connection diagrams.

These diagrams may use a code-type origin and destination code similar to the connection diagrams. Manufacturers of electrical equipment generally use the coding method to indicate interconnection between elements of their equipment.

Complete interconnection diagrams showing all electrical connections are a very useful drawing for tracing the routing of circuits when used with elementary diagrams. They often prove invaluable for troubleshooting.

7-3 Preparing Conduit and Wire Schedules

Preparation of conduit and wire schedules can generally start only after the one-line diagrams and elementary diagrams have been developed. It is necessary to work up these schedules in a preliminary way before conduit and other raceway drawings are started because the cable information is required in order to size the raceway system properly.

The work of *roughing out* schedules is generally done by an experienced designer who is meticulous and imaginative because he must *see* the circuits as they will be run and be careful to include all circuits required. Each element of the one-line diagram and control data from the elementaries must be translated into wire and cable requirements, which includes the number of conductors required, size, and type. In many instances the designer will have to decide whether or not he can include several circuits in a multiconductor cable, etc. The wire and cable information in turn will determine the size of the conduit or other raceway.

It is fairly common to divide the schedules into *service* categories and preface each conduit or cable number under that service with a characteristic letter. As an example, all alternating-current circuits concerned with the main generator one line described in Chapter 4 would have conduit or cable numbers prefaced with the letter G. The first number would be G1, and numbers would follow consecutively. The designer would *block out* a set of G numbers, say from G1 to G50, and start scheduling all the alternating-current circuits relating directly to the main generator. In central power plant design the conduit or cable characteristic letters might be as follows:

Preface Letter	*Service*
G	Generator
P	Power
AN	Annunciator
L	Lighting
DC	Direct-current circuit
M	Miscellaneous
A	Ash handling
SB	Soot blowers
IC	Interconnection
W	Welding
V	Valves
CB	Circuit breaker

In addition to identifying the service of all conduits or cables, this method also overcomes the problems of a long list of consecutive numbers that tend to be confusing especially if some attempt is made to keep services grouped. If a conduit or cable number has to be added, it may have to be inserted out of sequence in the grouping. With the prefix characteristic, this is less of a problem because at least the number is identified with its own group by reason of the unique prefix.

The scheduling designer will generally have sepias (paper reproducibles) or some other reproducible blank conduit and wire schedule forms made to start his work. Working with the one-line and elementary diagrams, he will note conduit and cable information onto the form for each circuit or combination of circuits.

CONDUITS
Al – Aluminum
EMT – Elect. Met. Tubing
F – Fiber
FX – Flexible
PV – Polyvinylchloride
T – Transite
X – X-Trey

SEE N.E. CODE FOR
A, AA, AI, AIA, AVA, AVB, AVL
MI
RH, RHH, RW, RHW, RUH, RUW
SA
T, TW, THHN, THW, THWN
UF
V
XHHW

OTHER INSULATION
B – Butyl
K – Oil base
P – Paper
PE – Polyethylene
-S – Shielded
SI – Silicone

OTHER FINISH (COVERING OR JACKET)
A – Armor
Al – Aluminum
B – Braid
L – Lead
NE – Neoprene
PE – Polyethylene
PV – Polyvinylchloride
T – Tape

Service	CONDUIT*			CONDUCTOR										
Service	Conduit or Cable Number	Size and Material	Length	Total Number	No. of Spares	Normal Current	A-C or D-C	Working Voltage	Area in MCM or Size AWG	Type of Insulation	Type of Finish	FROM	TO	FOR
DEAERATOR PUMP NO. A1	P1	3/4"		1-3/c	1		DC	125	#12	SI	PVC	MOTOR CONTROL CENTER A	PRESSURE SWITCH PS-21	CONTROL
	P2	1"		1-7/c	2		DC	125	#12	SI	PVC	MOTOR CONTROL CENTER A	PUSH BUTTON STATION AT MOTOR	CONTROL
	P3	1 1/4"		1-10/c	1		DC	125	#12	SI	PVC	MOTOR CONTROL CENTER A	CONTROL SWITCH ON CONTROL CONSOLE	CONTROL
	P4	2 1/2"		3	0		AC	480	250	RH	NONE	MOTOR CONTROL CENTER A	MOTOR	POWER SUPPLY

*Rigid steel unless otherwise noted.

Throughout design of the project the schedules in Fig. 7-3 will be used for assignment of conduit or cable numbers onto layout drawings where the conduits and other raceways are physically located. The schedules will be used by the electricians, in the field, for installation of most of the project wire and cables into the raceway system.

Figure 7-3 Section of Typical Conduit and Wire Schedule

Typical scheduling for circuits to a deaerator pump in a central power plant is shown in Fig. 7-3. Throughout design of the project the schedules in Fig. 7-3 will be used for assignment of conduit or cable numbers onto layout drawings where the conduits and other raceways are physically located. The schedules will be used by the electricians, in the field, for installation of most of the project wire and cables into the raceway system.

7-4 Preparing Cable and Interconnection Sheets for Wire and Cable Information (see also Sec. 7-7)

The cable and interconnection sheets are prepared in much the same way as the conduit and wire schedules as far as conduit numbers and wire and cable information are concerned. The principal difference, which is immediately evident, is that four sheets would be required to cover the data for wires and cables in the Fig. 7-3 example, i.e., a sheet for each circuit. Later in the development of the project, however, these same sheets will be used to show the connections of each cable at both ends. An example of the information that would be shown for the wire and cable information of the P4 conduit in Fig. 7-3 is shown in Fig. 7-4.

CABLE & INTERCONNECTION SHEET

Cable No. P4 For: DEAERATOR PUMP NO. A1-MOTOR

Size conductors 3-250MCM RH AC ☑ DC ☐

No. spares 0

Conduit no. P4 Size 2 1/2 Tray only ☐

From MOTOR CONTROL CENTER A

To MOTOR

Routing via

Figure 7-4 Cable Data Portion of Cable and Interconnection Sheet

Generally these sheets are kept bound in a loose-leaf notebook during the course of design work. Some convenient form of indexing these sheets is often helpful because several hundred sheets will be required on a large power plant project.

7-5 Preparing Connection Diagrams

The connection diagram of Fig. 7-2 showed the code *origin and destination* scheme for panel wiring in order to define this type of diagram. An explanation for the wiring determination will help to illustrate how these diagrams are prepared. Figure 7-5 shows a one-line diagram, its associated three-line diagram and the elementary of the relaying and metering associated with the panel

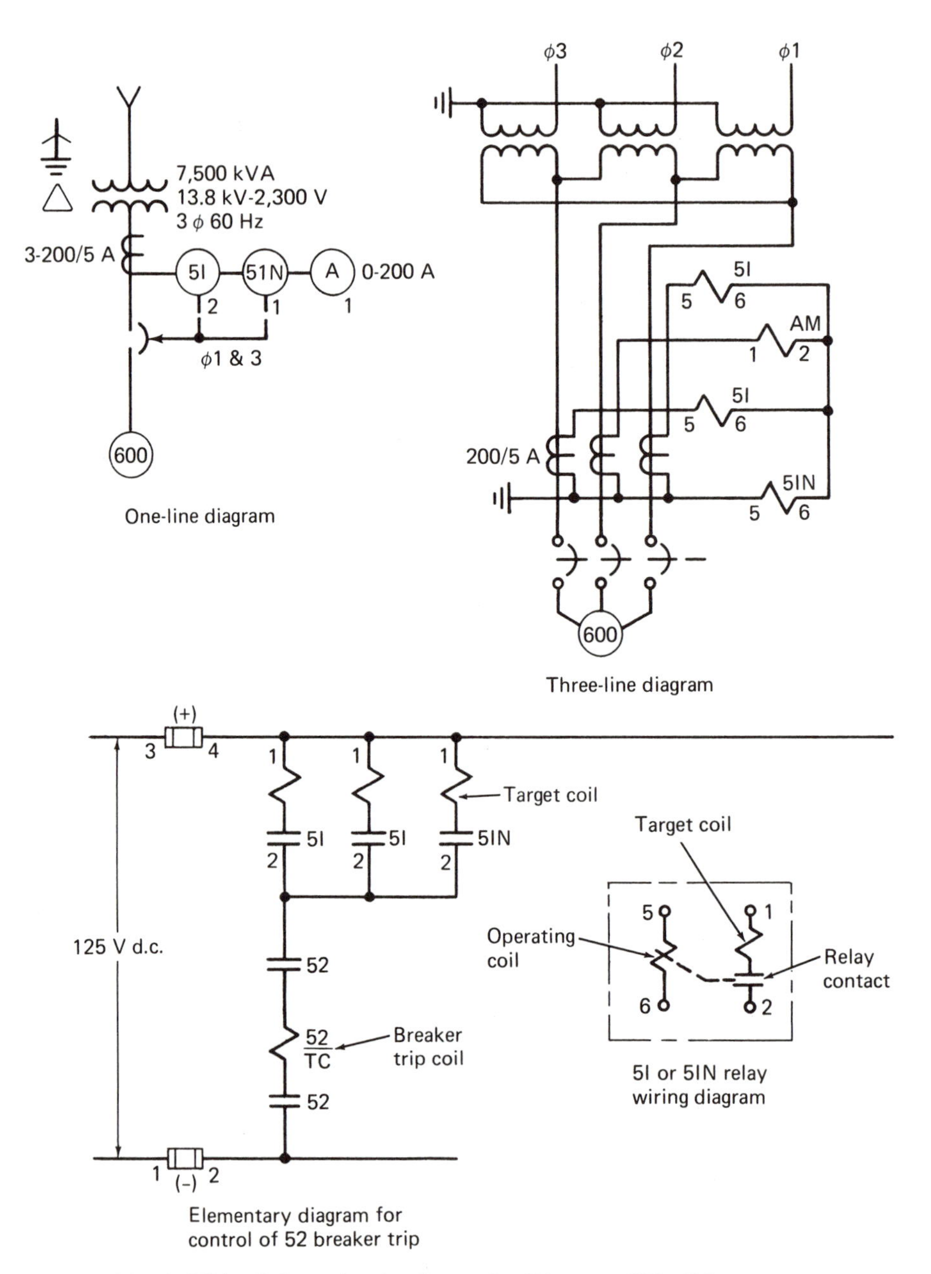

Figure 7-5 Basic Wiring Information for Connection Diagram of Fig. 7-2

indicated in Fig. 7-2. From this basic information the panel wiring can be developed.

In order to understand how panel wiring is developed from the diagrams of Fig. 7-5, each diagram should be studied starting with the one line.

The one-line diagram indicates a circuit for a 600-hp, 2,300-V, three-phase motor that utilizes an air circuit breaker as a motor starter. The motor is protected from phase overcurrents by overcurrent relays (51) and from ground overcurrents by a ground overcurrent relay (51N). These relays will trip the circuit breaker on abnormal conditions and the operator line (—— - —— - —) shows this action functions. An ammeter is provided to monitor current flow.

The one line is translated into a three line so that the operating coils of the relays and the coil of the ammeter can be shown as they are into the complete current transformer circuit. If it is known that a particular relay will be used, then the connection diagram for the relay will show the terminal numbers. In this case the operating coil is connected between terminals 5 and 6. The ammeter terminals are numbered 1 and 2. This terminal data is noted on the three-line diagram.

The elementary diagram shows the 125-V direct-current power supply, the fuse protection, and the parallel connection of the relays so that action by any relay will trip the circuit breaker by energizing the trip coil. (The 52, normally open, contacts on either side of the trip coil will be closed when the breaker is closed.)

Wiring information on the relay shows that the relay contact and target coil are connected between terminals 1 and 2. This is noted on the elementary. The target coil is energized when the relay contact closes. This causes a visual *flag* or *target* to show in the window section of the relay so the operator can tell which relay shut the motor down.

It is also necessary to have knowledge of the physical arrangement of the relays and meters on the panel before starting the connection diagram. In the example being used (Fig. 7-2) the front of the panel would appear as shown in Fig. 7-6.

Showing this layout may seem insignificant but many wiring mistakes are made by failing to remember that back view wiring diagrams *flip* or reverse device locations; i.e., the *right*-hand device in a front view becomes the *left*-hand device in a rear or back view wiring diagram!

With the foregoing data the connection diagram can be started. It is now essential to bear in mind the *location* in the circuit of several elements. The meters and relays are on the panel. The current transformers are located in the main conductors connecting to the line side of the circuit breaker. The trip coil of the circuit breaker is on the breaker mechanism. With these elements

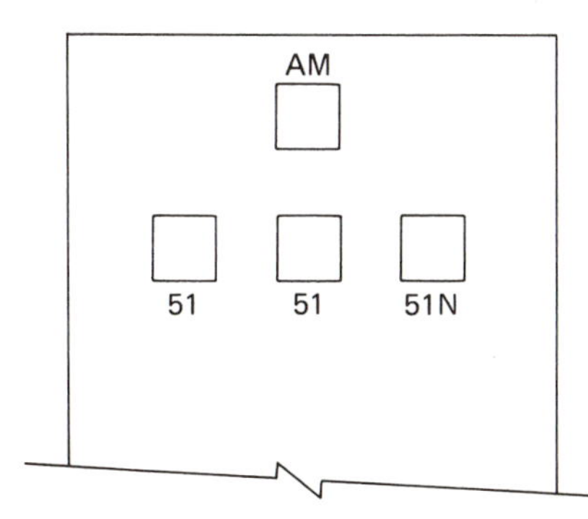

Figure 7-6
Front View of Panel for Connection Diagram of Fig. 7-2

separated physically it becomes obvious that a common *collecting point* for the conductors that must be connected to the several elements is required. The terminal block is made for this type of service and can be handily located on the back of the panel. The missing link in the connection data actually falls into the category of an interconnection diagram and is shown in Fig. 7-7.

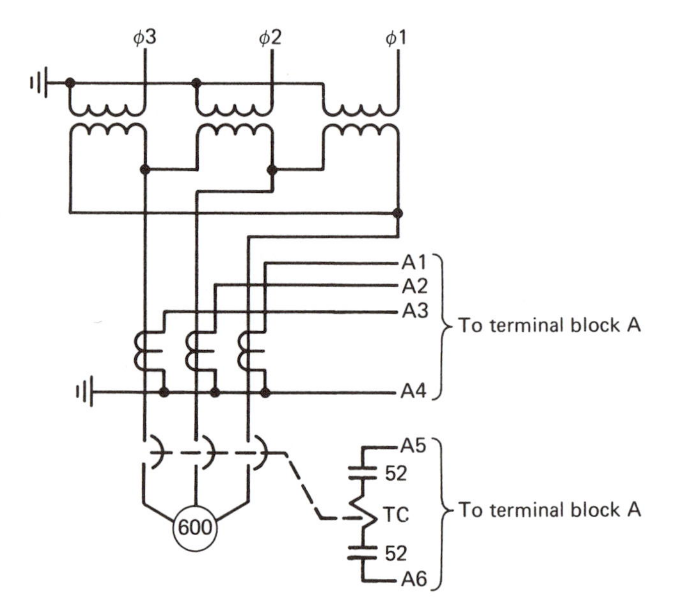

Figure 7-7 Connection Designations for Current Transformers and Circuit Breaker Trip Coil Conductors to Terminal Block of Fig. 7-2

Using information from Fig. 7-5 and 7-7 it can be seen that the connection diagram of Fig. 7-2 directs wiring of the panel in accordance with the information given. As an example, conductor A1 (Fig. 7-7) connects to terminal No. 1 on block A. From terminal No. 1, a conductor is directed to run to F5, which is one side of the operating coil of a 51 relay. This agrees with the three-line information of Fig. 7-5. At the relay, the other side of the coil, which is terminal 6, conductors are directed to run to C2 and E6. The C2 conductor makes a connection to terminal No. 2 of the ammeter and the E6 conductor as part of the common connection for all the devices by running to terminal No. 6 of the other phase overcurrent relay. At E6 this same common conductor is directed on to D6 to complete the common connection. All this agrees with the three-line information of Fig. 7-5.

By following the wiring designations further it can be seen that the circuit requirements are completed.

These are *not* simple diagrams to prepare. They need careful attention to accurate notation of wiring origin and destination directions and back-checking to make certain that both ends agree.

7-6 Preparing Interconnection Diagrams

As defined, these diagrams show complete connections between equipment units or unit assemblies and associated apparatus. On a project of any appreciable size this requirement can amount to the detailing of connections for a great many interrelated pieces of equipment and devices in the electrical systems. To reduce the design and drafting requirement, consideration should be given to showing interconnection for only those systems that are too complex in nature to depend on the electrical contractor for wiring detail. As an example, enumerable circuits for simple start–stop push-button control of a motor certainly do not require an interconnection diagram for wiring in the field. On the other hand, control and relaying protection of involved systems are more readily developed by the design team and will save untold hours of the contractors' time if interconnections are provided for these systems.

The design team must also determine if the interconnection diagrams should be *system oriented* or *area oriented.* With the system diagram each system is shown as a separate interconnection diagram. It is conceivable that any one diagram could cover several areas and possibly several floors in a building. This method makes troubleshooting of an individual system simpler. It increases the number of drawings required to show an equivalent amount of interconnection wiring as would be shown in the *area-oriented* diagrams.

The area-oriented system blocks out all the electrical equipment and devices, which are to be included in the interconnections, into areas and shows all the wiring requirements for that area. As an example, all connections to terminal blocks or other devices in a lineup of switchgear would be shown on a diagram. This method quite often makes it necessary to trace through several diagrams in order to follow any one particular electrical system or circuit.

There are pros and cons for both methods; however, on larger projects the area-oriented methods seems to be more commonly employed.

Just as with the instruction given for preparing connection diagrams (Sec. 7-5) it is necessary to collect basic data before attempting to start the diagrams. Such data will usually include

1. One-line diagrams
2. Elementary diagrams
3. Riser diagrams (if used)
4. Conduit and wire schedules (if used)
5. Layout drawings
6. Manufacturers' drawings

The diagram should be roughed out before final drafting starts because shifting and changing the arrangement of the layout will usually be necessary before all the elements are included and properly interconnected.

There are no standard symbols for these diagrams as such. Symbols used for both one-line diagrams and elementary diagrams are often used. In addition the following lines and symbols are useful:

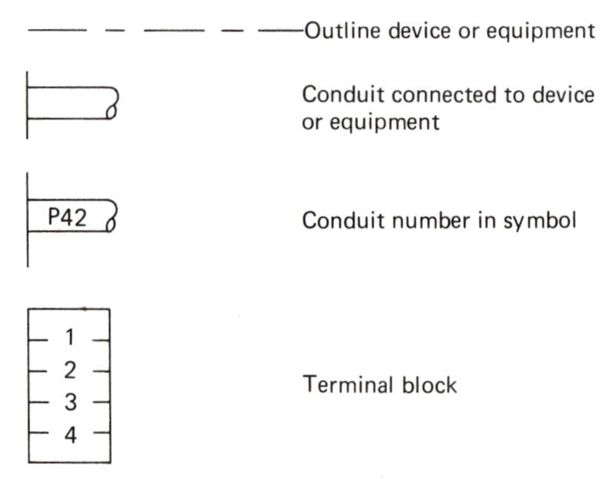

Conductors may be grouped into a *trunk* line to reduce the number of interconnecting lines; however, care must be exercised to identify each conductor as it leaves the trunk line. An example is shown in Fig. 7-8.

Figure 7-9 shows a portion of an interconnection diagram.

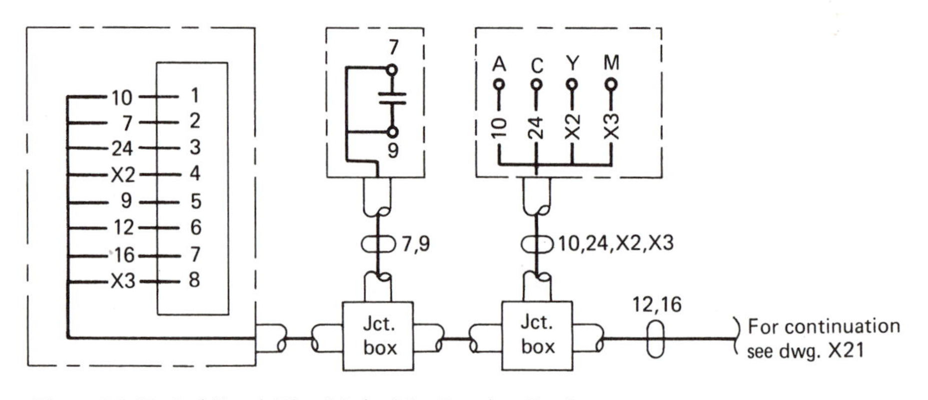

Figure 7-8 Typical Trunk Line Method for Routing Conductors

7-7 Preparing Cable and Interconnection Sheets for Interconnection Information (also see Sec. 7-4)

After these sheets have been set up to provide the wire and cable installation information, they are completed by adding the interconnection data. The form is prepared for only one cable on each sheet and the interconnection data for both ends of the cable. The sheet includes standard color coding of the conductors in a multiconductor cable. There is also a space provided for noting the number of the drawing upon which the terminating connections are shown. This is helpful to both the design team and the field electricians during checking and testing operations. An example of the same interonnection data supplied in Fig. 7-9 for the cable in conduit No. P121 is shown in the form of Fig. 7-10.

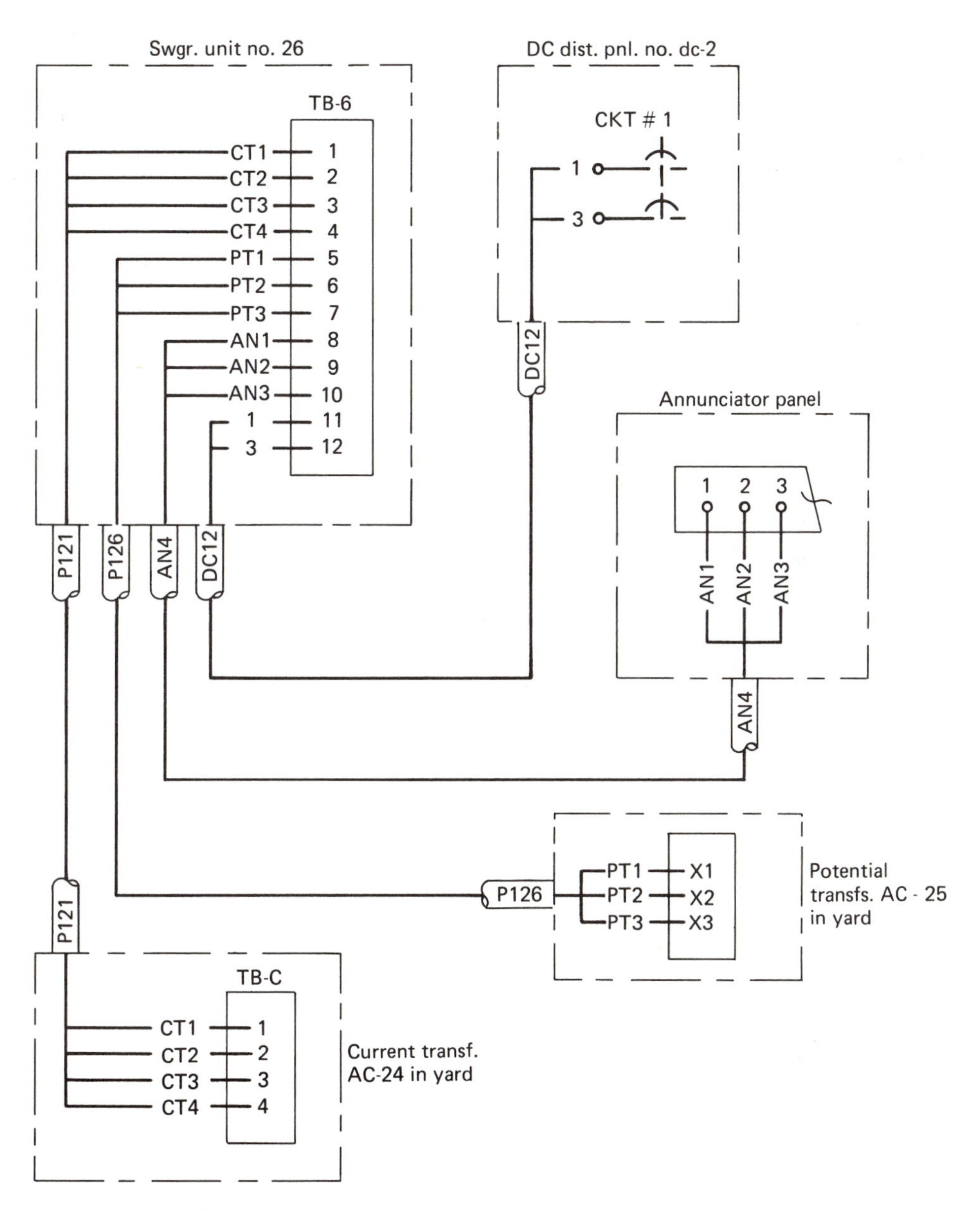

Figure 7-9 Typical Interconnection Diagram

7-8 Computer Program Methods for Wire and Cable Schedules and Interconnection Information

Computer programs have been developed that printout wire and cable schedules and also interconnection data. These programs are more or less an outgrowth of the cable and interconnection sheet. They have been developed

Cable & interconnection sheet

Cable No. P121 For: CURRENT TRANSFORMER-AC-24

Size Conductors 4/C#10 AC ☑ DC ☐

No. spares 0

Conduit No. P121 Size 1¼ Tray only ☐

From SWITCHGEAR UNIT NO. 26

To CURRENT TRANSFORMER TERM. BOX AT AC-24 IN YARD

Routing via

Ref. dwg. AX&YCO. 2173421 Term. blk. TB-6	Wire No.	Color	Wire No.	Ref. dwg. AX&YCO. 2175826 Term. blk. TB-C
1	CT1	Black	CT1	1
2	CT2	White	CT2	2
3	CT3	Red	CT3	3
4	CT4	Green	CT4	4
		Orange		
		Blue		
		White Black		
		Red Black		
		Green Black		
		Orange Black		
		Blue Black		
		Black White		
		Red White		
		Green White		
		Blue White		
		Black Red		
		White Red		
		Orange Red		
		Blue Red		
		Red Green		
		Orange Green		
		Black White Red		
		White Black Red		
		Red Black White		

Remarks:

Figure 7-10 Example of Finished Cable and Interconnection Sheet Showing Terminal and Wire Numbers

further to provide information on the composition of multiconductor cables; i.e., if several conductors must run between the same exact locations (terminal board to terminal board), the computer will combine conductors into cables in accordance with the paramenters built into the program.

The necessary information required by the program is noted on the appropriate form, keypunched onto computer cards and fed to the computer. The output data is used by both the design team and the field for installing wires and cables and making connections. Space is provided on the output sheets for drafting, wiring, or connection details that may be required.

Needless to say, the computer program system is designed for large projects only.

Chapter Review Topics and Exercises

7-1 What is the primary purpose of a conduit and wire schedule?

7-2 What information does the interconnection diagram provide?

7-3 Why would the number of spare conductors in a cable be useful information?

7-4 What advantage does a cable and interconnection sheet have?

7-5 Explain how the *origin* and *destination* system for making connection diagrams works.

7-6 How does an interconnection diagram differ from a connection diagram?

7-7 What basic data does a designer need in order to start roughing out conduit and wire schedules?

7-8 Why are preface letters helpful in setting up conduit and wire schedules?

7-9 How do conduit and wire schedules relate to physical layout drawings that will show where the conduit is to be installed?

7-10 How many cables does a single cable and interconnection sheet cover?

7-11 Referring to Fig. 7-5, what are the terminal numbers for the following devices?

(a) Control power fuse in positive line (+)
(b) Operating coil of the phase overcurrent relay (51)
(c) Operating coil of the ground overcurrent relay (51N)
(d) Target coil and relay contact circuit of the 51 relay
(e) Ammeter

7-12 What is significant about showing the front view of a panel?

7-13 Referring to Fig. 7-7, what wire designation has been assigned to the common connection for the current transformers?

7-14 Why do connection diagrams require special attention to detail wiring notations?

7-15 How can interconnection wiring diagram data be minimized?

7-16 Explain the ideas behind system-oriented and area-oriented interconnection diagrams.

7-17 What sort of data does the designer need in order to develop interconnection diagrams?

7-18 What is the trunk line method of grouping conductors and what precautions are required in its application?

7-19 There seems to be a decided advantage of interconnection diagrams over several sheets of cable and interconnections. What is the apparent advantage?

7-20 Referring to Fig. 7-10, what wire number connects to terminal 3 on terminal block C and what is the color of the wire?

8

GROUNDING

8-1 General

The word *grounding*, as applied to electrical power systems, is commonly used to cover both *system grounding* and *equipment grounding*. System grounding is a connection to ground from one of the current-carrying conductors of a distribution system or an interior wiring system. The grounding of the center, common connection of a wye- (Y) connected transformer, is an example of system grounding. Equipment grounding is a connection to ground from one or more of the noncurrent-carrying metal parts of the wiring system or of apparatus connected to the system. As used in this sense, the term *equipment* includes all such metal parts as metal conduits, metal trays, metal wireways, outlet boxes, cabinets, motor frames, lighting fixture housing, and switchgear.

System grounding depends on many engineering considerations too numerous and complex to include in this text. The decision to provide a grounded system and at what location to make the grounds, the method of grounding, and how many grounds should be made will be determined through engineering study. The results of the study will generally appear on the rough one-line diagrams prepared by the project engineer.

Equipment grounding is generally the responsibility of the design team and therefore this chapter will deal primarily with equipment grounding requirements.

8-2 Purpose of Equipment Grounding

Equipment is grounded for the following reasons:

1. To ensure against dangerous electric shock voltage exposure to any person who may come in contact with electrical equipment.
2. To provide a current carrying path of sufficient capability to adequately accept the ground fault currents that may be allowed to flow before the fault is cleared by the circuit overcurrent protective devices. This requirement must be met without danger of creating a fire or an explosive hazard.
3. To make the electrical system function properly when a grounding con-

dition develops, i.e., the protective devices will function because the ground path is properly designed.

8-3 Definitions

Some definitions for designing and understanding grounding systems are as follows:

Bonding Jumper: A reliable conductor to assure the required electrical conductivity between metal parts required to be electrically connected.

Ground: A ground is a conducting connection, whether intentional or accidental, between an electrical circuit or equipment and earth or to some conducting body that serves in place of earth.

Grounded: Grounded means connected to earth or to some conducting body that serves in place of earth.

Grounded conductor: A system or circuit conductor that is intentionally grounded.

Grounding conductor: A conductor used to connect equipment or the grounded circuit of a wiring system to a grounding electrode or electrodes.

Grounding conductor—equipment: The conductor used to connect noncurrent-carrying metal parts of equipment, raceway, and other enclosures to the system grounded conductor at the service and/or the grounding electrode conductor.

Grounding electrode conductor: The conductor used to connect the grounding electrode to the equipment grounding conductor and/or to the grounded conductor of the circuit at the service.

Voltage to ground: In grounded circuits the voltage between the given conductor and that point or conductor of the circuit that is grounded; in ungrounded circuits, the greatest voltage between the given conductor and any other conductors of the circuit.

Existing ground electrode: An underground metallic piping system, metallic building framework, well casings, steel piling, and other underground metal structures installed for purposes other than grounding are all existing ground electrodes.

Made ground electrodes: Made electrodes include driven steel or iron rods, driven pipes, steel plates, and bare copper conductor encased in concrete and located in concrete foundation footing that is in direct contact with earth.

Ground bus: A section of copper or aluminum bar connected directly to the ground electrode(s) and accessible for connection of grounding conductors from individual pieces of equipment.

A graphic representation of some of the definitions listed above is shown in Fig. 8-1.

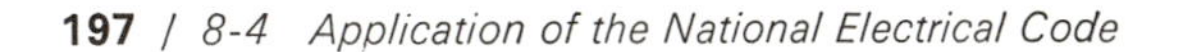

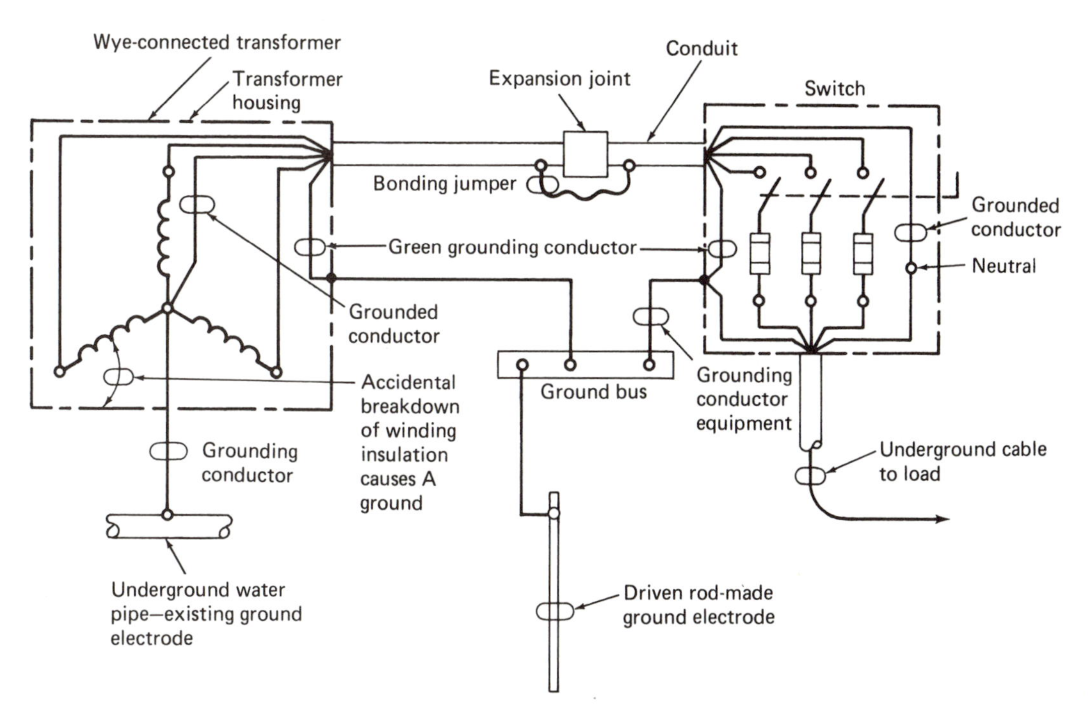

Figure 8-1 Graphic Representation of Some Grounding Terms

8-4 Application of the National Electrical Code

All grounding systems must be installed in strict accordance with the latest edition of the National Electrical Code unless the code does not apply (see the introduction to the latest code).

The article on grounding in the NEC covers general requirements for grounding and bonding of electrical installations and specific requirements for the following:

1. Systems, circuits, and equipment required, permitted, or not permitted to be grounded
2. Circuit conductors to be grounded on grounded systems
3. Location of grounding connections
4. Types and sizes of grounding and bonding conductors and electrodes
5. Methods of grounding and bonding
6. Connections for lightning arresters

In other articles in the code, applying to particular cases of installation of conductors and equipment, there are requirements that are in addition to those in the article on grounding. The NEC lists the related articles. Considerations for grounding must take into account these regulations as well.

8-5 Grounding Symbols

There are no published standards for symbols to use on grounding drawings. Symbols that show a reasonably graphic representation of the physical grounding wiring are as follows:

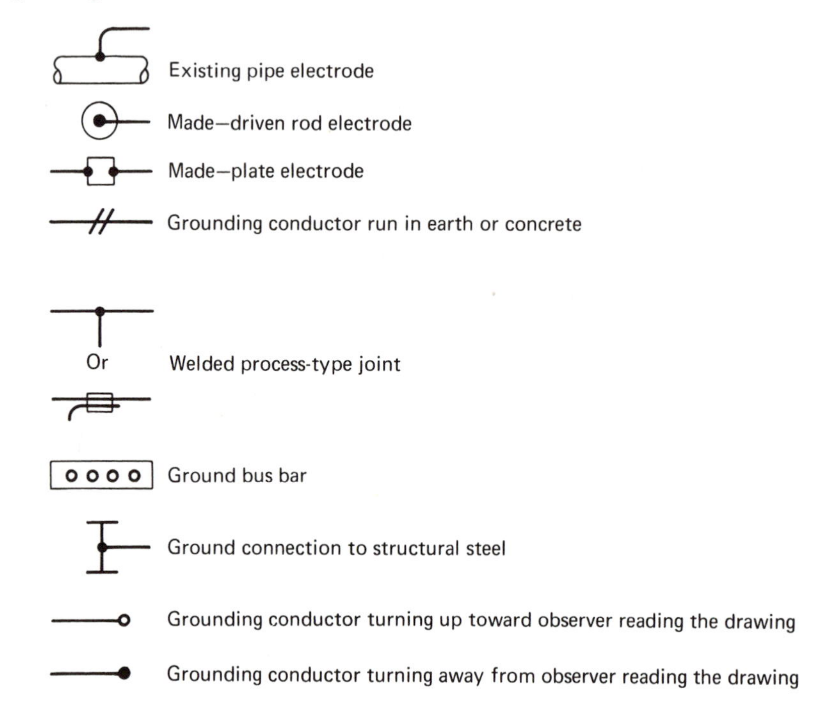

A legend may also be used with the symbols to indicate conductor size rather than spelling out the size on the drawing. An example is

8-6 Grounding against Electric Shock

One of the stated purposes of equipment grounding is to limit the potential (voltage) between noncurrent-carrying parts of electrical equipment to a safe value so that a person who may come in contact with the equipment will not receive an electrical shock if the noncurrent-carrying part becomes accidentally energized. This can in general be accomplished by connecting a grounding conductor (equipment) between the metal parts and a ground electrode. If the earth or other possible ground area that a person is standing on is at zero volts and the piece of equipment is connected directly to the zero-volt potential, it follows that no hazardous condition exists. It is not always quite this simple and many papers have been written on the problems associated with equipment grounding under various circumstances. As a general practice, however, in industrial and utility design and in commercial design to some lesser degree, the frames of motors, switchgear, distribution equipment, etc., are connected by grounding conductors (equipment) to ground electrodes to guard against electric shock. More detailed methods are outlined in other sections of this chapter.

8-7 Grounding for Conducting Ground Fault Currents

The grounding connections to guard against electric shock outlined above seldom are "seen" by currents feeding into a ground fault as the path of least impedance back to the source of power feeding the fault. It has been proved theoretically and through experimentation that even though a ground path has been provided directly from the ground fault to a ground electrode, the fault current will take a conducting path back along the route *closest* to the conductors supplying current to the fault.

The results of the test prove that by displacing a ground return path physically away from the supply conductors, the return path impedance is greatly increased. Naturally the return current will return along the path of least impedance, which is any conductor close to the supply conductors. A properly installed conduit system, with close attention given to good solid connections at couplings, boxes, and other fittings, provides the next best solution to the return path requirement because the conduit itself is physically close to the supply conductors.

8-8 Connections to Ground (Earth)

The word *ground* in electrical language does not always mean *earth*. Automobiles and aircraft have *ground* systems. A *ground bus* may be present several stories above the earth in the electrical system of a multistoried building and to the electrician it may mean the same as *earth* if he were asked to define it. For these reasons the connections to ground in these paragraphs will relate to connection into earth itself.

Earth is not a very good conductor. Its resistivity is said to be about 1 billion times that of copper. An 8-ft long, $\frac{3}{4}$-in. diameter ground rod driven into earth might test out at about 25 Ω resistance (depending upon reasonably good organic soil). Because the driven rod may have resistance greater than 25 Ω several rods may have to be driven into the soil to reduce the total resistance (the NEC requires that made electrode resistance shall not exceed 25 Ω). Continuous metallic underground water or gas piping systems in general have a resistance to ground of less than 3 Ω. Water pipes are usually installed at depths below the frost line where the moisture is relatively permanent and where resistance is barely affected by seasonal variations. Metal frames of buildings, local metallic underground piping systems, metal well casings, and similar electrodes have in general a resistance less than 25 Ω. Where available on the premises, a metal underground water pipe shall always be used as the grounding electrode (NEC requirment).

Where artificial grounding devices must be used, driven rods or pipes have proved most economical. The pipes used are generally galvanized steel whereas the rods are most often made with a steel core surrounded by a relatively thick sheath of pure copper. These rods are called *copper-clad*.

It has been determined that about 93% of the entire voltage drop occurs within a 6-ft radius of a driven rod and 82% of the total voltage drop within a 1-ft radius. Because about 90% of the total potential drop takes place within 2 ft of the pipe or rod electrode, multiple rods should be kept at least 6 ft apart wherever possible. A commonly used ground cluster arrangement uses three rods driven 10 ft apart in a triangular arrangement as shown in Fig. 8-2.

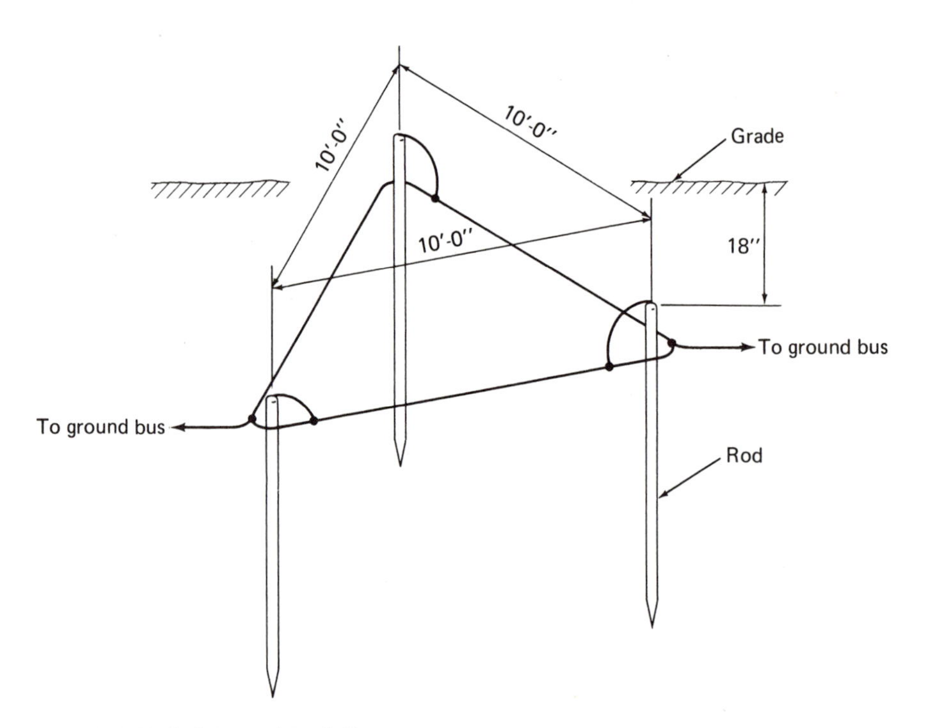

Figure 8-2 Typical Ground Rod Cluster

Made electrodes should be tested periodically to be sure resistance to ground is less than 25 Ω and therefore in compliance with the National Electrical Code. A relatively simple and inexpensive installation that allows ready access to the top of a driven rod or pipe electrode is shown in Fig. 8-3.

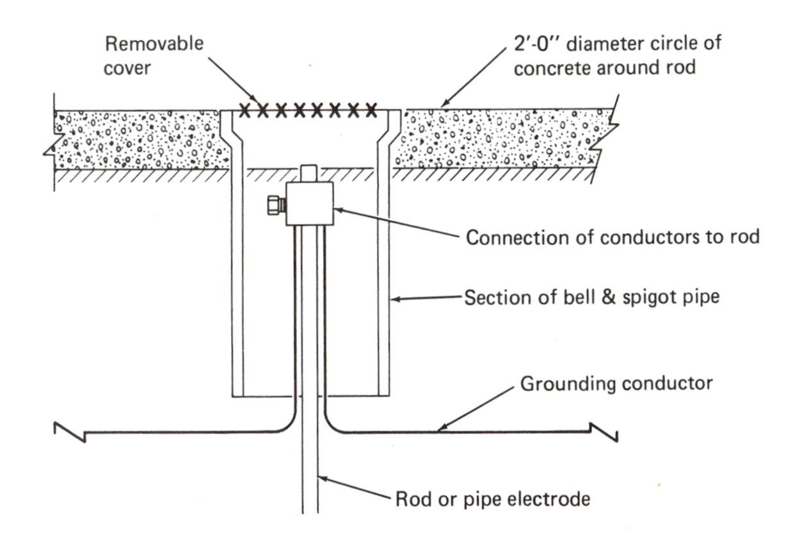

Figure 8-3 Method for Installing Electrode to Allow for Testing

8-9 Grounding of Outdoor, Open-Frame Substation

Open-frame substations (and switchyards) are constructed with various pieces of major electrical apparatus laid out in an open area on foundation *islands*. The incoming and outgoing overhead lines terminate on tower-like structural frames that are set up on their foundation islands. Bare conductors, supported on insulators, run overhead in the framework in order to interconnect the various pieces of apparatus. The total layout presents a grounding problem to bring the metal parts that a person could touch to zero potential. Voltages on different pieces of apparatus vary and the magnitude of electric-shock exposure will depend on the voltage stress imposed on the equipment insulation and the voltage difference between the metal parts of the equipment and the ground surface on which the person stands when he touches the equipment. The fence enclosing the substation also presents a possible hazard to the unauthorized person or an animal that may contact it. The fence and the adjacent ground surface must maintain a near-zero potential between them. The grounding problems can be difficult and careful engineering is required.

A fairly common solution to the problem is to provide a *frame grounding system* and a *surface grounding system* with interconnection between the two.

Bare, copper grounding conductors, ranging in size from 2/0, AWG for smaller stations to 500 MCM for larger stations, are run underground to interconnect all station equipment and structures in the frame system. The conductors in turn are connected to electrodes distributed throughout the station. Structural steel tower baseplates are bonded to the reinforcing steel in the tower foundations. The reinforcing steel in the concrete functions as an electrode. Any overhead line ground conductors terminating at towers at the outer boundary of the substation are continued overhead across the station and down to the frame ground following as close a parallel route with the phase conductors as possible (see Sec. 8-7 for reasons). Careful attention is given routing of grounding conductors and usually any junction is joined together by brazing or thermite welding. This grounding conductor network constitutes the frame ground.

Because the frame system results in spots of relatively low resistance paths where the electrodes are located and the grounding conductors run, there is a good chance that some areas adjacent to the frame system will permit hazardous potential differences to exist under ground fault conditions. To bring the entire surface area within the substation to a reasonably uniform potential level with the frame system, a grid mesh is installed slightly below grade. The grid, called a *counterpoise*, is constructed of relatively small bare conductors laid at 90° to each other in a square pattern with each junction brazed or thermite welded. The grid is connected to the frame system and although it does not reduce the overall station grounding resistance by very much, it functions to bring all parts of the substation surface, above the grid, to nearly the same potential as the frame system. If the apparatus casing, the structural frames, and the ground surface are all at the same potential, the chances that a person will be shocked by an energized piece of equipment (which has an insulation breakdown and is therefore energized) is greatly reduced. There is no potential difference to give a shock. At specific locations in the station where operators must perform switching operations, grounded metal platforms are provided to put the operator at the same potential as the equipment being operated. (A careful study of using a grounded metal platform versus a completely insulated platform may be necessary under some circumstances. There are differences in opinion in this matter.) Crushed rock fill used as a yard surface helps in this regard. The frame and surface grounding systems are diagrammatically shown in Fig. 8-4.

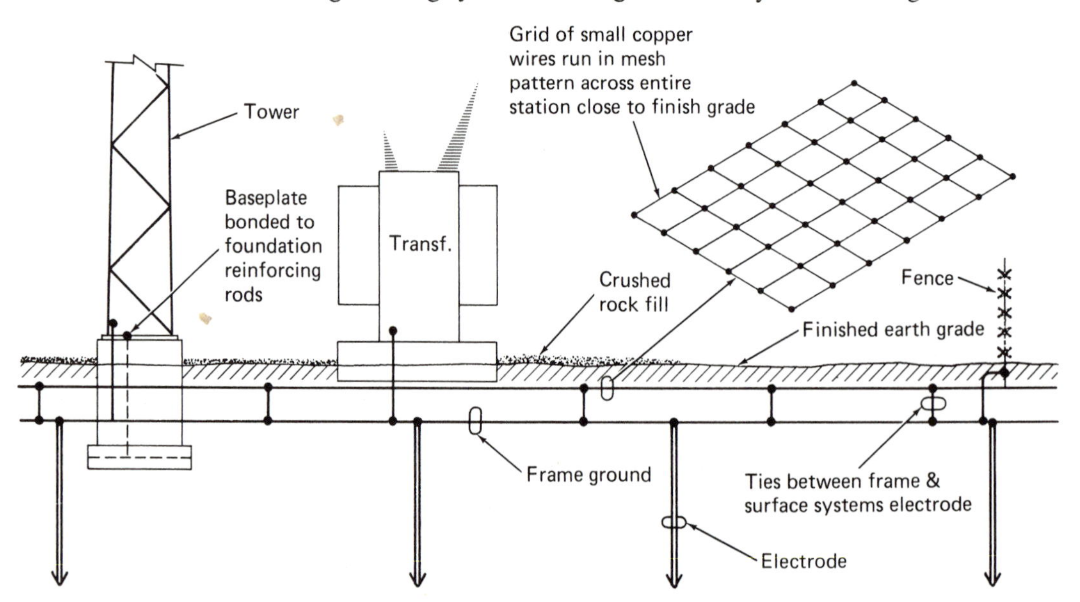

Figure 8-4 Section Showing Frame and Surface Ground Systems in Substation

In general, the practice seems to be to bond the fence to the station grid and extend the grid beyond the fence far enough to protect a person or animal touching the fence on the outside. It is very important to avoid an underground metallic conductor such as a water or air line or messenger cable from extending some distance beyond the fence. Such an extension conveys the station ground mat potential to the end of the metal extension and because surface potential drops off fairly rapidly, it is possible that a person standing on earth and touching the pipe or wire might receive a shock. A potential *difference* could exist between the metallic extension and the surrounding earth that does not have the grid under it.

8-10 Grounding of Outdoor Unit Substation

Unit substations combine the apparatus items into a single metal-enclosed package that greatly simplifies the ground problem. The objectives for grounding remain unchanged from those stated previously. Usually the unit is looped with a grounding conductor run about 18 in. below grade with electrodes driven in clusters at diagonal corners of the unit. An incoming overhead ground line on a tower presents no significant problem and is grounded to the grounding conductor. Figure 8-5 shows a typical arrangement.

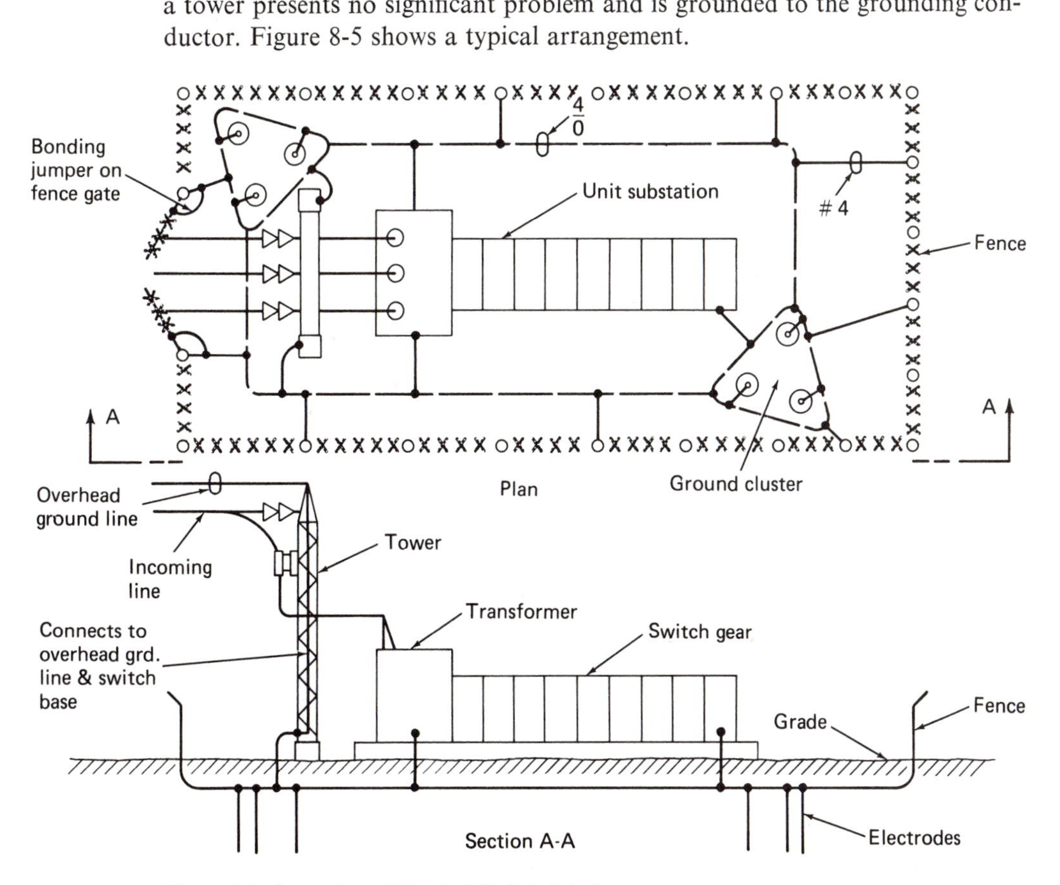

Figure 8-5 Grounding of Typical Unit Substation

8-11 Grounding of Building Service Entrance Equipment

Service entrance equipment is made up of the switch or circuit breaker or other protective device that is installed in the electrical supply entering a building. The National Electrical Code sets forth the mandatory rules governing such equipment and the manner in which connections can be made to it. Many systems serving buildings are served by systems that come from external power sources that are *grounded.* Buildings with *in-house* substations generally distribute power via *grounded* systems. More effective control of voltage stresses in an electrical power system can be realized if one of the power conductors is grounded. The NEC requires that an alternating-current electrical power system supplying interior wiring and interior wiring systems be grounded if by so doing the voltage on each ungrounded conductor can be set at a value not exceeding 150 V to ground. Such systems to be solidly grounded are shown in Fig. 8-6.

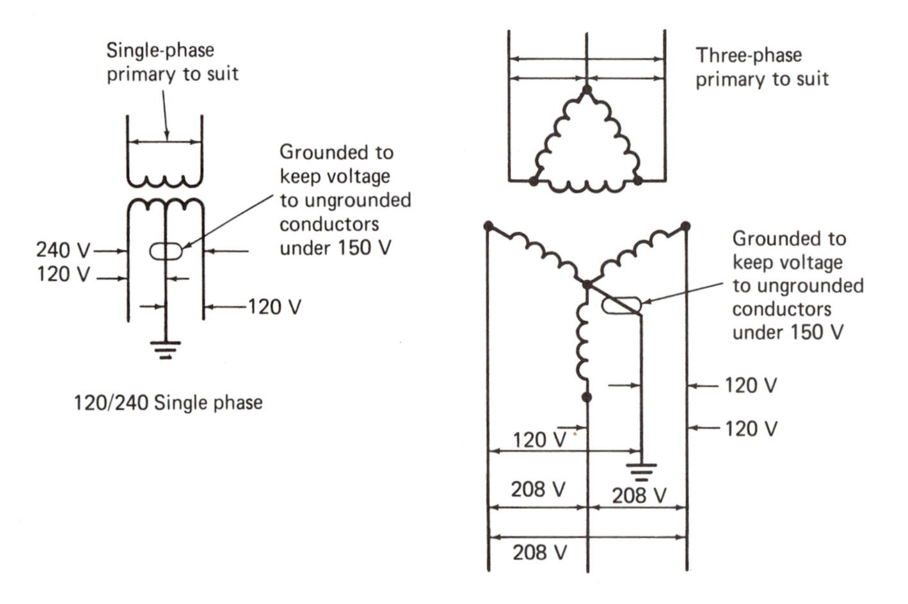

Figure 8-6 Systems that Must Be Solidly Grounded per NEC

The NEC also requires that systems nominally rated 480/277 V, three phase, four wire, in which the neutral is used as a circuit conductor be grounded. This is a wye-connected system similar to the 120/208-V system in Fig. 8-6 except the phase-to-ground voltage is 277 V and phase-to-phase voltage is 480 V. The latest edition of the code should be used in determining grounding requirements for all systems.

The NEC requires that any secondary alternating-current electric power system that operates with a grounded conductor shall have the grounded

conductor extend to the grounding junction at each service being supplied. The grounded conductor must be no smaller than prescribed by the NEC. This rule provides for making sure that the *service* grounded conductor is brought to the service entrance equipment from the power supply (usually a transformer). This conductor at the service entrance and beyond in all branch-circuit wiring is identified with white or gray surface-colored wire.

The NEC also requires that the grounded conductor must be connected to an earth electrode *at* the service entrance equipment. This is usually the service water pipe in accordance with the code.

Finally, the code requires that equipment grounding is continuous throughout the system. This may take the form of a well-bonded metallic raceway system such as metal conduit or grounding conductors run with the power conductors in the same enclosure. No matter which method is used, all metal boxes, cabinets, motor starters, switches, etc., *must* be connected back to the service entrance equipment with a grounding conducting path. The service entrance requirements are diagrammatically summarized in Fig. 8-7.

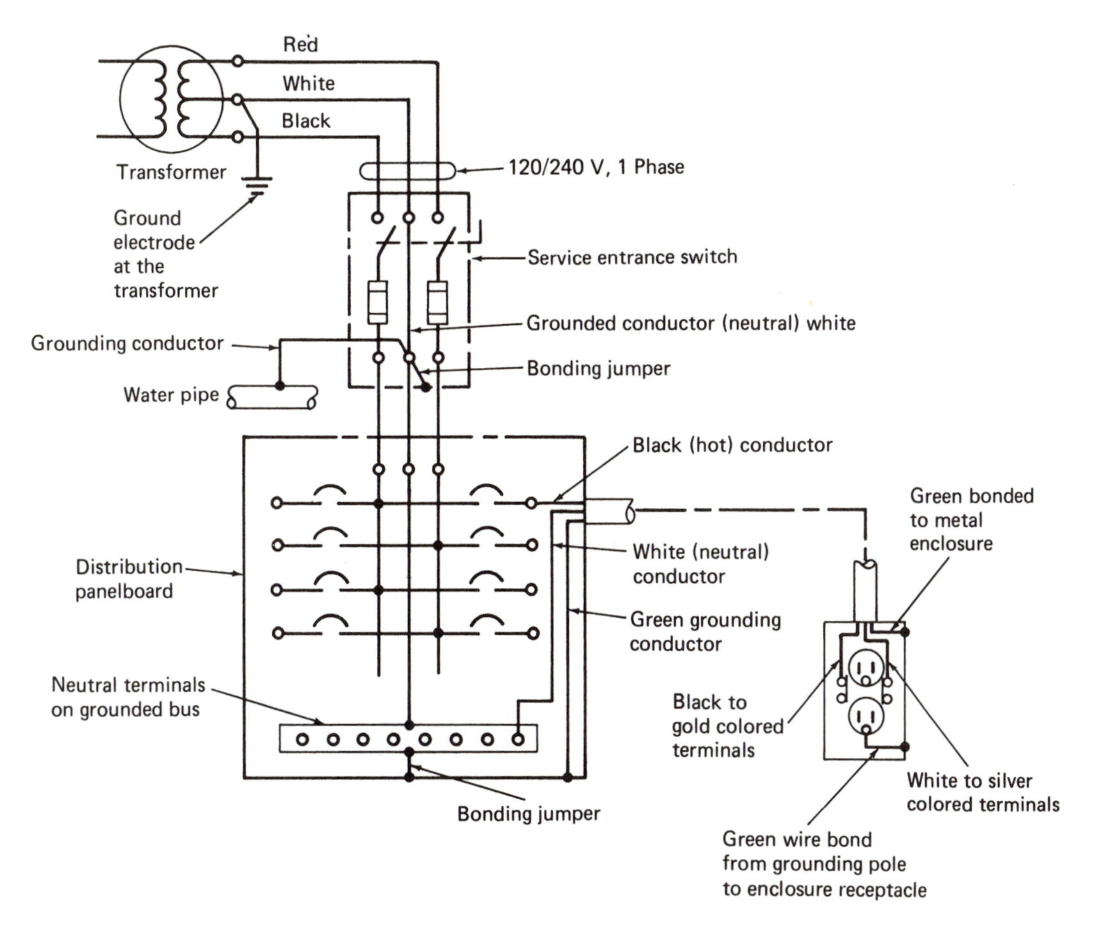

Figure 8-7 Grounding at Service Entrance

8-12 Grounding Requirements for Interior Wiring Systems

Figure 8-7 shows extension of the power system to the building interior from the service entrance equipment via a branch circuit to a receptacle. Note that in this example a green grounding conductor is run with the power conductors to bond the receptacle box solidly to the distribution panelboard. This complies with the code and with the result of the experiment described in Sec. 8-7 to control ground fault currents. The raceway itself could be used instead of the green wire. A properly installed raceway system will provide essentially the same grounding path as the wire. Many firms have adopted the green wire requirement as standard design because it is a more reliable grounding path.

The NEC specifies the size of the green grounding conductor as it relates to the size of the power conductor being used.

Under normal conditions no current flows in the green grounding conductor and therefore a zero potential exists between all the metal parts that are bonded by this conductor (or raceway system) and earth (or the equivalent of earth). When unplanned, unwanted fault currents flow along this path, there will be a voltage difference present. With a direct, low impedance path such as the green wire, the protective devices will operate to clear the fault.

8-13 Grounding of Interior Unit Substations and Switching Centers

The unit substation is usually a factory assembled unit complete with transformer (or transformers), switching equipment, and distribution devices. The entire unit is metal enclosed and internal grounding conductors protect all the equipment elements when properly connected in turn to earth. The interior grounding conductors may be either wire or bus bar and generally terminate in a horizontal grounding bus bar that runs the complete length of the unit in the bottom section. Grounding conductors from earth electrodes or equivalent are connected to this bus bar. At least two grounding connections should be made to the bar as protection against electric shock.

Because a transformer is part of a unit substation, it must be recognized that the secondary of the transformer is a new electrical system in terms of grounding for ground fault protection. The grounding for fault currents in the primary circuits must relate back to the source of power for the primary windings. Similarly the secondary winding represents the source of power beyond this point, therefore grounding conductors (green wire or raceway) must run back to the grounding bus associated with the secondary. Careful attention needs to be paid to this distinction between primary and secondary grounding buses. One is not so good as the other!

Actually the two grounding buses may be interconnected but they should be respected as separate grounding systems.

Conduits or other raceways that enter unit substations through the bottom section may terminate completely clear of any contact with the metal structure itself because very often there is no *floor* in the equipment. Careful design will call for bonding jumpers between the raceway and the enclosure grounding bus.

Metal raceways joining hard into the steel sheets of the unit enclosure are usually required to be made up tight with double locknuts (check code for specific requirements). Bonding jumpers or grounding-type bushings provided with grounding screws may be required if the raceway itself is to be utilized as the grounding conductor.

Removal of paint where grounding bonds are to be made and bonding across removal plates to which conduits are attached are important details to care for in good design.

8-14 Grounding at Utilization Equipment and Devices

Ultimately the power terminates in equipment or devices that provide for use of the power. Such apparatus as motors, electric heaters, lighting fixtures, and receptacles for further extension to household appliances must be properly bonded to the grounding conductor. The size and terminations of the grounding conductor must be in accordance with the code. Figure 8-8 displays the grounding requirements in general and shows the distinctions between the groun*ded* conductor and the ground*ing* conductor. (The green wire method is used.)

Note that the white wire is connected to the shell of the lamp holder in the lighting fixture, which will usually have a silver colored terminal screw. The center terminal has a gold colored screw for the *hot* wire. By placing the hot terminal inside the lamp holder, the contact to the lamp is guarded from accidental contact by a person and against accidental grounds. The shell is grounded by the white wire.

Ground terminals are mandatory for all receptacles. Portable cords to electric hand tools have their own green wire that extends the grounding conductor to the case of all portable tools as a safeguard against shock.

Bonding connections to the building columns, which in turn are connected to grounding electrodes, is appropriate for the fixed equipment such as motors and panelboards even though the green wire is provided. These additional grounds may be redundant but there is always the chance that the green wire may become interrupted.

8-15 Lightning Protection Grounding

Lightning is the discharge of electrical energy between clouds or clouds and earth. Charges of one polarity develop and build up in the clouds; the opposite polarity builds up in the earth. When the charge becomes so great that the insulation of the air masses can no longer contain it, a discharge takes place. The damage can be very extensive. Heat and mechanical forces produced by the current flow through the natural resistance elements in its path tears trees apart, sets fires, destroys electrical equipment, and causes interruption of power service. If a path of direct low resistance is provided, the chances of damage

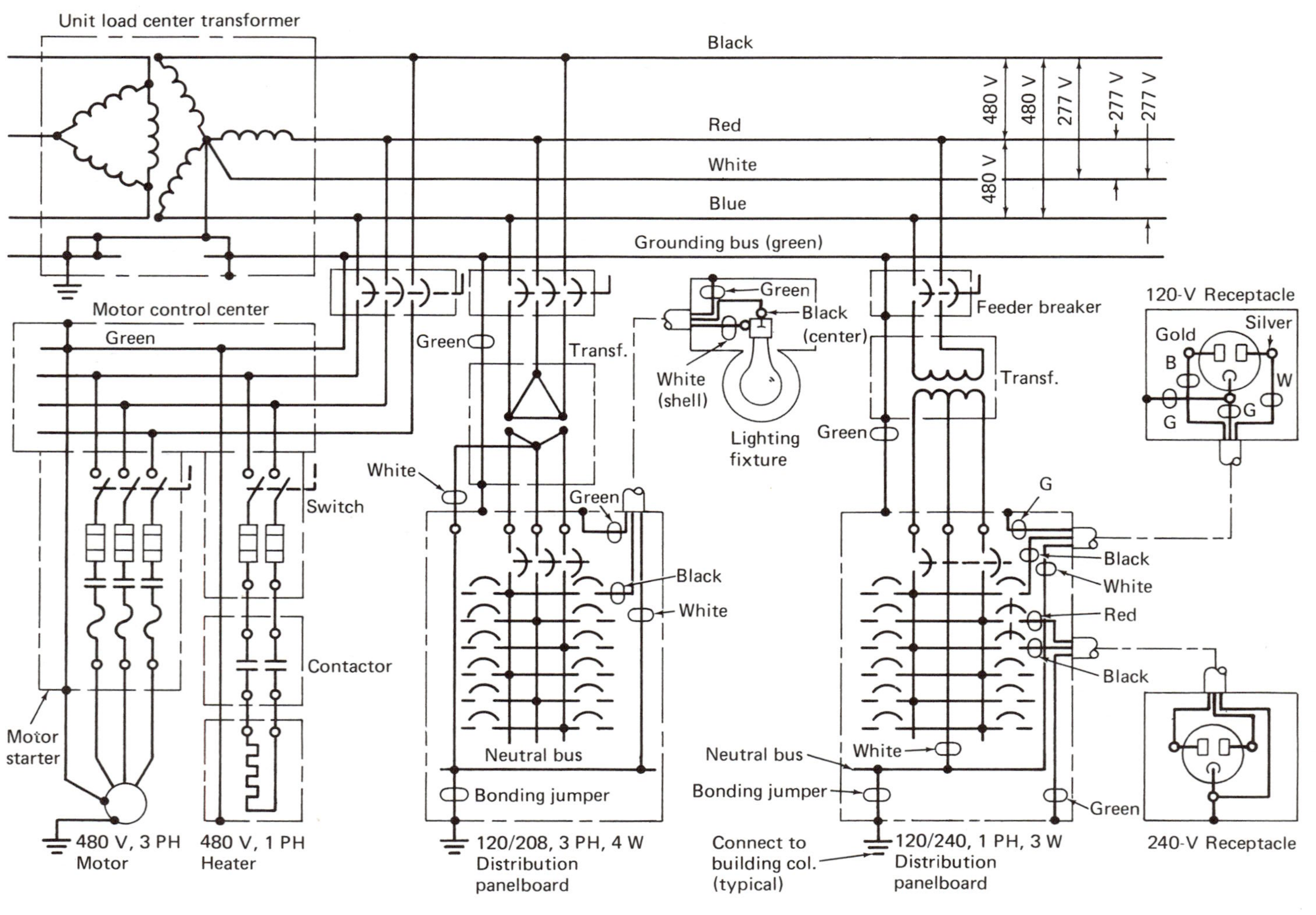

Figure 8-8 Typical Power Supply Showing Grounded (White) and Grounding (Green) Conductors

are reduced. This is the fundamental theory of lightning protection. High-voltage transmission lines are protected by running ground wires above the power conductor from tower to tower. Figure 8-9 shows the location of these wires.

The ground wire is clamped solidly to the tower structure and the structure itself is connected to earth electrodes or other earth systems such as a local counterpoise. The power conductors are protected from lightning strokes within the general area of a *cone of protection* as shown in Fig. 8-9 because the stroke is attracted to the ground wire and conducted to earth.

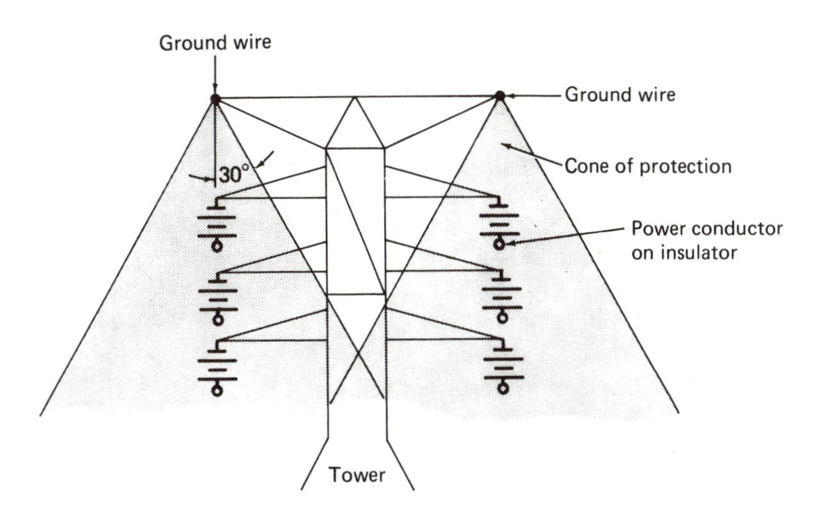

Figure 8-9 Transmission Line Static Wires for Lightning Protection

In substations and switchyard areas, lightning masts are often erected either from the ground surface up (similar to a flagpole) or as vertical rods extending up from the tops of structures. From the tops of these masts a cone of protection (like an umbrella) shields the equipment below from lightning damage.

Where ground wires and masts can be effectively used, they provide a good deal of protection; however, lightning protection of power stations, substations, distribution systems, and switchyards must include protection of equipment by means of lightning arresters. The elementary principle behind a lightning arrester is to connect a simple horn gap in series with a resistance as shown in Fig. 8-10.

The horn gap prevents current leakage from the line when voltage on the

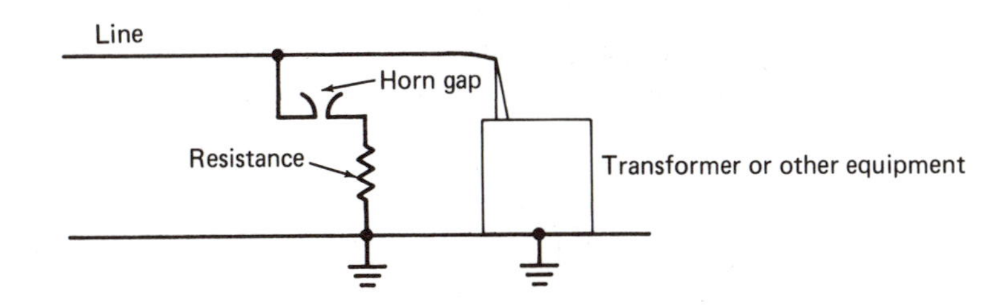

Figure 8-10 Elementary Lightning Arrester

line is normal. If lightning strikes the line, the path of least resistance is across the gap and through the resistance rather than through the windings of the apparatus. Without an arrester the high-voltage stress from a stroke would probably puncture the windings insulation. While the high voltage is being drained off to earth, the resistance in series with the horn gap limits the current flow to earth. When the discharge has passed, the heat of the arc at the horn gap causes an upward draft of hot air that carries the arc upward, stretches it, and causes it to break. Note that the horn gap configuration leads to promoting a stretching effect. Also the magnetic field on the inside of the V helps to drive the arc upward. This arresting action takes place without interrupting the normal load circuit.

This elementary form of arrester is not used in modern practice; however, the principle remains the same in manufactured arresters. There are many different types to suit the wide variety of applications and manufacturers publish detailed information for use by engineers and designers.

The important considerations in terms of grounding is to connect the grounding terminal of the arrester directly to a ground electrode in as short and straight a path as possible. In addition, the grounding conductor should be connected into the common station ground bus.

8-16 Grounding Connectors and Connections

Grounding connections are made in two principal manners. The first method uses mechanical-type connectors with bolted fittings. These have been used for years with satisfactory results. The second method employs an exothermic welding process where no outside source of power or heat is required. The reaction in this process is the reduction of copper oxide by aluminum producing molten, superheated copper. A weld is made between copper conductors from the welding heat obtained from the superheat in the molten copper. This method provides a permanent connection; therefore, if disconnecting is required for testing, it must be provided at some other proper location by means of a bolted connection.

Mechanical connectors for grounding are available in many shapes and configurations to fit driven rods and pipes and for connection to water or other service pipes. Clamps and fittings can be bought to accommodate a range of cable sizes as well as multiple cables on one connector. Loop arrangements can be made so that the grounding conductor can be run unbroken. Fittings are available for clamping grounding conductors to flat plates such as building columns.

Other fittings are especially designed for fence and fence gate grounding. Flexible braid-type bonding jumpers provide for connections where movement is expected.

Thermit connectors accommodate a wide range of grounding connections but here the configuration of the connection is in the mold required to make the joint. Figure 8-11 shows a typical mold for a cable-to-rod connection with an illustration of the finished joint.

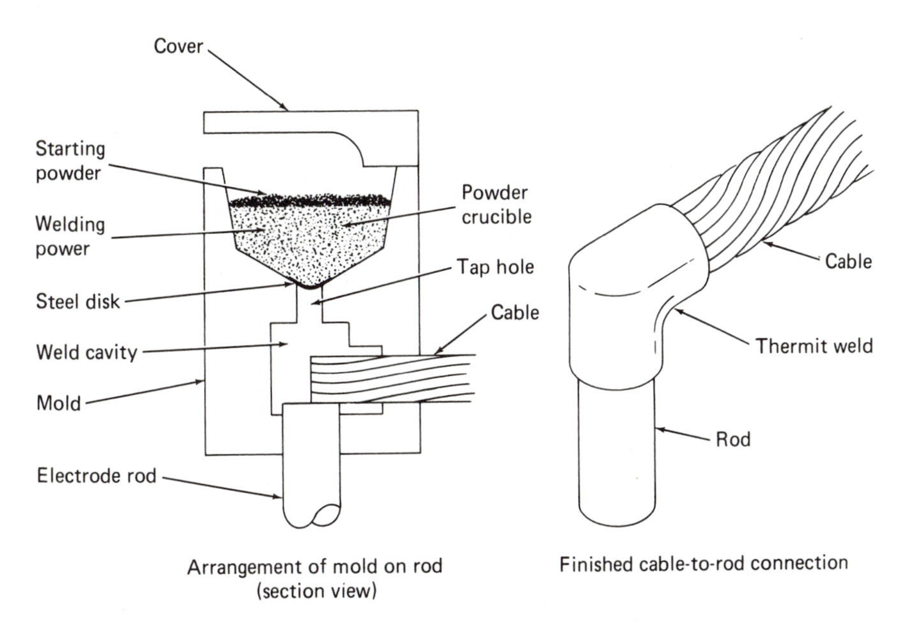

Figure 8-11 Thermit Method of Making Electrical Connection

To make a Thermit connection similar to that shown in Fig. 8-11 the proper mold must be used and set up as shown. The steel disk is placed in the bottom of the powder crucible and then the welding powder is emptied in (cover removed). The powder is packaged so that the starting powder is in the bottom of the powder carton and therefore comes out last and ends up on top. The cover is put on and a flint gun is used, through the cover opening, to ignite the starting powder. The heat generated through burning of the powder causes the welding powder to burn and become molten and burn through the steel disk. The molten material flows down around the cable and rod and generates enough heat so the rod, cable, and molten material becomes homogeneous. Heat dissipates sufficiently in about 10 sec so that the mold can be opened and slag cleaned from the connection. Molds are available as one-shot or throwaway after about 100 welds or in the form shown in Fig. 8-11, which can be cleaned and used many times. This connection method can be made in several configurations for

1. Cable to cable
2. Conductor to electrode rod (bus or cable)
3. Cable to flat surface
4. Cable to terminal lug, bus, and tube
5. Bus to bus
6. Bus to tube
7. Cable to reinforcing bars

Chapter Review Topics and Exercises

8-1 Explain the difference between system grounding and equipment grounding.

8-2 Does a *grounded* conductor relate to system grounding or to equipment grounding?

8-3 Why is grounding important?

8-4 How should grounding conductors relate physically to the conductors' feeding power?

8-5 Can the metallic path provided by a conduit carrying power conductors be used as a grounding conductor? How does it compare as a return ground path to the grounding conductor run with the power conductors?

8-6 What special care must be taken in using metallic conduit as a grounding conductor?

8-7 Why are water pipes usually good electrodes?

8-8 How are man-made electrodes provided?

8-9 Describe a ground rod cluster and why it is made in that form.

8-10 What is the value of providing access to driven ground rod electrodes?

8-11 Explain the *frame grounding system* for a typical substation.

8-12 What does the *surface grounding system* accomplish in a substation?

8-13 Why should the counterpoise ground extend beyond the substation fence line on the outside?

8-14 Why should grounding connections to underground pipes that might run through a substation site be avoided?

8-15 Sketch the electrical distribution (low-voltage) systems that *must* be grounded to comply with NEC rulings.

8-16 What color *must* the *grounded* conductor be to comply with the NEC?

8-17 What color *must* the *grounding* conductor be to comply with the NEC?

8-18 Why should careful attention be given to grounding both primary and secondary systems when transformers are part of the circuit?

8-19 Where should bonding jumpers be required for switchgear installations?

8-20 Describe the *cone of protection* as it relates to lightning protection.

8-21 How should grounding conductors be run from the ground terminal of a lightning arrester?

8-22 Why should mechanical-type ground connectors be used in some cases as opposed to a permanent welded connection?

8-23 What kind of a connection does a Thermit process provide?

9

RACEWAY LAYOUTS

9-1 General

The National Electrical Code defines a *raceway*, as *any channel for holding wires, cables, or bus bars, which is designed expressly for and used solely for this purpose.* Raceways may be of metal or insulating material and the term includes

1. Rigid metal conduit
2. Rigid nonmetallic conduit
3. Electrical metallic tubing (EMT)
4. Flexible metallic conduit
5. Surface raceways
6. Cellular concrete floor raceways
7. Cellular metal floor raceways
8. Underfloor raceways
9. Wireways
10. Busways

In addition, *cable trays*, which are defined in the code as *a unit or assembly of units or sections and associated fittings, made of metal or other noncombustible materials forming a rigid structural system used to support cables*, appear on drawings covering raceway layouts. Cable trays include ladders, troughs, channels, solid bottom trays, and other similar structures.

Raceway layouts are physical-type drawings, generally showing plan views and necessary elevations and sections of any one or a combination of several of the raceway systems above. As a rule the word *raceway* does not find common usage in the title of these drawings. They are more likely to be called *conduit and cable tray layout* or *conduit and busway layout*, etc. *Conduit* is used so extensively that few layouts on a major job will be without it. Installation of any raceway, covered by the National Electrical Code, must be made in accordance with the rules of the latest edition of the code.

9-2 Descriptions of Raceways

A brief description of the raceways named in Sec. 9-1 follows. Because manufacturers make raceways in several designs, the examples given are shown as typical for each type.

Rigid Metal Conduit

Rigid metal conduit is really nothing more than pipe except that it is made expressly for electrical work and therefore has a smooth interior so wires or cables can be pulled into it without injury to the conductor insulation. Rigid metal conduit is made in 10-ft lengths and threaded on both ends with a coupling furnished on one end of each length. Standard trade sizes (inside diameter) are $\frac{1}{2}''$, $\frac{3}{4}''$, 1″, $1\frac{1}{4}''$, $1\frac{1}{2}''$, 2″, $2\frac{1}{2}''$, 3″, $3\frac{1}{2}''$, 4″, $4\frac{1}{2}''$, 5″, and 6″. Conduit can be bent by hand in sizes up through 1″ by using a hand conduit bender, commonly called a *hickey*. The larger sizes are generally bent by using hydraulic bending machines. In addition, standard 45 and 90° elbows and large-radius elbows are manufactured that often prove less expensive than field bends. Rigid conduit is joined into pressed steel boxes, sheet steel boxes, cabinets, etc., by using threaded locknuts and bushings. There are enumerable fittings made with threaded or threadless connections to accommodate the changes in direction that the wide range of layout conditions may dictate. Rigid metal conduit is made in both steel and aluminum. Figure 9-1 illustrates conduit construction,

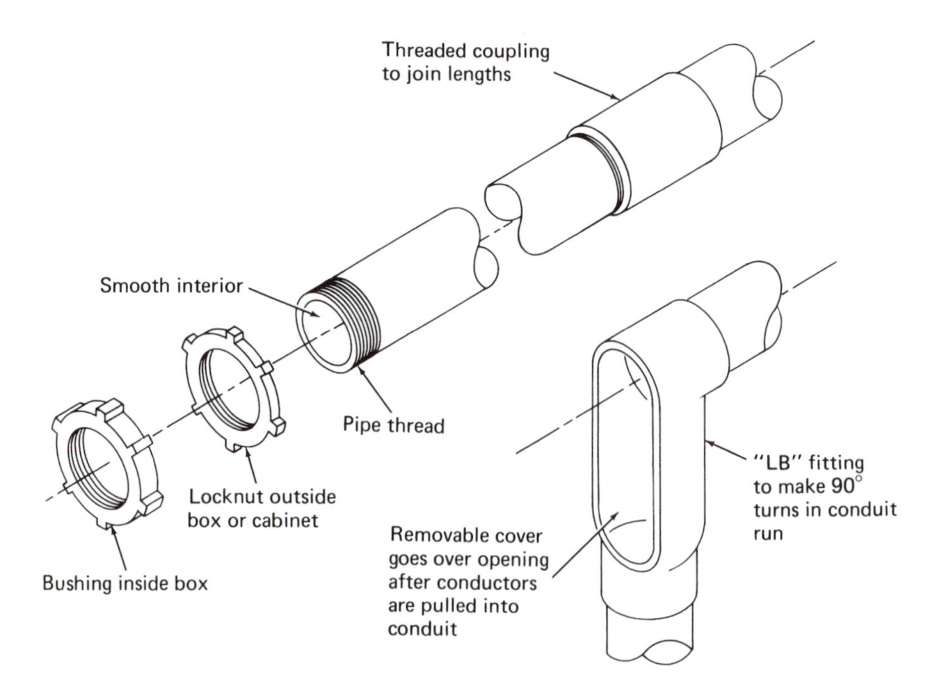

Figure 9-1 Typical Rigid Metal Conduit and Common Fittings

a locknut and bushing, and a typical *condulet* fitting for making a right-angle bend with access to the conduit for pulling conductors.

Rigid Nonmetallic Conduit

Rigid nonmetallic conduit and fittings must be made of a suitable nonmetallic material that is resistive to moisture and chemical atmospheres in order to meet NEC requirements. In addition, for use above ground, it must be flame retardant and resistant to impact and crushing, to distortion due to heat under conditions likely to be encountered in service, and to low temperature and sunlight effects. For underground use, the material must be acceptably resistant to moisture and corrosive agents and of sufficient strength to withstand abuse, such as impact and crushing, in handling and during installation.

Materials that have been recognized as having suitable physical characteristics when properly formed and treated include

1. Fiber
2. Asbestos cement
3. Rigid polyvinyl chloride (PVC)
4. High-density polyethylene

Rigid nonmetallic conduit is used primarily in underground installations and the conduits are generally encased in concrete. Figure 9-2 shows a section view of such a *duct bank* of underground conduits.

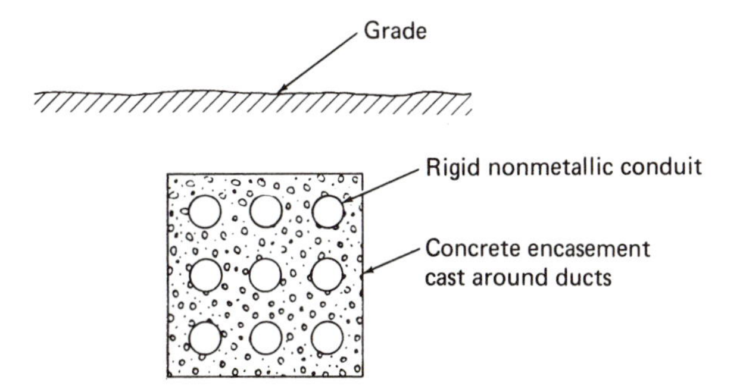

Figure 9-2 Typical Duct Bank Using Rigid Nonmetallic Conduit

Electrical Metallic Tubing (EMT)

Electrical metallic tubing is a thin wall conduit that does not permit threading. Fittings, called *connectors*, are made to connect the tubing to boxes, panelboards, etc. Tubing couplings are also made for joining sections of tubing together. Both the connectors and couplings slip onto the tubing and are made tight either by setscrews or by compression rings in the fitting. The connectors are threaded for joining into boxes, etc., by a locknut furnished with the connector. EMT is furnished in standard trade sizes of $\frac{1}{2}''$, $\frac{3}{4}''$, 1″, $1\frac{1}{4}''$, $1\frac{1}{2}''$, 2″, $2\frac{1}{2}''$, 3″, and 4″ in 10-ft lengths. Unlike rigid metal conduit, couplings are not

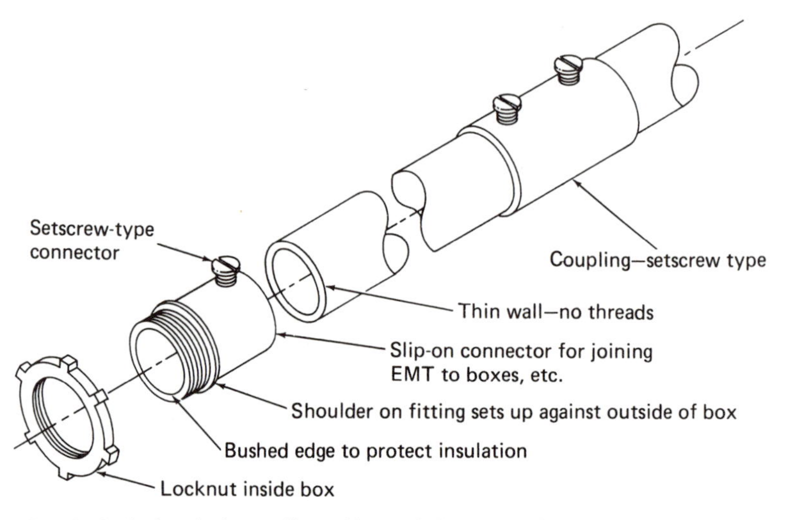

Figure 9-3 Typical Electrical Metallic Tubing and Common Fittings

furnished with each length. Tubing may be used for both exposed and concealed work. Figure 9-3 shows EMT with typical fittings.

Flexible Metal Conduit

Flexible metal conduit is a corrugated tubing made by wrapping a formed ribbon of metal into an interlocking spiral resulting in a flexible metal tubing. BX cable is actually flexible metal conduit with the conductors prefabricated into the conduit to form the cable. Flexible metal conduit without conductors is sometimes called *Greenfield.* Standard sizes are $\frac{3}{8}$″ (code restrictions on this size), $\frac{1}{2}$″, $\frac{3}{4}$″, 1″, $1\frac{1}{4}$″, $1\frac{1}{2}$″, 2″, $2\frac{1}{2}$″, and 3″. Threaded connectors are used to join the flexible metal conduit to boxes, etc. It finds a wide variety of uses where rigid connections are not feasible or where vibration becomes a factor.

Flexible metal conduit is also made in a liquid-tight construction. Special fittings are required for this type in order to make the entire assembly liquid-tight. Figure 9-4 shows a section of flexible metal conduit.

Surface Raceways

Surface raceways are made in several shapes and forms in both metal and nonmetallic construction. They may only be installed in dry locations. Fittings are made to accommodate almost any conceivable layout. This type of raceway is installed on floors, walls, ceilings, and any other surface where a concealed system is too costly or impractical. Sizes and configuration vary with each manufacturer. Figure 9-5 shows some typical types.

Cellular Concrete Floor Raceways

Cellular concrete floor raceways are the hollow spaces in floors constructed of precast cellular concrete slabs together with suitable metal fittings designed to provide access to the floor cells. The *cell* is defined as *a single, enclosed*

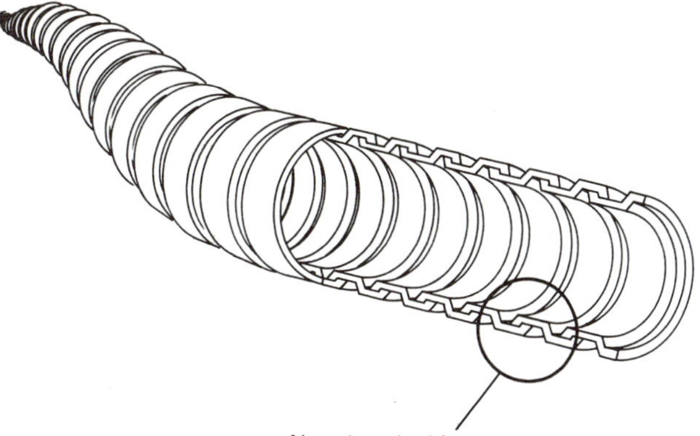

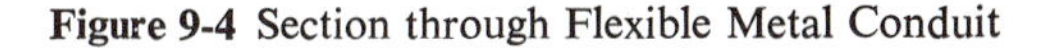
Figure 9-4 Section through Flexible Metal Conduit

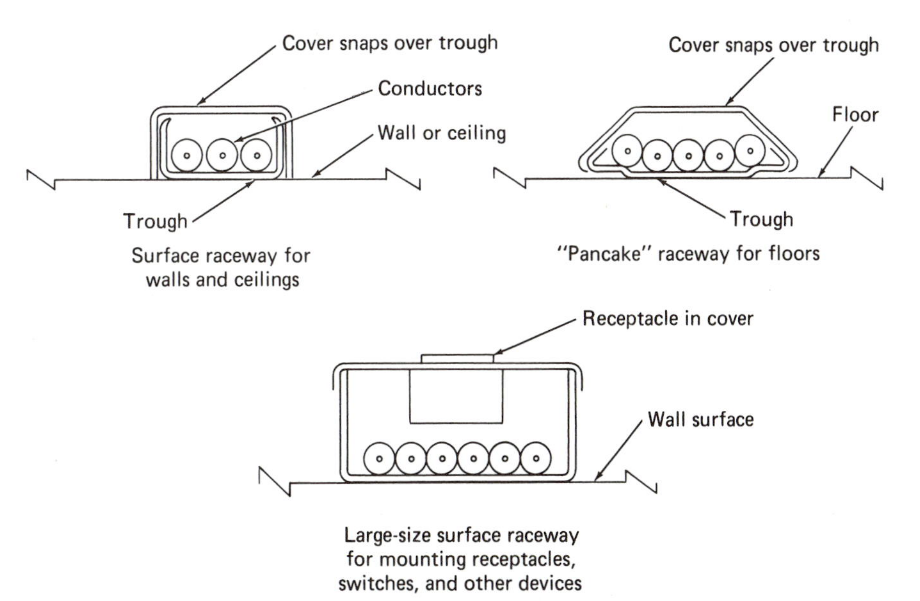

Figure 9-5 Typical Surface Raceways

tubular space in a floor made of precast concrete slabs, the direction of the cell being parallel to the direction of the floor member. *Header ducts* are installed at 90° to the cells in order to furnish access to the cells. These header ducts are generally made of metal that is cast in place in the concrete fill covering the precast slabs. Figure 9-6 shows typical construction.

Cellular Metal Floor Raceways

Cellular metal floor raceways are the hollow spaces of cellular metal floors. Construction is similar to the cellular concrete floor raceway system except

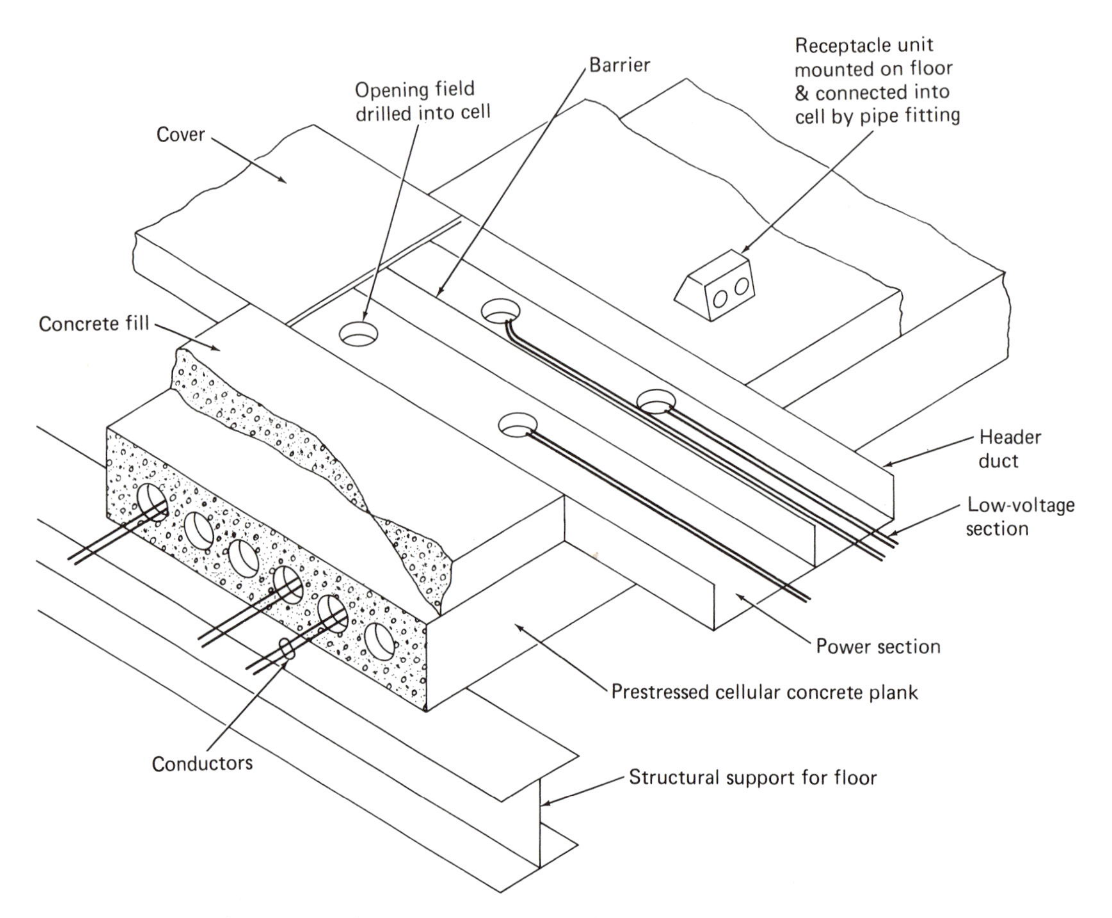

Figure 9-6 Typical Construction of Cellular Concrete Floor Raceway

that the structural floor system is constructed of prefabricated metal *planks* having a cell-type construction. The metal planks or decking is secured by welding to the structural steel framing members in the building and then concrete is cast in place over the decking to form the structural floor. When finished, the floor has continuous cells running through it that are used for raceways. Header ducts join the cells into the above-floor electrical system. Figure 9-7 shows typical construction of this system.

Under-floor Raceways

Under-floor raceways are metal ducts of various forms and shapes that are laid onto a rough concrete slab and then covered with finished concrete. Like the cellular systems the result is a raceway made integral with the floor so electrical service is available by tapping into the in-floor cells. Header ducts connect the runs of underfloor raceways together and to the panelboards that supply power to the under-floor system. Figure 9-8 shows a typical double-duct system.

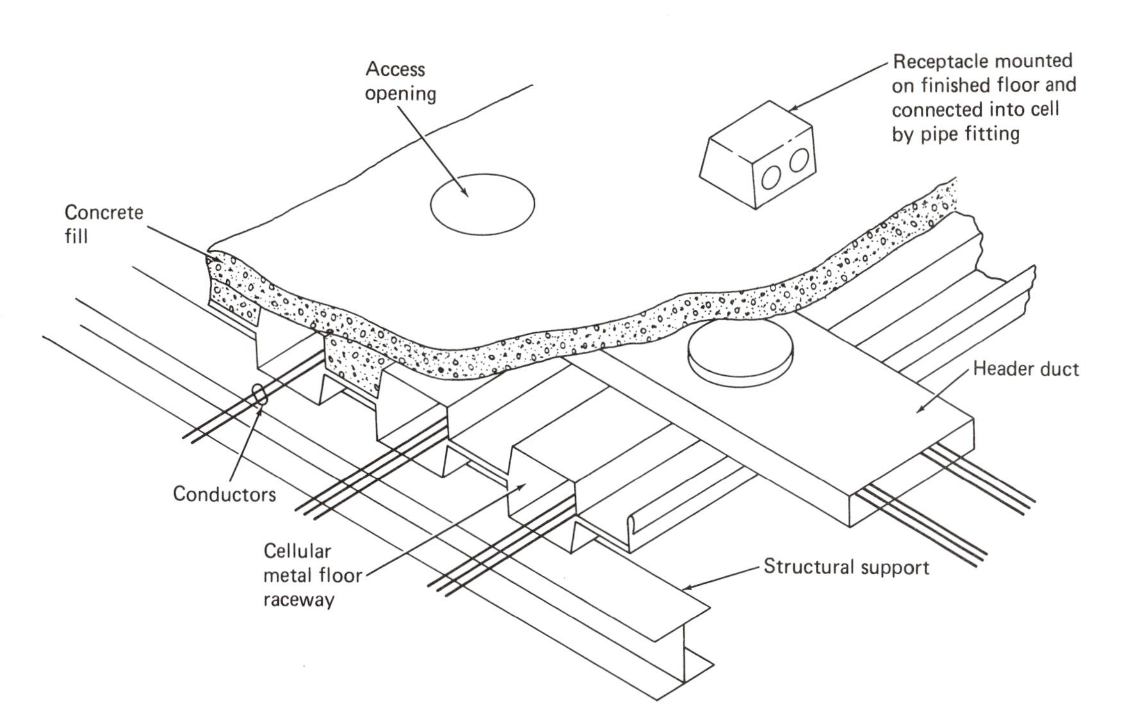

Figure 9-7 Typical Construction of Cellular Metal Floor Raceway

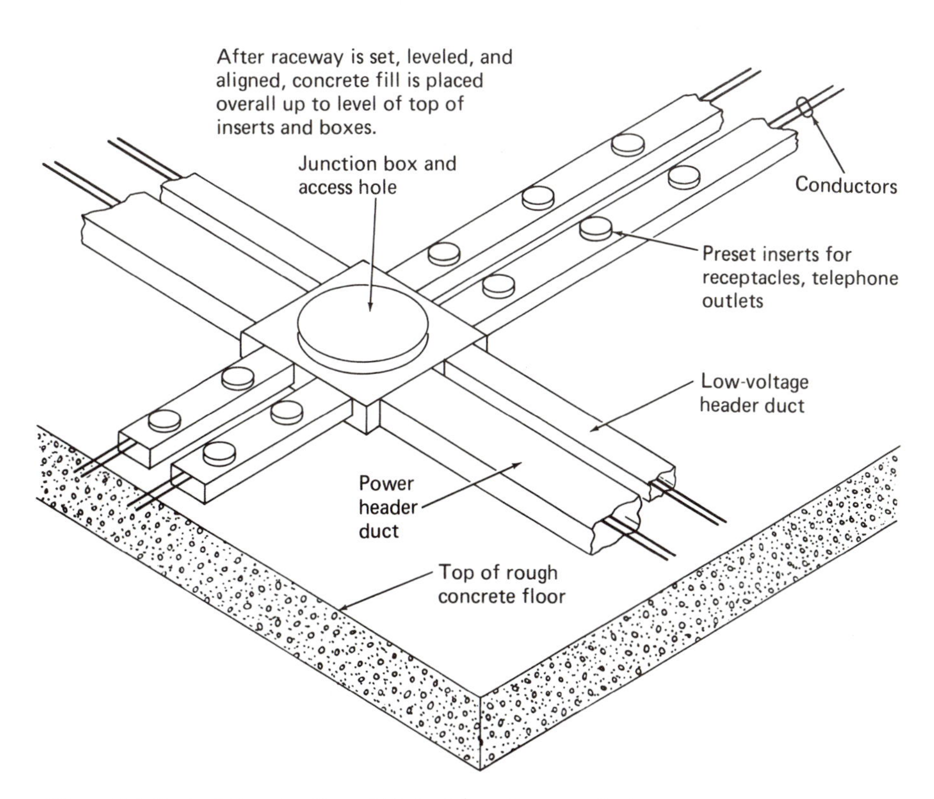

Figure 9-8 Typical Construction of Under-Floor Raceway System

Wireways

Wireways are sheet metal troughs with hinged or removable covers for housing and protecting wires and cables. The conductors are laid in place after the wireway is installed as a complete system. Wireways are made for exposed work only. They are available as standard products for both dry and wet locations. For wet locations, the wireway is properly gasketed and approved for such installations. Wireways come in standard sizes of 4″ × 4″, 6″ × 6″, and 8″ × 8″ and standard lengths such as 60″, 24″, and 12″ with prefabricated elbows, tee fittings, junction and pull boxes, and hangers. Figure 9-9 shows a section of wireway with an elbow and tee fitting.

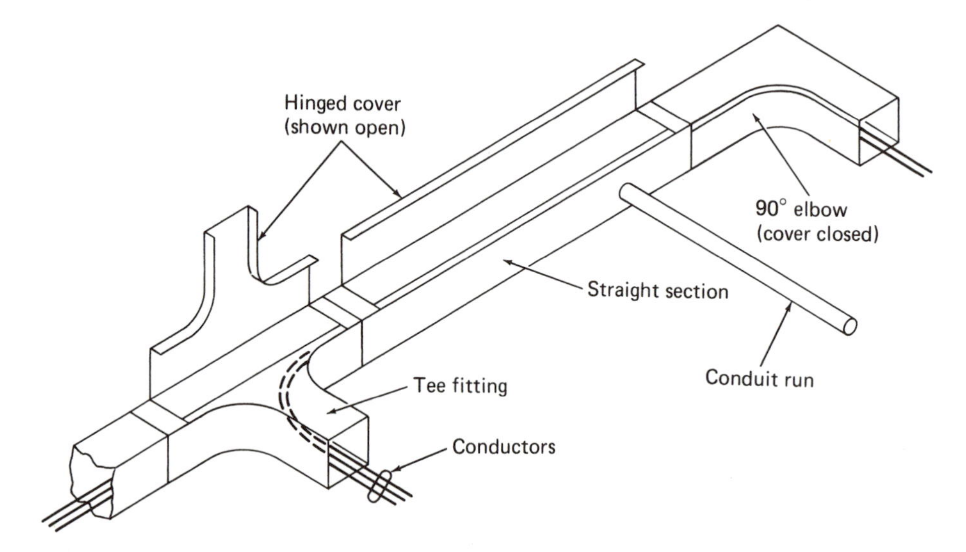

Figure 9-9 Typical Construction of Wireway

Busways

Busways are a prefabricated assembly of bus bars, insulators, and a metal enclosure that are used in various forms for power distribution. Busways are used only in exposed work. They are available in various forms and ratings. Standard length is generally 10 ft and fittings are available to accommodate most layouts. Custom-made busways may be required for many installations. Figure 9-10 illustrates one typical type of busway called *plug-in bus* that is made so that power may be taken from the bus at several points along a run.

Cable Trays

Cable trays or *cable troughs* are prefabricated assemblies of straight sections and fittings that can be joined together to form a raceway system. In general, three types of cable trays are available:

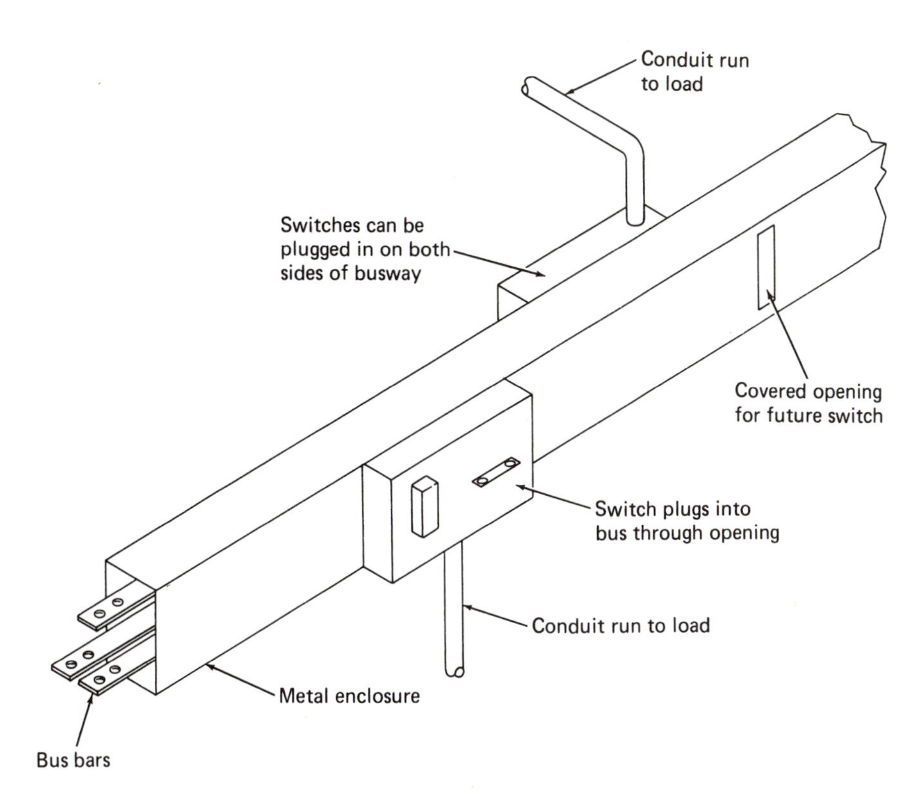

Figure 9-10 Typical Plug-In Busway Construction

Trough type: Has a continuous bottom either ventilated or nonventilated. Standard widths are 6″, 9″, 12″, 18″, and 24″. This type is commonly used when conductors are small and need full support. Figure 9-11 shows this type.

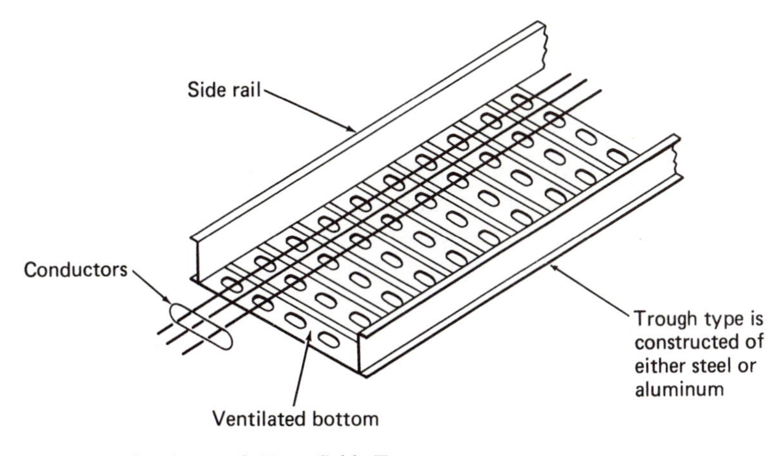

Figure 9-11 Ventilated Trough-Type Cable Tray

Ladder type: Consists of two longitudinal side members connected by individual rungs. Generally used for power cable support. Rung spacing varies. Standard widths are 6″, 9″, 12″, 18″, 24″, and 30″. The ladder type is made in both steel and aluminum. Figure 9-12 illustrates this type.

Channel type: Is a one-piece ventilated channel section. These are generally used to support either a single power cable or several small multiconductor control cables. Standard widths are 3″ and 4″ and the channel is made of either steel or aluminum. Figure 9-13 shows a typical channel tray.

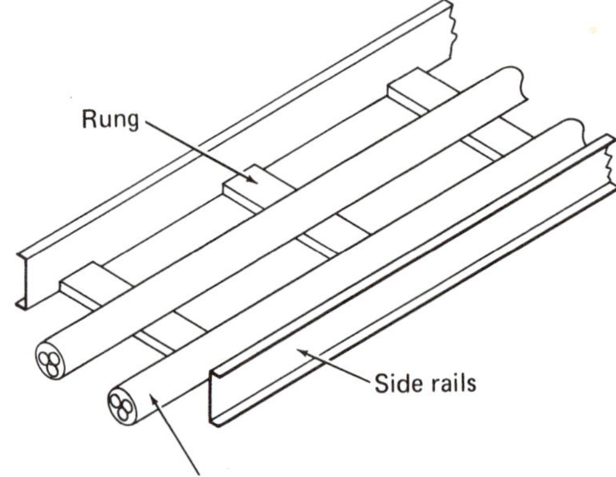

Figure 9-12 Ladder-Type Cable Tray

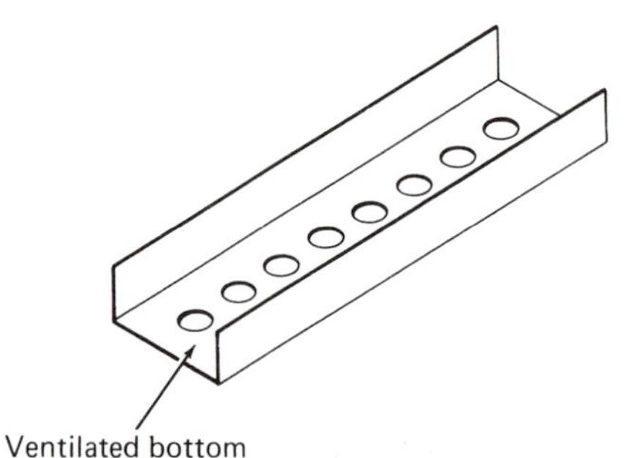

Figure 9-13
Channel-Type Cable Tray

A wide range of fittings are made to provide 90° bends (both horizontal and vertical), offsets, dropouts for running cables out through the bottom of the trough type, reducing sections, tee sections, etc. In addition, supporting pieces and other hardware are available to hang or otherwise mount the tray system.

9-3 The Out-of-Function Part of a Layout

In order to show how conduit or other raceways are to be installed in a building or elsewhere, a relating framework of physical walls, columns, stairways, etc., must be drafted before the electrical elements of the drawing can be started. This related framework is often called the *out-of-function* or *shell.*

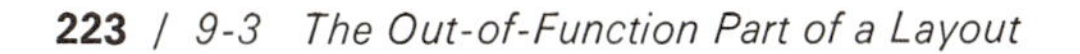

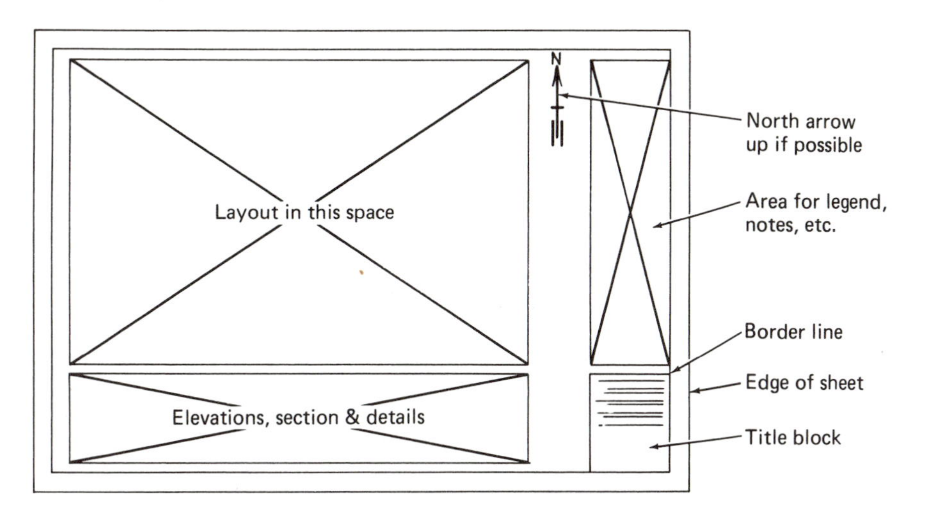

Figure 9-14 Typical Arrangement for Drawing Layouts

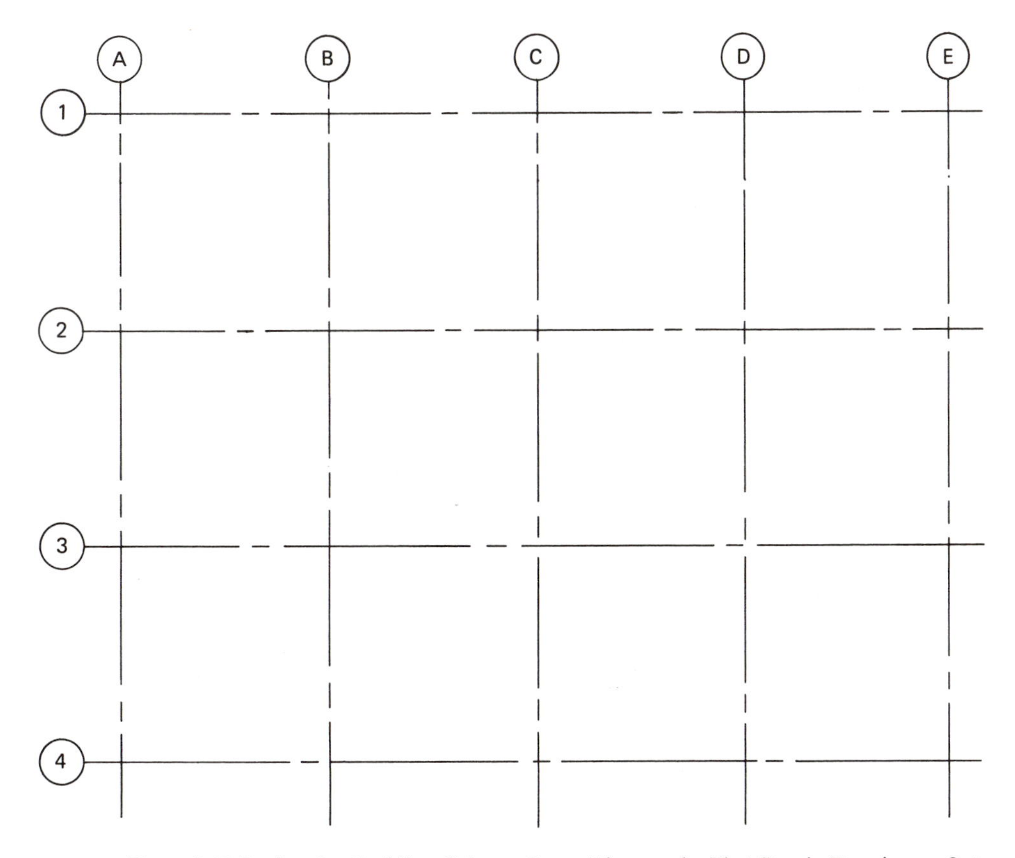

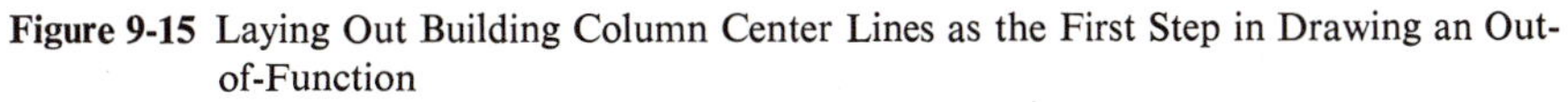

Figure 9-15 Laying Out Building Column Center Lines as the First Step in Drawing an Out-of-Function

Information for this work is acquired from the architectural and structural drawings. It is important to understand how the building is to be constructed in order to work the electrical system into it with a minimum of interference. It is also important that the electrical discipline identify space requirements for electrical equipment early in the building design before building layouts become too solidly fixed.

Layout drawings are generally drawn to scale so choosing the best scale to use is the first decision to be made. The size of the building, whether the entire floor plan can be shown on one drawing or whether it must be spread over several sheets, must be considered. Generally raceway layouts should be drawn at the scale of $\frac{1}{4}'' = 1' - 0''$. A scale of $\frac{1}{8}'' = 1' - 0''$ is used but if the layout is at all complicated, this scale will be hard to draw and harder to read. Every attempt should be made to make the scale $\frac{1}{4}'' = 1' - 0''$ or larger, if at all possible.

The layout should be started up into the upper left-hand corner of the drawing sheet with the northwest corner of the building. *Do not start any layout in*

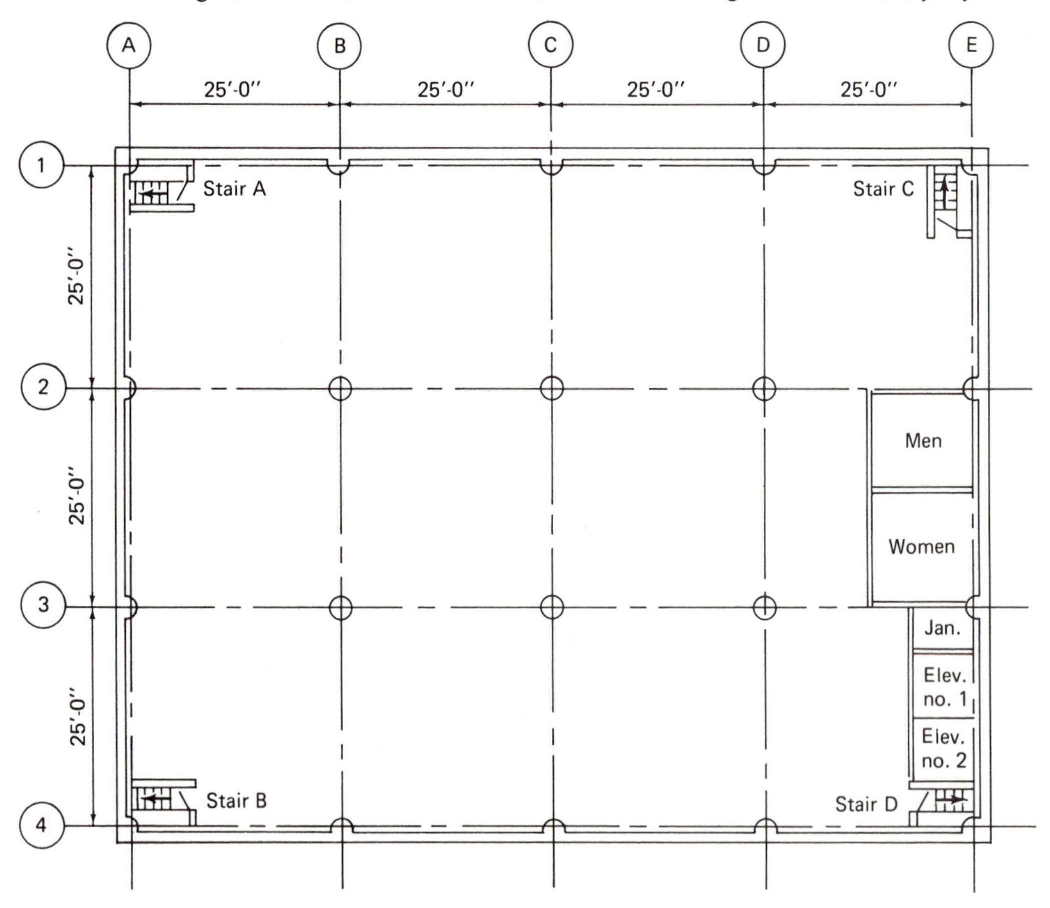

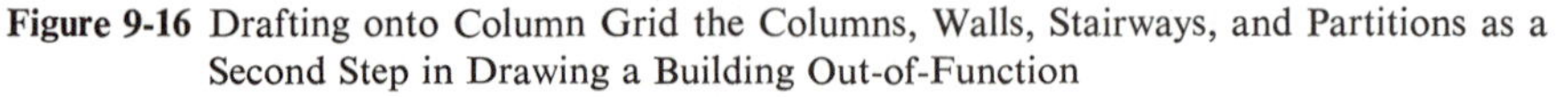

Figure 9-16 Drafting onto Column Grid the Columns, Walls, Stairways, and Partitions as a Second Step in Drawing a Building Out-of-Function

the center of the sheet because valuable drawing space will be wasted. Allow room along the right-hand side of the sheet for legend, notes, etc. Allow room on the bottom of the sheet for elevations, sections, and details if at all possible. Naturally the size of the layout, the scale to be used, and the size of the drawing sheet itself will have a bearing on the arrangement of the drawing. It may be necessary to put the legend, notes, details, etc., on separate sheets. A typical ideal arrangement of the drawing is shown in Fig. 9-14.

The out-of-function part should be started by drawing the column center lines onto the sheet first. "Balloons" are used to identify the column line letter or number and these are, of course, the same letters and numbers shown on the architectural and structural drawings. Figure 9-15 shows an example. Linework should be made with a fairly hard pencil (2H or 3H) to keep it sharp and fairly light.

The second step in developing the out-of-function part is to draft in onto the column line grid the columns, walls, stairways, doors, etc. This further development of Fig. 9-15 is shown in Fig. 9-16.

The building of Fig. 9-16 is constructed of reinforced concrete with round columns topped by a dropped panel that supports the flat concrete slab of the roof above. This layout shows the third floor so stairways in each corner go down to the street level. Other partitioning includes toilets, a janitor's closet, and passenger elevators. After this drafting is done, the out-of-function part is ready for the electric layout work called the *in-function.*

9-4 Symbols for Raceway Layouts

There are no accepted standard symbols for use on raceway layouts. Usually, however, the symbols used by various firms are shown in a legend on the drawings so symbols used are readily identified. The following symbols are representative of those commonly used:

———o	Conduit run exposed and turning up or toward observer
– – – –●	Conduit run concealed and turning down or away from observer
✕✕✕✕✕✕✕✕✕	Flexible conduit—if liquid-tight, an "LT" is noted above and along the symbol; if explosion-proof, an "EP" is used
(J)	Junction box; T = Thermostat, F = Flow switch, etc.
[T]	Transformer (size & voltage to be noted)
[F]	Fused disconnect switch
(15)	Motor—numeral denotes horsepower
▨	Power distribution panel

Lighting distribution panel

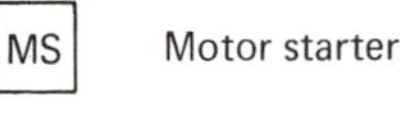

O Push-button station

Major equipment such as switchgear, and motor control centers are generally indicated to scale. Note that only a single line is used to show a conduit. Underfloor raceway systems are usually shown using single lines to represent each raceway and each header duct. An example of this is shown below.

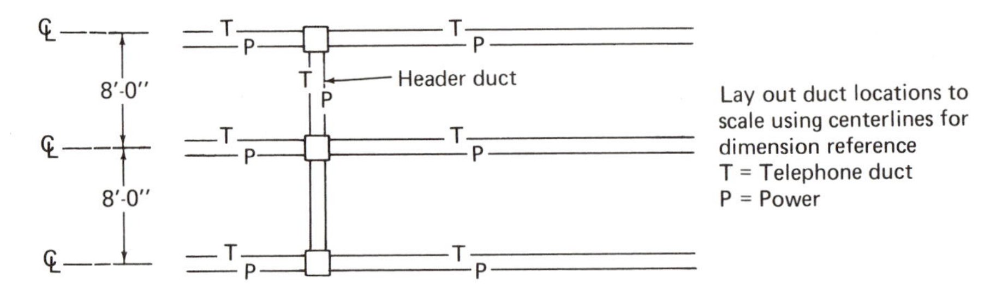

Cable trays are always laid out to scale because they are usually fairly good sized and room must be allowed to run them in the space provided. Interference of the cable tray with other equipment and other trades, such as ventilating duct work, must be considered and the tray located to clear all interferences. A typical cable tray layout is shown in Fig. 9-17.

In Fig. 9-17 the trays are laid out to scale on the drawing out-of-function. The X numbers identify the tray route similar to route numbers on a road map. These numbers are used to schedule the routing of cables in the trays (see Chapter 7).

9-5 The In-Function Part of a Layout

The elements of the electrical system to be physically installed in the building is the in-function part of the layout. It is really what the drawing is all about—its purpose.

The first steps should be the location of the major equipment such as switchgear, motor control centers, and panelboards. Busways and cable trays should be laid out to scale. Conduit runs are made next but running them to scale is next to impossible so diagrammatic runs are generally acceptable. All this work should be done with a pencil lead soft enough to make the work stand out against the out-of-function. Notations should be made as the last step because if they are made as the work is developing, they will invariably get in the way of the in-function line work.

If concealed conduit runs are required and the conduits must turn up out of concrete in an exact location, then the location must be dimensioned to the building column lines. A fairly good test of the completeness of a raceway layout is to ask whether the run shows *from–to* and *where*. Obviously the run

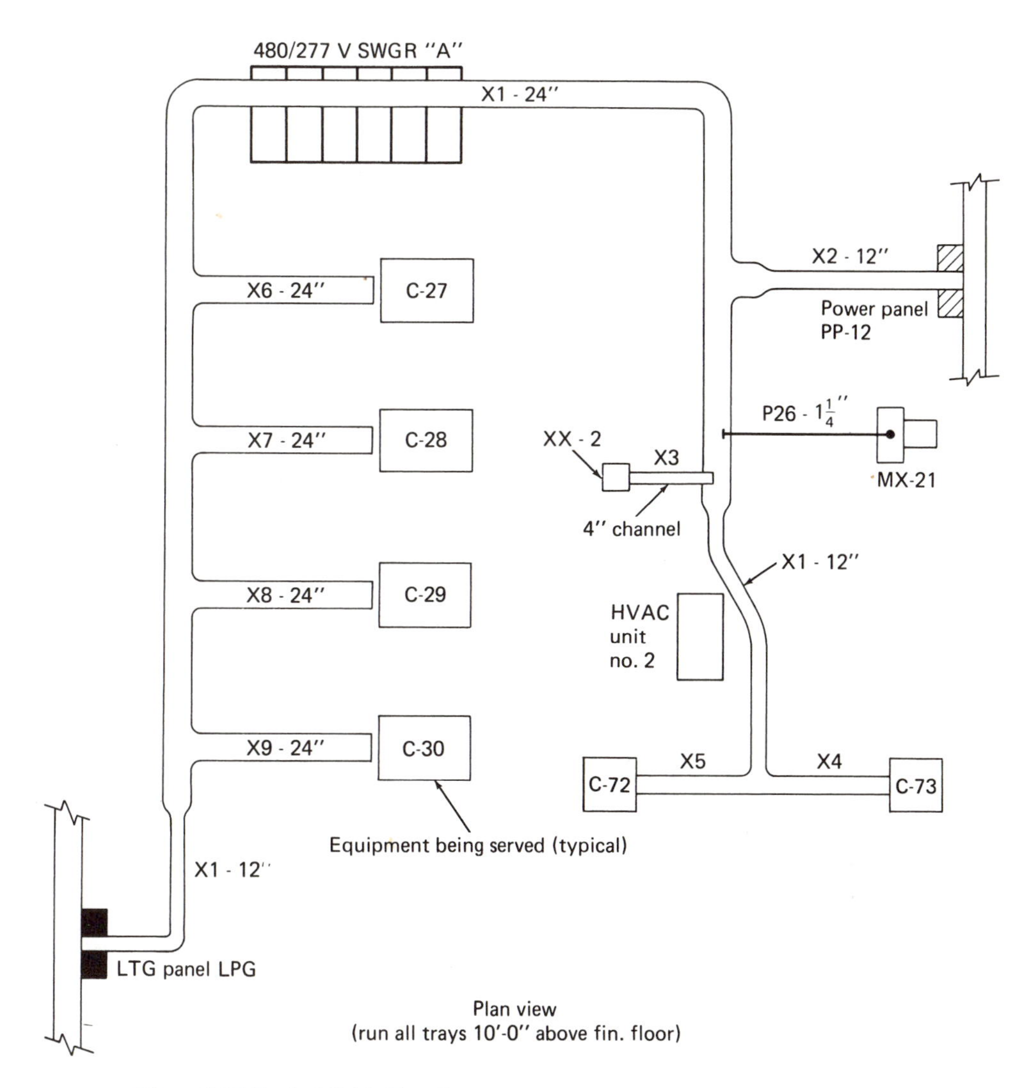

Figure 9-17 Typical Cable Tray Layout

should either show or note the route of the raceway as going *from* some piece of equipment or device *to* some other element and further information should either show or note *where* the run is made. As an example, a note was made in parentheses on Fig. 9-17 saying *run all trays 10'–0" above finished floor*. This tells *where*. Another similar note might be *run all conduits above hung ceiling*.

9-6 Concealed Conduit Runs

For the most part, concealed conduit runs are those made in concrete floors, walls, columns, etc. They provide an easy and direct way of interconnecting equipment that will probably remain fixed in place. There is no problem of support for the conduits as there is with exposed work, and because the runs cannot be seen, they can usually be made directly from one point to another. As an example of this, the layout of Fig. 9-18 shows conduits running from a motor control center to several fixed pieces of equipment and the runs are made point to point.

The runs shown in Fig. 9-18 are made as directly as possible with offsets being indicated to avoid interference. Note that a crossing of the conduits in one place was necessary. When this occurs, a check must be made to make certain that there is room in the concrete slab construction for a crossing to be made. The reinforcing rods in the slab construction very often limit the space available for conduit runs. Figure 9-19 shows a section through a typical slab and the space available for conduit.

If the slab shown in Fig. 9-19 were a 6-in. slab with $\frac{1}{2}$-in. rods set down from the top of the slab 1 in. and up from the bottom 1 in., then the approximate available space would be 2 in. This means that even 1-in. conduits could not cross because the outside diameter of the conduit is 1.315 in. and two conduits crossing would equal 2.630 in. or better than $2\frac{1}{2}$ in. Another routing for the conduits would need to be devised to avoid the crossing.

Concealed conduit runs should be dimensioned at points where they turn up from the concrete. Center line dimensions are generally given back to column center lines or to some fixed point such as the finished face of a wall.

Concealed runs can be made into banks for crossing floors, etc., with the runs fanning out as required. The National Electrical Code limits the number of quarter bends (90°) to four or a total of 360°. A good practice is to hold the limit to three bends for ease in pulling conductors. Figure 9-20 shows both dimensioning methods and the banking of runs.

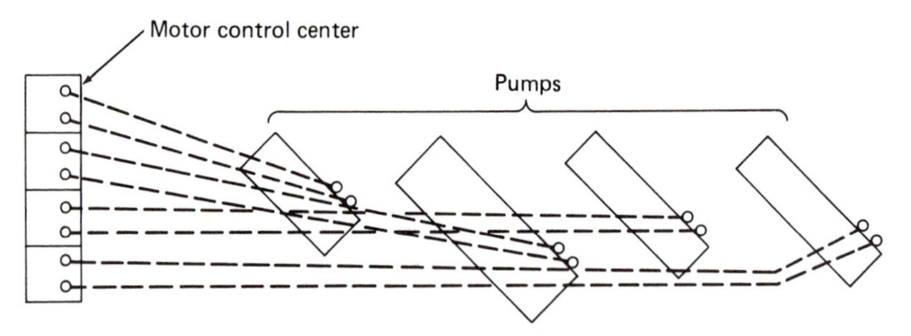

Figure 9-18 Example of Concealed Conduits Run in Concrete Floor from Point to Point

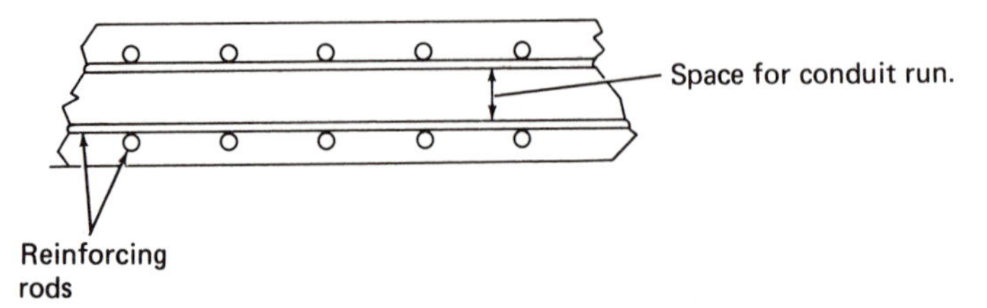

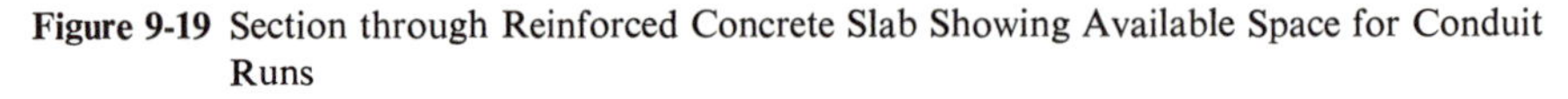

Figure 9-19 Section through Reinforced Concrete Slab Showing Available Space for Conduit Runs

The runs of conduit of Fig. 9-20 will have a 90° elbow at each end and part of a third 90° turn in the offset kicks so each run has a total bend of less than 270°. The example indicates a total run between columns of about 250 ft so with the bends shown this is a fairly good pull for the conductors.

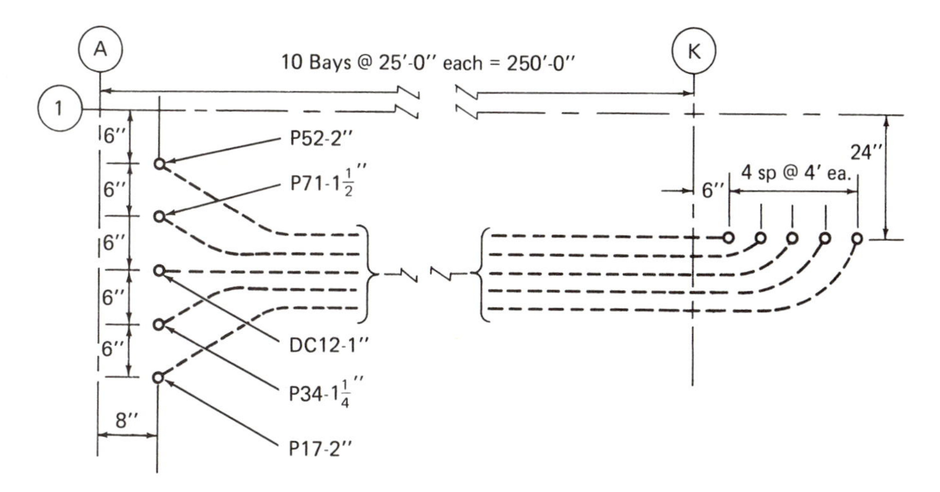

Figure 9-20 Banking and Dimensioning Concealed Conduit Runs in Concrete Slab

9-7 Exposed Conduit Runs

As with many design problems, organization of the objective to be met is important to exposed conduit runs. As an example, by determining the number of runs that will follow the same general route, a system of common supports for the conduits can be developed that is an economy in design. Perhaps a better arrangement of the layout of electrical equipment will result in fewer crossing of conduits, etc. A neat routing of the runs will tend to promote the same in actual installation and improve the appearance of the finished installation. Figure 9-21 shows the exact same conduit problem of Fig. 9-18 except the conduits are run exposed overhead.

The same problem of a crossing pair of conduits is shown in Fig. 9-21 as occurred in the concealed run of Fig. 9-18. If this were enough of a problem, perhaps the second and third sections of the motor control center layout could be swapped to avoid the crossing. Note that in Fig. 9-21 the layout indicates all conduits angling off to the motor control center at the same point. This tends to direct the same idea to the installing electrician.

Another solution to the crossing problem is to make the supports shorter and two level. The crossing then would go above or below the other conduits. Still a third solution is to make use of a length of wireway above the motor control center so that only wires within the wireway cross as shown in Fig. 9-22.

Some designers frown upon combining several circuits in a common enclosure such as the wireway of Fig. 9-22; however, if installed in accordance with the code, it is a perfectly practical solution to be considered.

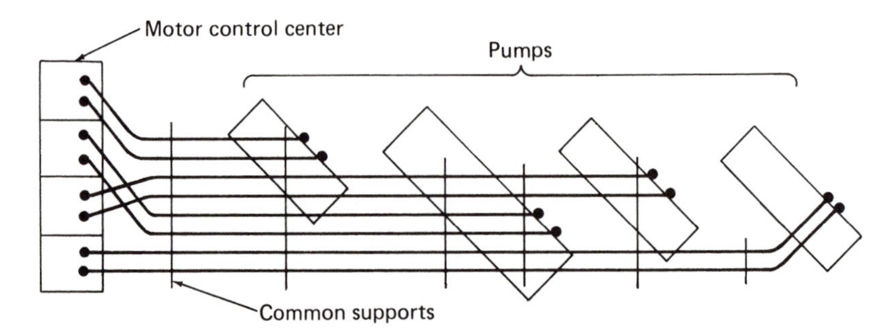

Figure 9-21 Exposed Conduit Run Using Common Supports

Exposed runs should be limited to three 90° bends as good practice and of course never more than four as allowed by the code. Pull boxes can be easily installed in exposed runs either for changing direction with a pull point or in the middle of a run strictly as a pull point. Of course most conduit fittings, such as the LB fitting illustrated in Fig. 9-1, are pull points in themselves because they have removable covers that allow access to the conduit. In this regard, caution should be exercised in using large-sized conduit fittings for runs with large conductors because bending the conductors to fit into the fitting can cause injury to the cable insulation.

Designers and draftsmen should be careful in laying out conduit runs to see that the design can be installed without undue problems in the field. For instance, a design showing a requirement for turning a conduit elbow onto a run of conduit where there is not room to swing the elbow is a common error in design. This is especially true in the larger conduit sizes (3″ and up) where in order to swing the elbow into place a space of 3 ft or better would be required.

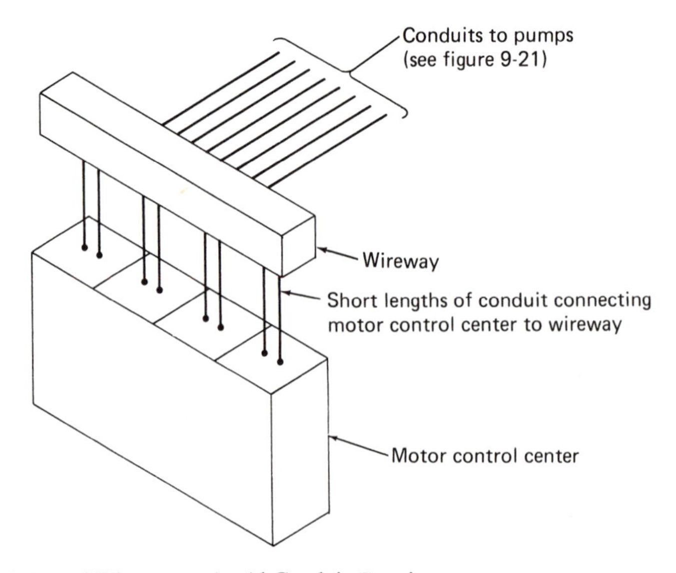

Figure 9-22 Use of Wireway to Avoid Conduit Crossings

9-8 Underground Duct Lines

The installation of electrical distribution systems underground is gaining preference rapidly. A great many new housing developments have underground utilities, which certainly enhances the appearance of these areas. There are several ways of burying the electrical conductors. This discussion will be confined to underground raceway systems using rigid nonmetallic conduits.

Because underground duct lines generally cover large areas, the layout drawings are made at scales of 1″ = 20′ – 0″ or 1″ = 30′ – 0″ or 1″ = 40′ – 0″. As with conduit layout drawings, smaller scales are harder to draw and harder to read.

Conduit sizes are generally not smaller than 3 in. with 4 in. being a very common size in use. Choice of the type depends on some engineering considerations and on some personal preferences. These conduits are generally laid in tiers in a trench prepared for them. Forms may or not be required to contain the concrete encasement which forms the finished duct bank.

An out-of-function is prepared showing the buildings, roads, etc. Location of the service entrance point of the buildings or other termination point of the underground lines is noted and a study routing is made to determine how the distribution needs can be met. This study will point up the need for manholes or handholes and the number of ducts required to meet the present and future needs.

Topographical data showing how the ground lays or how it may be modified for the new construction is needed to determine depth of manholes and the profile of the underground duct line.

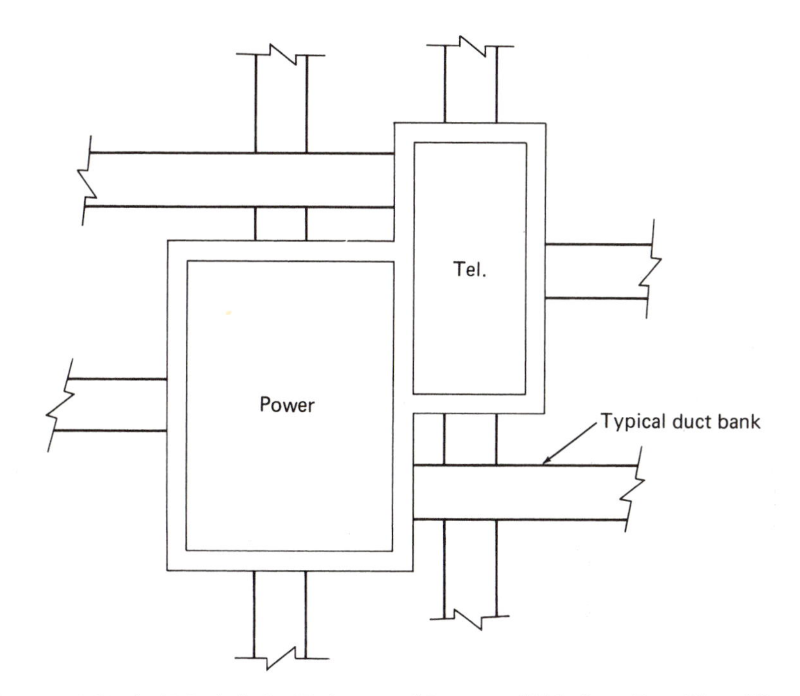

Figure 9-23 Typical Manhole for Underground Power and Telephone Duct Lines (Plan view)

Manholes are available in prefabricated form or they can be constructed in place. Some underground systems use a type with one space for power and another space for telephone designed with a common wall between the spaces. This type is shown in Fig. 9-23.

Note that the manhole allows for entrance of duct lines from all four directions into each section.

The underground duct line layout will generally show the approximate routing of the duct line, the manhole locations, and sections through the duct lines to show how many ducts are to be installed. Profiles may also be included if required that show the location of the duct lines below grade. A typical layout plan is shown in Fig. 9-24.

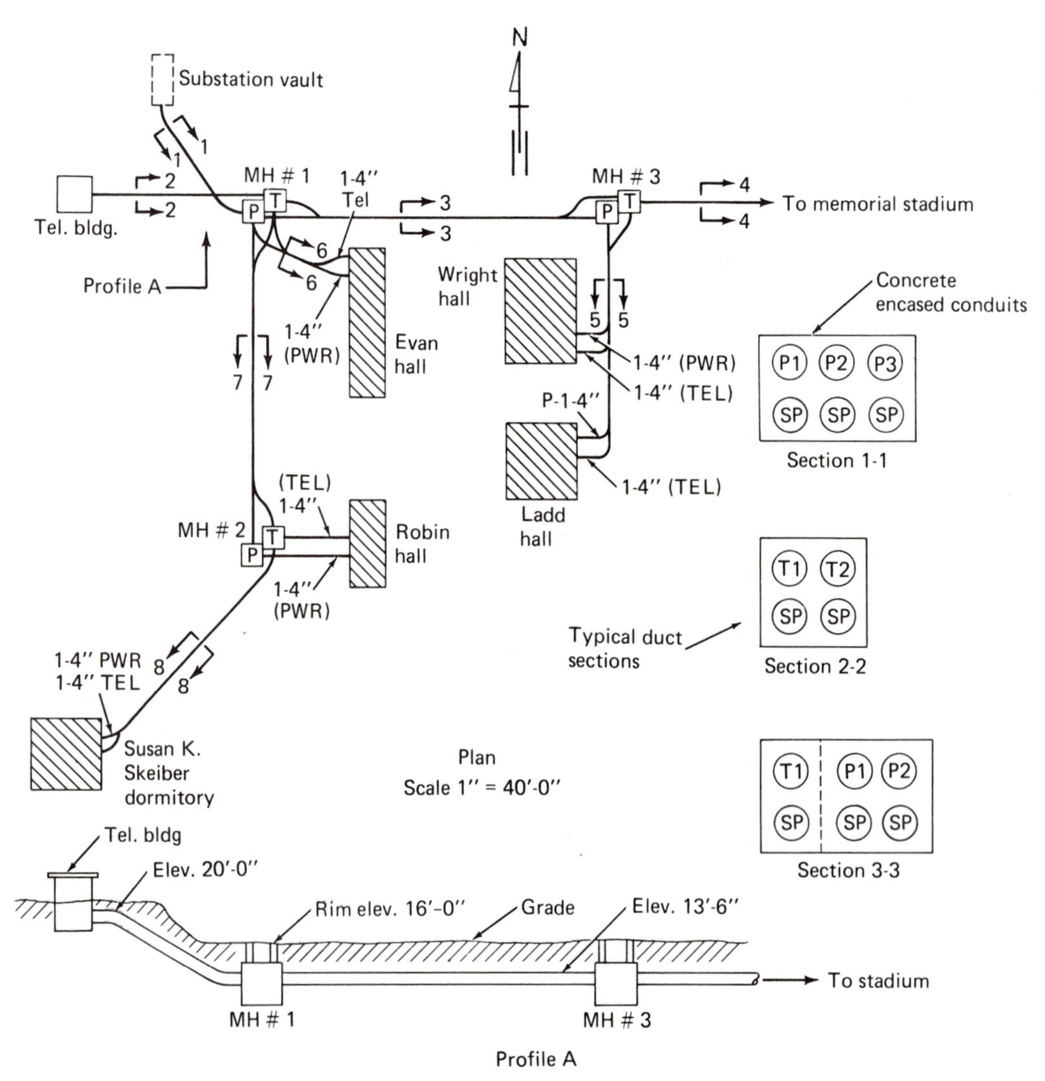

Figure 9-24 Typical Layout for Underground Duct Lines

Chapter Review Topics and Exercises

9-1 Why does conduit make a particularly good raceway?

9-2 Describe how conduit is joined into electrical boxes, cabinets, etc.

9-3 Why are conduit fittings made with removable covers?

9-4 Why is nonmetallic conduit usually encased in concrete when run underground?

9-5 How does EMT differ from rigid steel conduit?

9-6 Where is flexible metal conduit used?

9-7 Why are surface raceways used and why are they restricted to dry location installations?

9-8 What is the principal advantage of a cellular floor raceway system?

9-9 How do conductors run from above the floor panelboards, etc., to the under-floor cells?

9-10 Describe how an under-floor raceway system is installed.

9-11 What appear to be some advantages to the use of wireways?

9-12 What is the principal advantage of a plug-in bus?

9-13 Describe how a continuous rigid cable support of the ventilated trough type is constructed.

9-14 What is an out-of-function or shell?

9-15 What should be the minimum scale for a raceway layout and why?

9-16 Describe the steps to developing an out-of-function.

9-17 How are cable trays shown on the layout?

9-18 What is an in-function?

9-19 Why are conduit crossings in reinforced concrete floors sometimes impossible?

9-20 What is the limit in the number of 90° bends in a conduit run for good design practice? Why?

9-21 Why is neatness in exposed conduit run layouts a good idea as relates to field construction?

9-22 How can wireways be used to avoid conduit crossings and congestion?

9-23 Describe an underground duct line.

9-24 What does a profile show as drawn for an underground duct line layout?

9-25 Lay out the out-of-function of the pump house shown in Fig. 10-16 at a scale of $\frac{1}{4}'' = 1' - 0''$ using the following building characteristics:

(a) Spacing of number bays is 15′ — 0″

(b) Spacing of lettered bays is 30′ — 0″

(c) Columns measure 8″ deep and 6″ across the flange.

(d) Concrete block wall is 8″ thick and is laid on a line butting up against and toward the outside of the column flange.

(e) The large equipment door is 12′ — 0″ wide and the personnel doors are 3′ — 0″ wide. Large door is centered on building center line. The south side of the personnel doors is 4′ — 0″ from the Ⓑ line.

(f) Roof is a flat concrete slab 15′ — 0″ above the finished, slab-on-grade floor.

The electrical equipment to be installed is shown in the following one-line diagram.

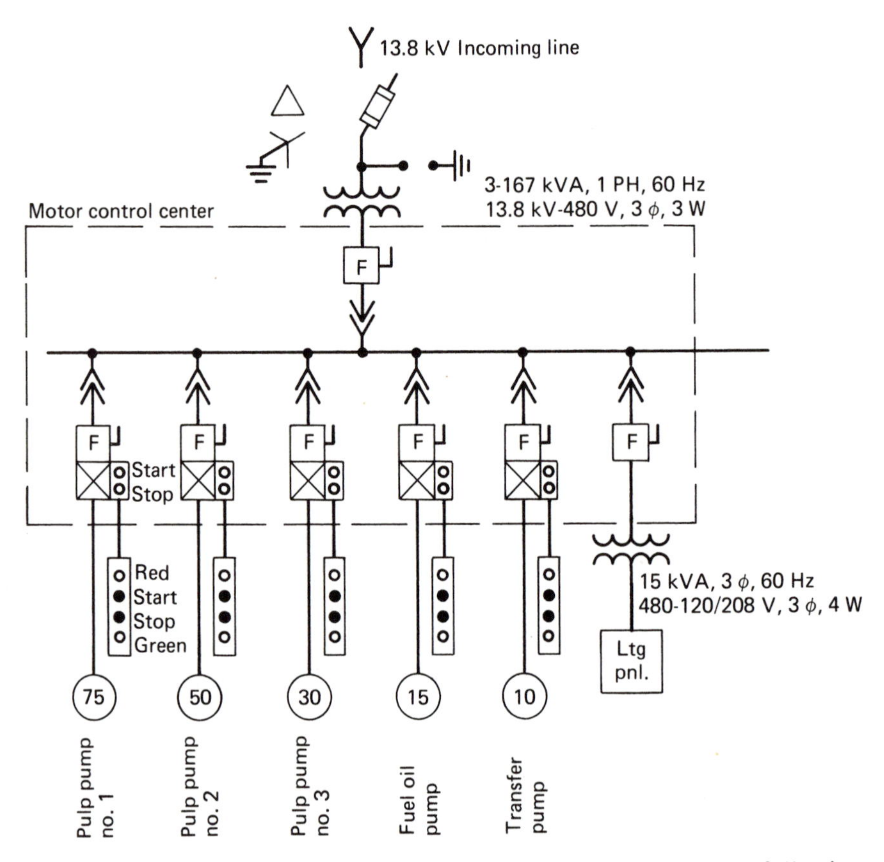

Raised concrete bases are provided for each pump with the following dimensions:

75 hp—3′ — 0″ × 8′ — 0″

50 hp—3′ — 0″ × 6′ — 0″

30 hp—2′ — 0″ × 6′ — 0″

15 hp—18″ × 4′ — 0″

10 hp—18″ × 4′ — 0″

The pump bases are set at a 45° angle to the center line of the building to make maximum use of the space. The 75 hp pump is the most westerly and located as shown in the following.

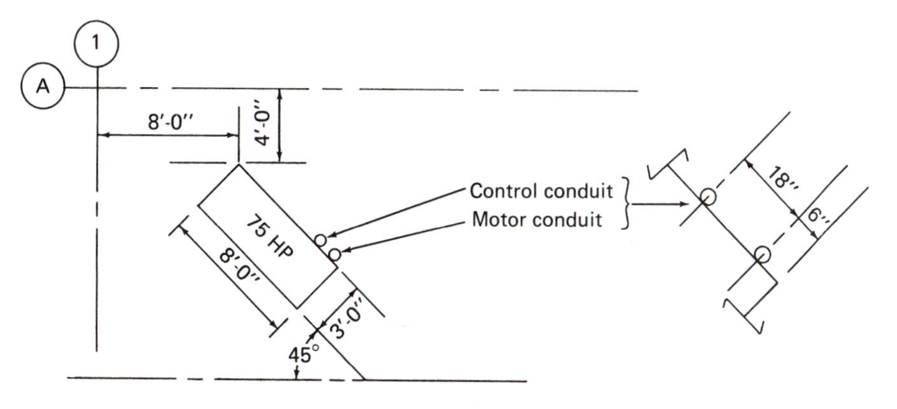

All pump bases are in line at 4′ — 0″ from the Ⓐ line. The 50 hp pump base is set 11′ — 0″ east of the 75 hp base; the 30 hp, 10′ — 0″; 15 hp, 9′ — 0″; and the 10 hp, 8′ — 0″, which results in a 14′ — 0″ dimension from the corner of the 10 hp base to the ⑤ line. All dimensions relate to the same point on the bases as is dimensioned for the 75 hp base illustrated above. Motor and control conduits are to be located as shown.

Front view of the motor control center is shown in the following.

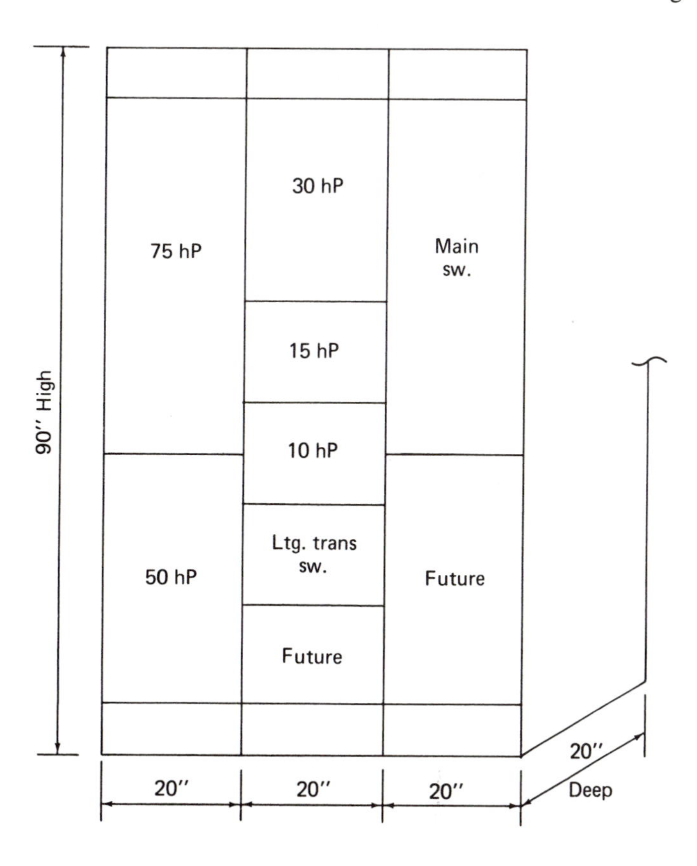

With the basic data above, locate the motor control center up against the west wall and run the motor and control conduits to each pump from the motor control center. Runs may be made in the floor slab or overhead as desired. Indicate the 15-kVA transformer as located on the east wall between the personnel door and the equipment door and run conduit to it from the motor control center.

10

LIGHTING LAYOUTS

10-1 General

The design of proper lighting is a study in itself. To become an expert in this field it is necessary to acquire knowledge and experience in the many types of light sources that are available and of the fixtures and other devices used to distribute the light. Light can be used in so many different ways to create pleasant working and relaxing atmospheres and those who study its uses find interesting and challenging work as they become the experts. The average engineer, designer, and draftsman, however, do not need the special talents for the broad field of general lighting. Knowledge of reasonably simple calculations and a basic knowledge of common lighting sources and fixtures will suffice.

10-2 Definitions and Descriptions of Lighting Terms

A knowledge of some fundamentals is basic to understanding the study of lighting. The more important terms and concepts are as follows:

Candlepower

The light giving intensity of a source of light is expressed in candlepower. One candlepower is equal to the light giving power of one candle that is made to certain standard specifications.

Footcandle

The intensity of illumination on a surface is measured in footcandles. One footcandle is the intensity of illumination on a surface one foot away from a light source of one candlepower. This is illustrated in Fig. 10-1. *The footcandle is the commonly used measure of illumination.*

Lumen

The amount of light flux that falls on a surface is measured in lumens. One lumen is equal to the amount of light flux that falls on one square foot of a surface on which the intensity of illumination is one footcandle. This is illustrated in Fig. 10-2. *The lumen is the commonly used measure of light source quantity,* and therefore manufacturers rate lamps in lumens.

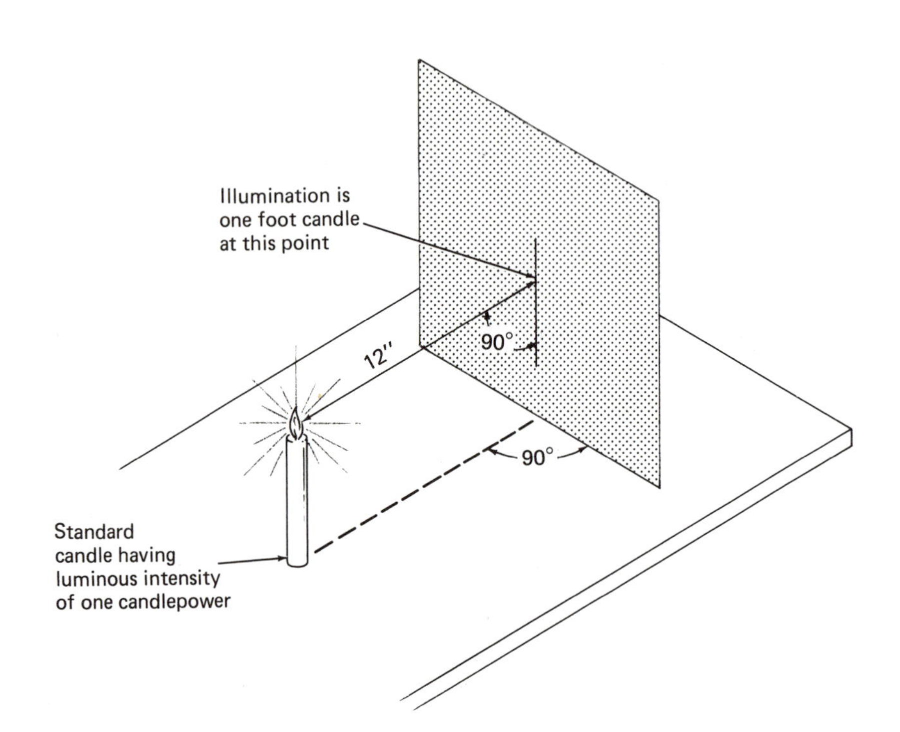

Figure 10-1 Basic Arrangement for Measurement of One Footcandle

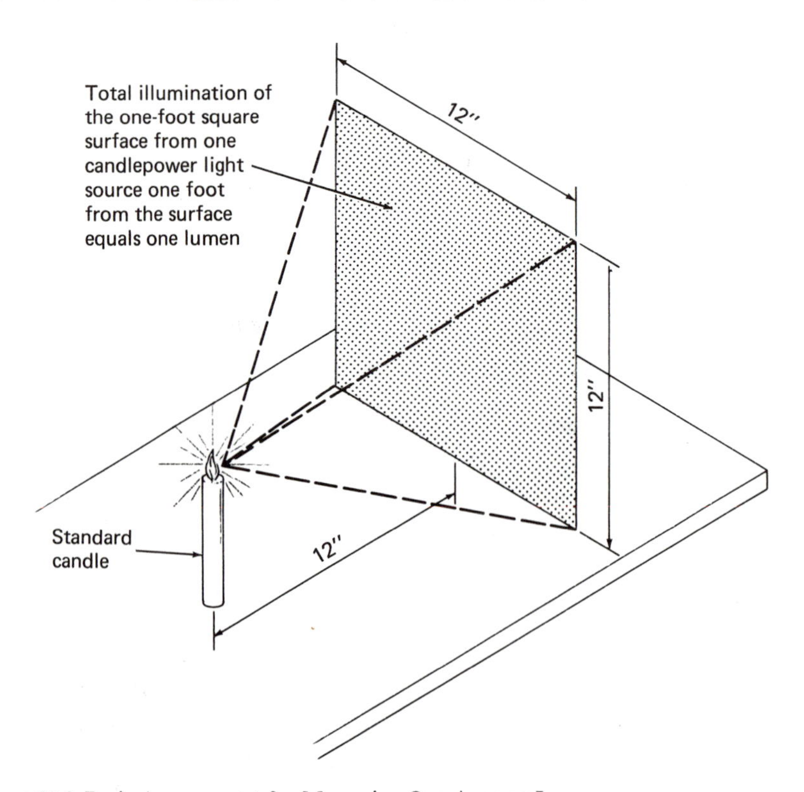

Figure 10-2 Basic Arrangement for Measuring One Average Lumen

Because a flat surface will not receive equal flux across it, the illustration of Fig. 10-2 shows an *average* lumen.

In order to measure the accurate lumen output of one candlepower, the source of light must be thought of as a point source at the center of an enclosed hollow sphere having a radius of one foot. With this arrangement, every point on the inside surface of the sphere will be equidistant from the light source and be illuminated to one footcandle. On every square foot of surface one lumen of light flux will fall. The interior surface of the sphere equals $4\pi r^2$ or 12.57 sq ft; therefore one candlepower equals 12.57 lumens. More important, illumination expressed in footcandles is really *lumens per square foot* and therefore, based on this analysis, lumens required to meet a desired level of illumination should be equal to footcandles times the area to be lighted. This analysis is not entirely true, however, because consideration must be given to the factor that effects the utilization of all the flux from the light source. This factor is called the *coefficient of utilization.*

Coefficient of Utilization

The coefficient of utilization is the ratio of the lumens reaching the working area to be lighted to the total lumens generated from the light source. This factor takes into account the following:

1. *Efficiency and distribution* of the light by the lighting fixture, i.e., what percentage of the light is absorbed by the fixture and therefore does not reach the area to be lighted?
2. *The mounting height* of the fixture above the area to be lighted. Distance is an important consideration. Light varies inversely (decreases) as the square of the distance from the light source; i.e., twice the distance will reduce the illumination to one-fourth. This is illustrated in Fig. 10-3. One lumen of light flux falls on the one square foot of area set up one foot away from the light bulb. This same flux will illuminate an area measuring two feet by two feet or four square feet at a distance of two feet from the bulb. Because there is only one lumen from the source, it must be spread across the entire four-foot area; therefore, each square foot will receive only one-quarter lumen!
3. *Size and shape of the room to be lighted.* The width and length of the room and the mounting height of the fixture (Fig. 10-4) are all important factors in the effectiveness of the light from the source in illuminating the working area. As an example, if the room is square and has dark walls and a dark ceiling so that reflection is negligible and if the lamp is mounted in the center of the room, the floor will be quite evenly lighted as shown in Fig. 10-4(a). If the room length is increased by 50%, the original floor space will be lighted as before but the added floor space will receive light at less intensity because of the greater slant of the light rays and the average intensity of illumination over the whole room will be less than with the square room.
4. *Reflection of light* from the ceilings and walls is another factor in the

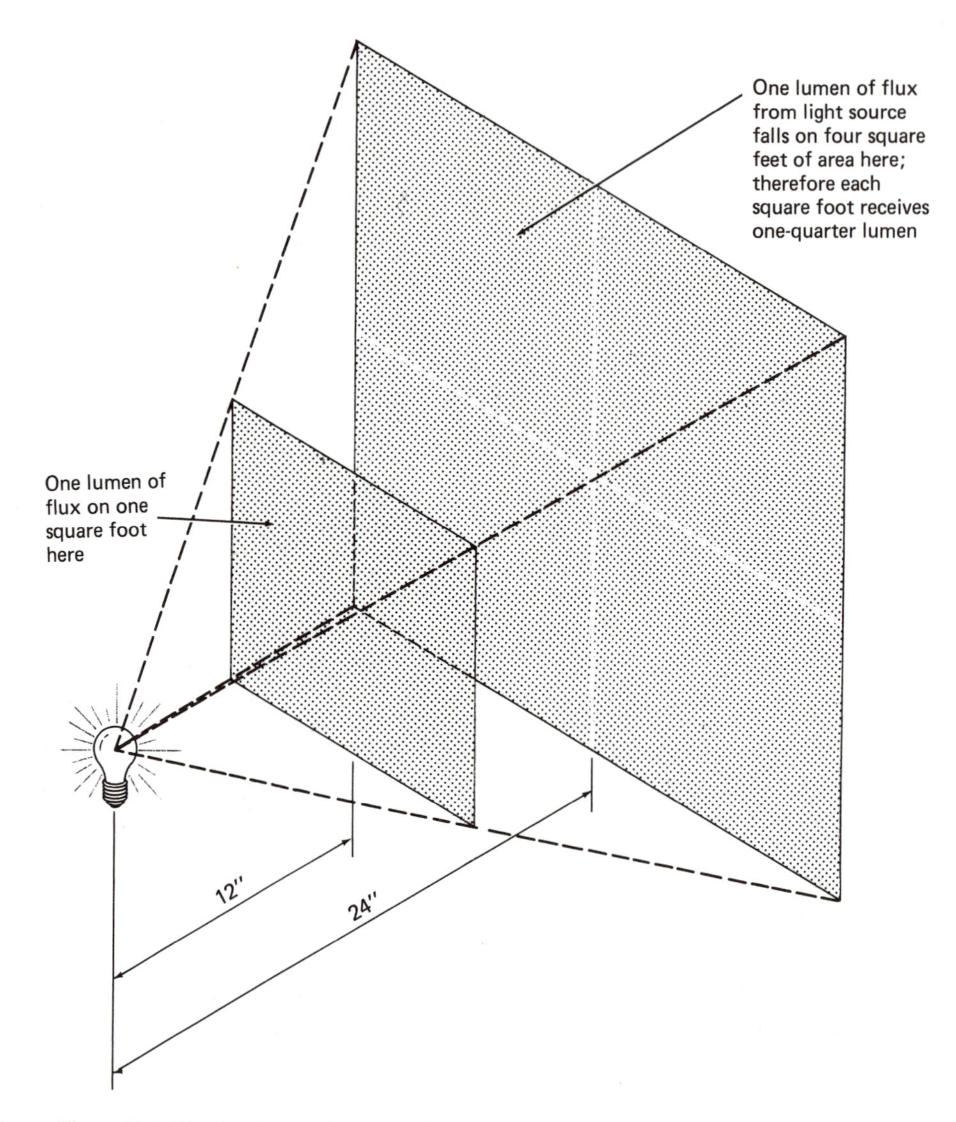

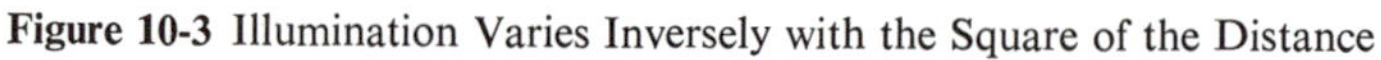
Figure 10-3 Illumination Varies Inversely with the Square of the Distance

utilization of light from the source. Light walls and ceilings are going to reflect more light than darker surfaces.

All these factors must be considered in lighting calculations so methods for determining lighting levels and data from manufacturers take them into account.

Maintenance Factor

All lighting calculations must result in a determination of the *in-service* footcandles, which is the level of illumination over the average span of time. To

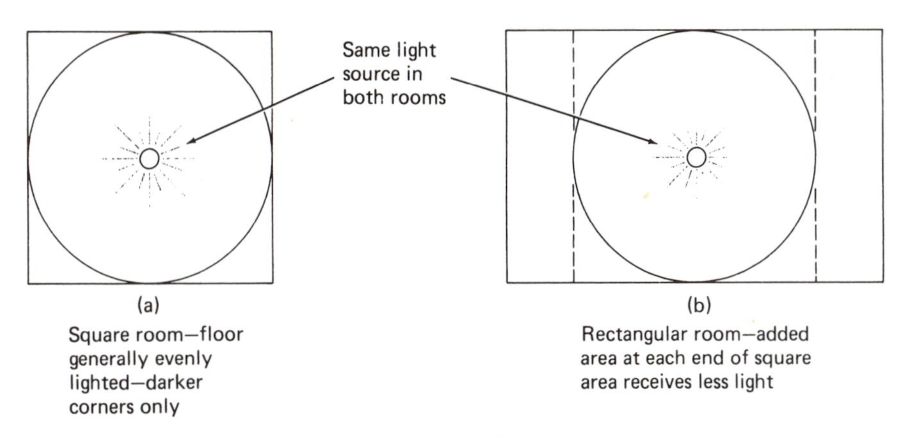

Figure 10-4 Effects of Room Shape on Lighting Efficiency

provide for the depreciation of lighting levels from the brand new installation to the in-service level, the following maintenance considerations must be made:

1. As lamps are used over a long period of time, their lumen output diminishes. Replacement of lamps *before* they burn out restores the output to the initial conditions.
2. Accumulation of dirt on the lamps and reflectors reduces the lumen output.
3. Accumulation of dirt on walls and ceilings reduces the reflectance capability of these surfaces.

Maintenance factors are categorized and manufacturers of fixtures provide numerical efficiency values for each category for different types of fixtures as follows:

1. *Good maintenance factor*: Where the atmospheric conditions are good, fixtures are cleaned frequently, and lamps are replaced in entire groups rather than one at a time as they burn out.
2. *Medium maintenance factor*: Where less clean (than "good") atmospheric conditions exist, fixture cleaning is fair, and lamps are replaced only after burnout.
3. *Poor maintenance factor*: Where the atmosphere is quite dirty and equipment is poorly maintained.

The designer must exercise careful judgement as to existing and anticipated conditons in order to arrive at a practical maintenance factor.

Spacing to Mounting Height Ratio

The designer needs to know how far apart he can space luminaires (fixtures), for a given mounting height, so as not to cause any undue variation in the lighting level across the working area. Generally speaking, even distribution is desired in a room regardless of the amount of light. Uneven distribution creates light and dark spots, shadows, and in many cases glare. Proper spacing of fixtures is therefore essential. General rules of thumb are as follows:

1. For fixtures with light primarily in the upward direction, the spacing between rows of fixtures can be from 1.2 to 1.5 times the ceiling height above the floor.
2. For fixtures with light distribution fairly well divided between its upward and downward component, the spacing can be from 0.9 to 1.1 times the mounting height above the floor.
3. For fixtures with the light distribution primarily downward, the spacing should not exceed from 0.7 to 0.9 times the mounting height above the floor.
4. Where continuous rows of fixtures are used, such as strips of fluorescent units, the spacing can be increased approximately 20% of those described above.
5. The distance from the wall to a fixture is approximately one-half the spacing between rows of fixtures.

10-3 Illumination Levels

The more difficult the seeing task, the more light required and the better comfort required regarding brightness and glare. Quality lighting goes beyond just providing the required footcandles. Careful consideration needs to be given to colors for walls, floors, and ceilings. Generally speaking, colors should be as light as practical for working and maintenance conditions.

The Illuminating Engineering Society (IES) publishes the minimum recommended levels for illumination for industrial, commercial, and institutional interiors that are keyed to the seeing task. This data should be consulted as a starting point to any lighting design layout. These are a guide for the designer. Further judgement needs to be made for the exact lighting level being designed. A representative listing of some of these levels is shown in the appendix.

10-4 Light Sources

A great many light sources are available in many sizes and shapes. The types most commonly used in general lighting design are the incandescent, fluorescent, and mercury arc.

Incandescent lamps are made with a filament and are a purely resistive load operating at unity power factor. They can be used almost anywhere if of the proper type and properly protected for weather conditions.

Fluorescent lamps are widely used in modern design. They have several advantages over the incandescent lamp with the principal one being a very large increase in lumens per watt. Light is also emitted over a larger lamp surface so that distribution of light is much better. These lamps operate on a current flow through mercury vapor that causes the phosphor coating on the inside of the tube to glow (fluoresce). Because autotransformers (called ballasts) are required to increase the voltage for the lamps, a magnetic factor is introduced that effects the power factor of this load type. Capacitors are used in the wiring, however, and they help to correct the power factor.

Mercury-arc lamps look somewhat like incandescent lamps but they too operate on a mercury vapor and require autotransformers. This type of lamp finds much application in medium-to-high mounting height installations.

Representative lamp data is given in Tables 5 and 6 in the appendix.

All data given should be used as a guide only. Ratings of lamps are being continually revised so the latest manufacturers' rating information should be used for any design layout.

10-5 Basic Formulas

The following basic formulas, used in conjunction with the method for making the lighting calculations described below, will generally serve the designer for most common lighting layouts.

$$\text{(1)}\quad \text{Number of fixtures required} = \frac{\text{footcandles} \times \text{area}}{\text{lamps per fixture} \times \text{lumens per lamp} \times \text{coefficient of utilization} \times \text{maintenance factor}}$$

$$\text{(2)}\quad \text{Number of footcandles in service} = \frac{\text{number of fixtures} \times \text{lamps per fixture} \times \text{lumens per lamp} \times \text{coefficient of utilization} \times \text{maintenance factor}}{\text{area}}$$

$$\text{(3)}\quad \text{Area per fixture} = \frac{\text{lamps per fixture} \times \text{lumens per lamp} \times \text{coefficient of utilization} \times \text{maintenance factor}}{\text{footcandles}}$$

10-6 The Zonal Cavity Method of Calculating Average Illumination Levels

The method of computing average interior illumination levels described here is based on the Illuminating Engineering Society (IES) zonal cavity method. This method improves older systems by providing increased flexibility in lighting calculations as well as greater accuracy but it does not change the basic concept that footcandles are equal to lumen flux over an area.

The zonal cavity method is founded upon the concept of considering a room to be made up of a series of cavities that have *effective* reflectances with respect to each other and to the work plane. Any room may generally be divided into three basic spaces or cavities, as shown in Fig. 10-5.

The space between the work plane and the floor is called the *floor cavity*; the space between the fixtures and the work plane is defined as the *room cavity*,

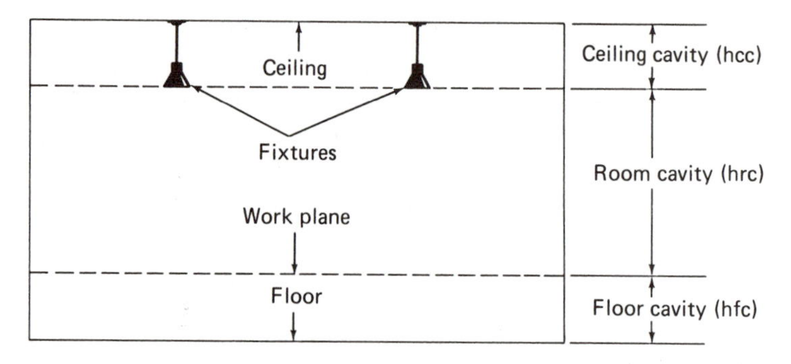

Figure 10-5 Section through a Room Showing Zonal Cavities

and the space above the bottom of the fixtures to the ceiling is the *ceiling cavity*.

By dividing the room in this way it is possible to calculate numerical relationships called *cavity ratios*, which in turn can be used to determine effective reflectance values of the ceiling and floor and from this to find the *coefficient of utilization*.

There are four basic steps to any calculation using the zonal cavity method. They are

Step 1: Determine cavity ratios.

Step 2: Determine cavity reflectances.

Step 3: Select coefficient of utilization using data from Step 2.

Step 4: Compute number of fixtures required or footcandle level using a basic formula (see Sec. 10-5).

These steps are worked as follows:

Step 1:

Cavity ratios may be determined in one of two ways. The most basic and accurate manner is by calculation using the following formulas (see Fig. 10-5):

$$\text{Ceiling cavity ratio (CCR)} = \frac{5\text{hcc}(\text{L} + \text{W})}{\text{L} \times \text{W}} \tag{1}$$

$$\text{Room cavity ratio (RCR)} = \frac{5\text{hrc}(\text{L} + \text{W})}{\text{L} \times \text{W}} \tag{2}$$

$$\text{Floor cavity ratio (FCR)} = \frac{5\text{hfc}(\text{L} + \text{W})}{\text{L} \times \text{W}} \tag{3}$$

where hcc = distance in feet from fixture to ceiling
hrc = distance in feet from fixture to work plane
hfc = distance in feet from work plane to floor
L = length in feet of room
W = width in feet of room

Cavity ratios may also be found in Table 1 in the appendix. This table covers typical-sized cavities in a wide range of room dimensions.

Step 2:

This step determines what the effective cavity reflectance is, as opposed to the actual reflectance that is estimated for the ceiling, walls, and floor. It is rare that a lighting designer has knowledge of the ultimate room reflectances when making calculations. The arbitrary use of, say, an 80% ceiling, 50% walls, and 30% floor is at best an estimate; however, with considered judgment such estimates will usually provide sufficient accuracy. The effective cavity reflectance is defined then as *the summation of actual reflectances within a given cavity into one hypothetical reflectance at the face of the cavity.* These reflectances are identified as follows:

Pcc = effective ceiling reflectance

Pw = effective wall reflectance

Pfc = effective floor reflectance

These are taken directly from prepared tables such as Table 2 in the appendix. They are located in the tables under the applicable cavity ratio and the actual reflectance of the ceiling, walls, and floor. Note that if the fixture is recessed into a hung ceiling or surface mounted on the room ceiling, then the ceiling cavity ratio (CCR) is 0 and if the work plane is to be the floor itself, then the floor cavity ratio (FCR) is 0 and in either or both cases the *actual reflectance* of the ceiling or floor will be the *effective reflectance.*

Step 3:

Step 2 will give values, from the reflectance table, for the effective reflectance of the ceiling and floor cavities. The wall reflectance remains as the *actual* reflectance value selected. Using these three values and the calculated or table-determined room cavity ratio (RCR), the coefficient of utilization can be determined from the appropriate table for the fixture being considered. An example of such a table and the fixture for which the table applies is shown in Fig. 10-6.

Step 4:

The coefficient of utilization, determined by Step 3, will have to be adjusted if the effective floor cavity reflectance is something other than 0.20. Tables 3 and 4 in the appendix are used for this purpose.

Step 5:

Enter the adjusted C.U. factor in the appropriate basic equation and solve for value sought.

Example 1: For an illumination calculation based on the zonal cavity method:

This example uses the semi-indirect lighting fixture shown in Fig. 10-6.

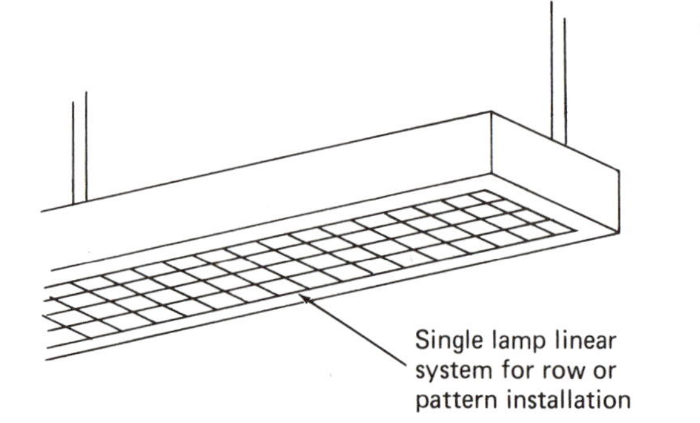

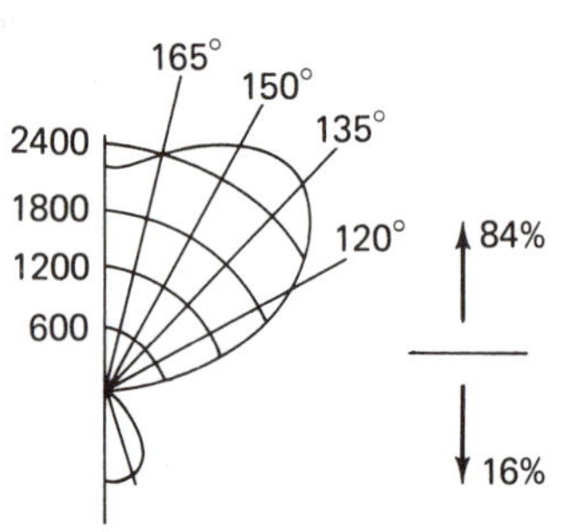

Semi-indirect

Pfc	20%			20%			20%		
Pcc	80%			70%			50%		
Pw	50	30	10	50	30	10	50	30	10
Room cavity ratio 1	0.70	0.67	0.64	0.61	0.59	0.57	0.46	0.44	0.43
2	0.61	0.57	0.53	0.54	0.50	0.47	0.41	0.38	0.36
3	0.54	0.49	0.44	0.47	0.43	0.39	0.36	0.33	0.30
4	0.48	0.42	0.37	0.42	0.37	0.34	0.32	0.29	0.26
5	0.42	0.36	0.32	0.37	0.32	0.29	0.28	0.25	0.23
6	0.38	0.32	0.28	0.34	0.29	0.25	0.26	0.22	0.19
7	0.34	0.28	0.24	0.30	0.25	0.21	0.23	0.19	0.17
8	0.30	0.25	0.20	0.27	0.22	0.18	0.21	0.17	0.15
9	0.27	0.22	0.18	0.24	0.19	0.16	0.18	0.15	0.13
10	0.25	0.19	0.16	0.22	0.17	0.14	0.17	0.13	0.11

Figure 10-6 Coefficients of Utilization for Semi-Indirect Fluorescent Lighting Fixture

Design Problem: Determine the number of fixtures (Fig. 10-6) required to light the work plane at 2′-6″ above the floor to a level of 100 fcs in a room measuring 24 ft wide, 70 ft long, and 12 ft in height. Reflectances are ceiling, 0.80; walls, 0.50; and floor, 0.10. The fixtures are to be suspended by stems so the bottom of the fixture is 30 in. below the ceiling. The fixture is to be lamped with single-tube, high-output fluorescent lamps rated 8,900 initial lumens per lamp. Judgment is that the maintenance will be medium with a factor of 0.80.

Step 1: Select ceiling (CCR), room (RCR), and floor (FCR) cavity ratios from Table 1. For the ceiling cavity ratio, follow the table in from the room width of 24 ft and length of 70 ft to the cavity depth column headed 2.5. This is 2′-6″, which is the depth of the cavity set by suspending the fixtures 30 in. below the ceiling. The same cavity ratio applies to the floor cavity because the work plane is 2′-6″ above the floor. The cavity ratio for both is 0.7, from the table.

For the room cavity ratio, subtract 5′-0″ (2′-6″ ceiling + 2′-6″ floor = 5′-0″) from the total room height of 12′-0″ to result in a cavity depth of 7′-0″. Following in on the table from the room width of 24 ft and length of 70 ft the cavity ratio is determined as 2.0.

In summary,

$$\text{CCR} = 0.7$$
$$\text{RCR} = 2.0$$
$$\text{FCR} = 0.7$$

Step 2: Select the effective ceiling (Pcc) and floor (Pfc) cavity reflectances from Table 2 using the cavity ratios found in Step 1 (the wall reflectance remains 0.50 as stated in the design problem).

For the ceiling, follow the table in from the ceiling or floor cavity ratio column at 0.7 to the column headed by an 80% ceiling or floor reflectance and a 50% wall reflectance and read the value of 70. For the floor, follow the same 0.7 level across to the column headed by a 10% ceiling or floor reflectance and a 50% wall reflectance and read the value 11.

In summary;

$$\text{Pcc} = 70\%$$
$$\text{Pfc} = 11\%$$
$$\text{Pw} = 50\%$$

Step 3: Using a room cavity ratio of 2.0 from Step 1 and the effective cavity reflectances from Step 2, select the proper C.U. from the table of Fig. 10-6. It is immediately seen that no 11% factor is indicated opposite the level for floor cavity reflectances. Step 4 modifies the C.U. to account for this. Using the column headed by 20% Pfc, 70% Pcc, and 50% Pw, opposite a room cavity ratio of 2, a value of 0.54 is determined.

Step 4: The effective floor cavity reflectance was found to be 0.11 so the C.U. of 0.54 must be modified by the factors in Table 3 or 4. In this example the difference between a 0.10 floor reflectance and a 0.11 is negligible so Table 3 is used and under the column headed by 70% Pcc and 50% Pw and opposite a room cavity ratio of 2 a multiplying factor of 0.949 appears. Therefore $0.54 \times 0.949 = 0.51$, which is the modified C.U. to be used in the basic formula.

Step 5: Compute the number of fixtures of the type to be installed to illuminate the area to an average maintained level of 100 fcs.

$$\text{Number of fixtures required} = \frac{\text{footcandles} \times \text{area}}{\text{lamps per fixture} \times \text{lumens per lamp} \times \text{C.U.} \times \text{maintenance factor}}$$

$$= \frac{100 \times (24 \times 70)}{1 \times 8{,}900 \times 0.51 \times 0.80}$$

$$= 46.3 \text{ (46 fixtures minimum)}$$

Example 2: For an illumination calculation based on the zonal cavity method.

This example uses a recessed lighting fixture in a suspended ceiling. It is a 2′ × 4′ fixture using four lamps.

Design Problem: The suspended ceiling pattern provides a balanced layout of 30 fixtures in a room 30 ft wide by 45 ft long and 8′-6″ in height from the floor to the suspended ceiling. Reflectances are ceiling, 0.70; walls, 0.50; and floor, 0.30. The work plane is 2′-6″ above the floor. What level of illumination in footcandles will the 30 fixtures provide on the working plane? The maintenance is judged as medium at 0.80. The fixtures C.U. table is shown in Fig. 10-7. Lamps are rated 3,100 initial lumens each.

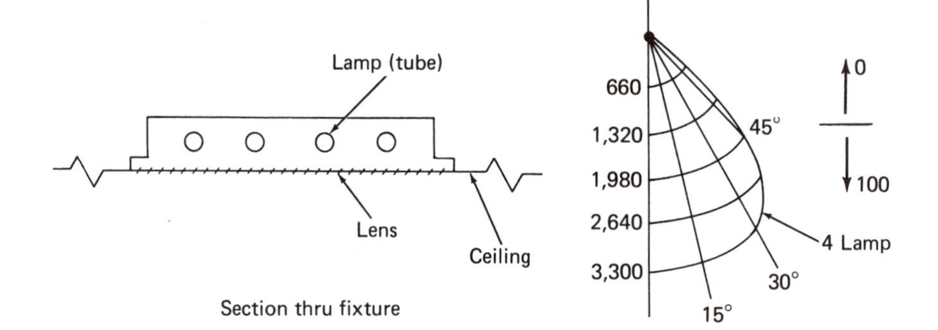

Pfc	20%			20%			20%			20%	
Pcc	80%			70%			50%			30%	
Pw	50	30	10	50	30	10	50	30	10	50	30
Room cavity ratio 1	0.68	0.66	0.64	0.67	0.65	0.63	0.64	0.62	0.61	0.61	0.60
2	0.61	0.58	0.55	0.60	0.57	0.54	0.58	0.55	0.53	0.56	0.54
3	0.55	0.51	0.47	0.54	0.50	0.47	0.52	0.49	0.46	0.51	0.48
4	0.50	0.45	0.41	0.49	0.44	0.41	0.47	0.43	0.40	0.46	0.43
5	0.45	0.39	0.36	0.44	0.39	0.36	0.43	0.38	0.35	0.41	0.38
6	0.40	0.35	0.32	0.40	0.35	0.31	0.39	0.34	0.31	0.38	0.34
7	0.37	0.32	0.28	0.36	0.31	0.28	0.35	0.31	0.28	0.34	0.30
8	0.33	0.28	0.25	0.33	0.28	0.24	0.32	0.27	0.24	0.31	0.27
9	0.30	0.25	0.21	0.29	0.25	0.21	0.29	0.24	0.21	0.28	0.24
10	0.27	0.22	0.19	0.27	0.22	0.19	0.26	0.22	0.19	0.26	0.22

Figure 10-7 Coefficients of Utilization for Four-Lamp Recessed Fluorescent Lighting Fixture

Step 1: Select ceiling cavity ratio, room cavity ratio, and floor cavity ratio from Table 1 or by formula.

CCR = 0 (when the fixtures are surface mounted or recessed, there is no ceiling cavity.)

RCR = 1.7 (cavity depth is 8′-6″ − 2′-6″ = 6′-0″)

FCR = 0.7 (cavity depth is 2′-6″ — 2.5 in Table 1)

Step 2: Determine the effective ceiling and floor reflectances from Table 2 using the cavity ratios from Step 1. The *effective* ceiling reflectance is the same as the *actual* reflectance of 0.70. The wall reflectance is also the actual reflectance of 0.50 and the floor reflectance from the table is 0.28. This value is found under the column headed by an actual floor reflectance of 0.30 and a wall reflectance of 0.50 opposite the floor cavity ratio of 0.7.

Step 3: Using the room cavity ratio from Step 1 and the effective cavity reflectances from Step 2, the C.U. for the fixture is selected from the table of Fig. 10-7. Using the column headed by a floor reflectance of 20 (0.28 is the value determined in Step 2 and this will require a modification in Step 4), a ceiling reflectance of 70, and a wall reflectance of 50, it can be seen that the C.U. for a room cavity ratio of 1.7 falls between 0.67 (RCR-1) and 0.60 (RCR-2). By interpolation the C.U. for 1.7 will equal approximately 0.62. (There are 7 percentage points between 0.60 and 0.67 and 1.7 is 0.7 greater than 1. To interpolate: $0.7 \times 7 = 4.9$ and $67 - 4.9 = 62.1$.)

Step 4: The effective floor reflectance was found to be 0.28 so the C.U. of Step 3 must be modified to account for the difference between 0.20 from the C.U. table and 0.28 from the reflectance table. Table 4 in the appendix is used for this adjustment. The procedure is as follows: Enter the table at a room cavity ratio of 1, a ceiling reflectance of 70%, and a wall reflectance of 50%. The figure is 1.070; at a room cavity ratio of 2, the figure is 1.057 with a difference between the figures of 0.013. The actual room cavity ratio is 1.7 so by interpolation the multiplying factor for 1.7 is 1.06 $[(0.3 \times 0.013) + 1.057]$. The actual effective floor reflectance is 0.28, however, so the factor must be modified further. The 1.06 figure is for 30%; therefore at 28% the figure will be less at 1.048, $[1.06 - (0.2 \times 0.06)] = 1.048$. For practical purposes, use 1.05. Adjusting the C.U. of Step 3, $0.62 \times 1.05 = 0.65$, and this is the proper C.U. to use in the basic formula.

Step 5: Compute the average maintained illumination level in footcandles using the basic formula.

$$\text{Number of footcandles in service} = \frac{\text{Number of fixtures} \times \text{lamps per fixture} \times \text{lumens per lamp} \times \text{C.U.} \times \text{maintenance factor}}{\text{area}}$$

$$= \frac{30 \times 4 \times 3{,}100 \times 0.65 \times 0.80}{(30 \times 45)}$$

$$= \frac{193{,}440}{1{,}350}$$

$$= 143 \text{ fc}$$

To simplify the steps necessary for zonal cavity calculations, a standard form is often used, which makes the work easier and serves as record of the calculation. A typical form is shown in Fig. 10-8.

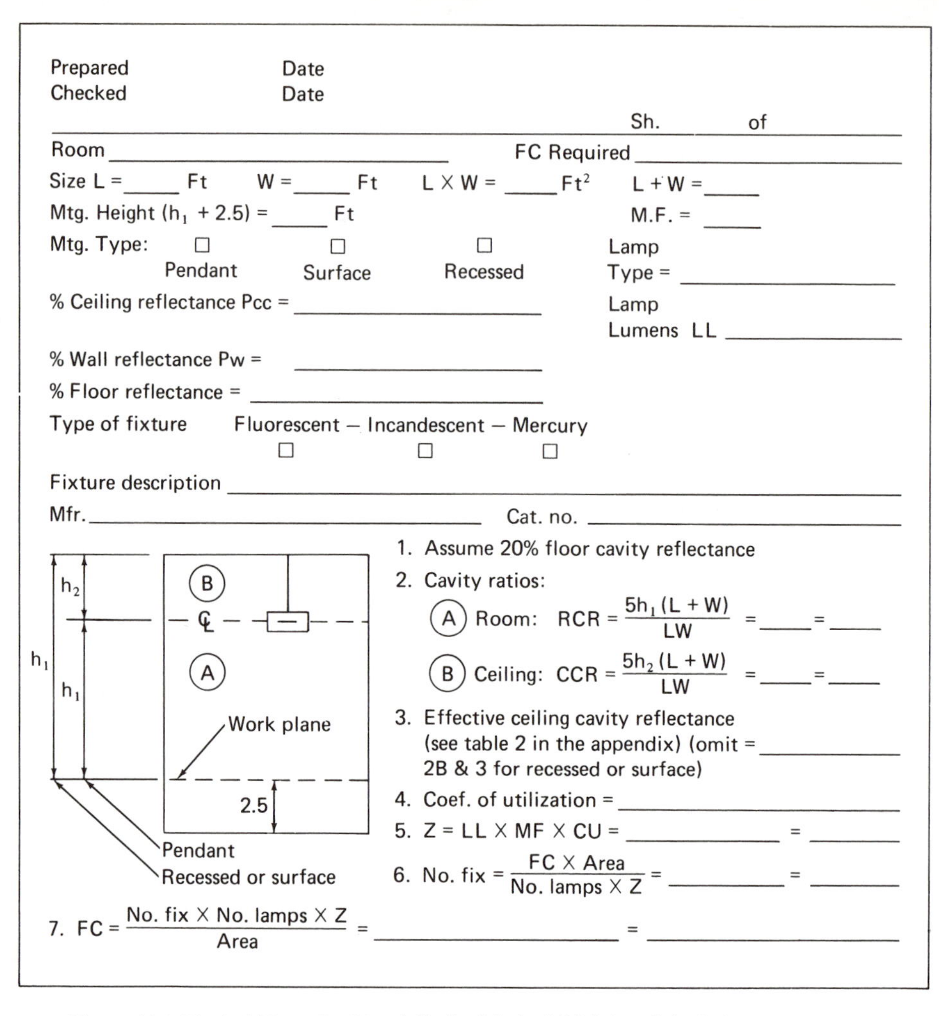

Prepared　　　Date
Checked　　　Date

Sh. ____ of ____

Room ________________ FC Required ________________

Size L = _____ Ft　W = _____ Ft　L X W = _____ Ft^2　L + W = _____

Mtg. Height (h_1 + 2.5) = _____ Ft　M.F. = _____

Mtg. Type: ☐ Pendant　☐ Surface　☐ Recessed

Lamp Type = ________________

% Ceiling reflectance Pcc = ________________

Lamp Lumens LL ________________

% Wall reflectance Pw = ________________

% Floor reflectance = ________________

Type of fixture　Fluorescent — Incandescent — Mercury
☐　☐　☐

Fixture description ________________

Mfr. ________________ Cat. no. ________________

1. Assume 20% floor cavity reflectance
2. Cavity ratios:

   (A) Room: $RCR = \frac{5h_1(L + W)}{LW}$ = _____ = _____

   (B) Ceiling: $CCR = \frac{5h_2(L + W)}{LW}$ = _____ = _____

3. Effective ceiling cavity reflectance (see table 2 in the appendix) (omit 2B & 3 for recessed or surface) = ________________
4. Coef. of utilization = ________________
5. Z = LL X MF X CU = ________________ = ________
6. No. fix = $\frac{\text{FC} \times \text{Area}}{\text{No. lamps} \times Z}$ = ________________ = ________
7. FC = $\frac{\text{No. fix} \times \text{No. lamps} \times Z}{\text{Area}}$ = ________________ = ________________

Figure 10-8 Typical Form for Zonal Cavity Method Lighting Calculations

Calculations by the zonal cavity method to determine coefficients of utilization can be made as accurate as the designer may wish to develop them. Reflectance factors are very important for an accurate calculation, which may require consultation with the architects for reflectance values of different colors and shades of colors.

The zonal cavity method can also be used with less (but often times sufficient) accuracy by rounding the room cavity ratio off to the nearest whole number and using it with the *actual* reflectance values to determine the coefficient of utilization. As an example of the rounding out idea and using the figures from example 2, the following calculation would result.

Reflectances are ceiling, 0.70; walls, 0.50; and floor, 0.30 (as given in the problem). The room cavity ratio from Table 1 is 1.7 and this would be rounded to 2.0. Referring to the table in Fig. 10-7, the C.U. opposite a room cavity ratio of 2 and under the reflectance columns of floor 20 (there is no value for

30 so this will have to be used), ceiling 70, and walls 50, a C.U. of 60 is determined. Solving the basic formula as in example 2,

$$\text{Number of footcandles} = \frac{30 \times 4 \times 3{,}100 \times 0.60 \times 0.80}{30 \times 45}$$
$$= 132$$

The more accurate method calculated a level of 143 fcs or 11 fcs more than determined by the rounding out method. The maintenance factor, being a judgment value, also has an effect on the calculation so the variance that may appear could be a result of this as well as from the less accurate calculation.

In summary, the zonal cavity method has a great potential where extreme accuracy is required and for unusual room conditions. Under many normal lighting conditions it may be the designer's judgment to use the less accurate method and the end results may not vary significantly. Some trial calculations and experience will help to make logical judgments.

10-7 Determining Lighting Fixture Arrangement and Spacing

After the number of fixtures are determined, to provide the level of illumination required, the arrangement of fixtures and spacing must be worked out. If lighting fixtures are to be installed in a suspended ceiling system, the arrangement of the ceiling tiles will usually be developed by the architect. He will design a *reflected ceiling plan* showing the tile layout. The electrical designer will have to work with the architect and the heating, ventilating, and air conditioning designers to develop a ceiling arrangement to suit all requirements.

Any lighting layout tends to be a cut-and-try problem. If the number of fixtures required and the spacing to mounting height requirements do not work out with one arrangement of rows and fixtures in each row, perhaps another row arrangement will work. There is no exact way of determining an acceptable layout so different layouts have to be tried to see if they will work.

As an example, consider the layout problem for installing 20 incandescent fixtures where the area to be lighted is 30 ft by 60 ft with a mounting height of 12 ft above the floor. Data on the fixture indicate that the spacing should not exceed 1.3 times the mounting height. With 20 fixtures a first consideration might be four rows of 5 fixtures each. This would lay out as shown in Fig. 10-9.

The layout meets all the requirements and is satisfactory; however, the spacing between fixtures is rather unbalanced so the lighting effect might show more *pools* of light than if the spacing could be more balanced. Consider adding one more fixture and laying out three rows of seven as in Fig. 10-10.

By using an arbitrary *circle of light* with the same diameter for the examples in Figs. 10-9 and 10-10, a comparison shows a better overlapping pattern with the three-row layout using one more additional fixture.

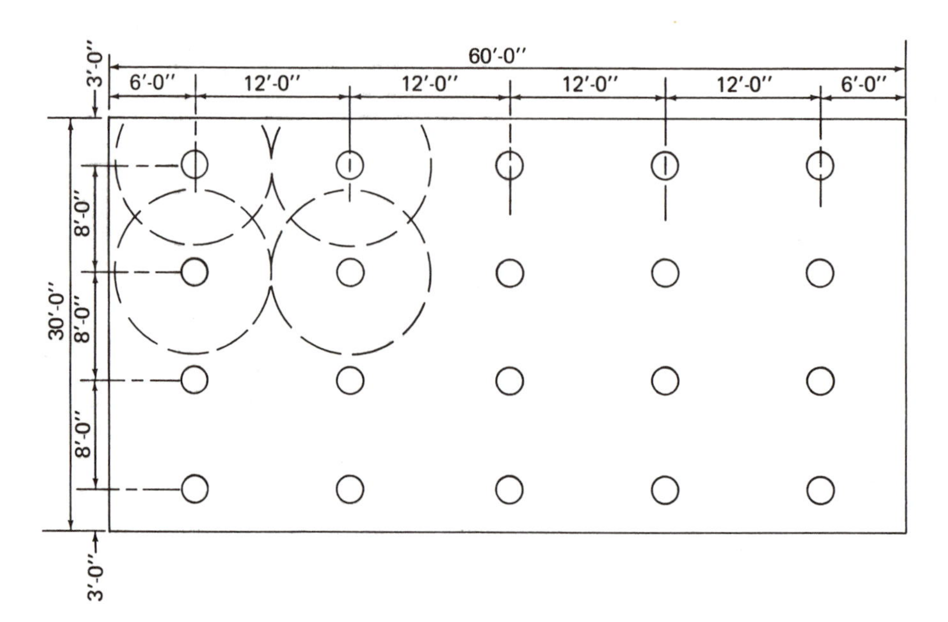

Figure 10-9 Fixture Layout for Room 30′ × 60′ using 20 RLM-300 W Incandescent Fixtures Mounted at 12′-0″ above Floor with Calculated Footcandle Level of 20 fc 30″ above Floor

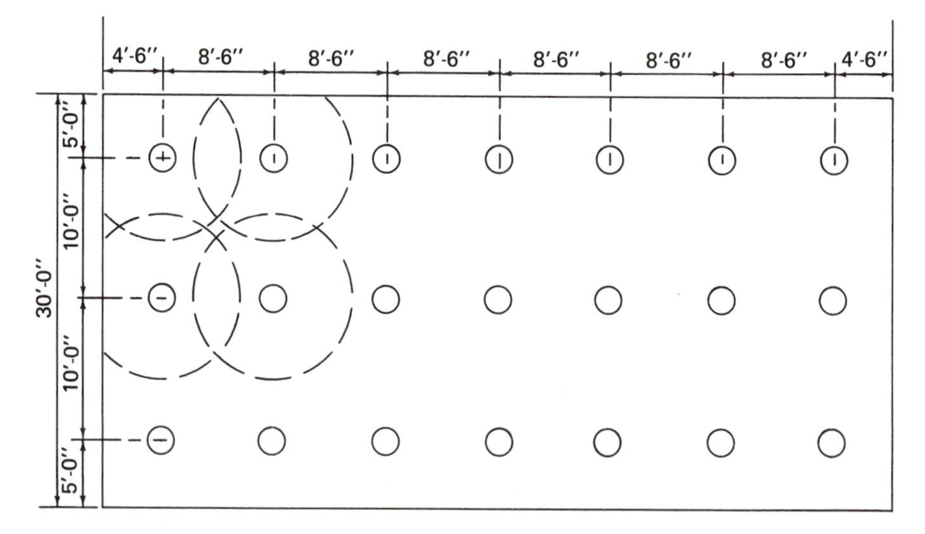

Figure 10-10 Fixture Layout for Room 30′ × 60′ Using 21 RLM-300 W Incandescent Fixtures Mounted at 12′-0″ above Floor with Calculated Footcandle Level of 20 fc 30″ above Floor

With fluorescent fixtures, the inclination is to run the length of the fixture parallel to the long dimension of the room. This may not always provide the best result, however, so a layout with the fixtures at 90° to the longest dimension may be worth trying.

Manufacturers' data and advice from representatives is always worthwhile where unique situations may exist and unusual patterns may be required.

10-8 Convenience Outlets (Receptacles) and Other Small Power Requirements in Lighting Layouts

In addition to lighting, the lighting layout drawings also show receptacles and small power requirements such as small fans, unit heaters, and drinking fountains that generally are rated for 120-V service. The National Electrical Code states specific rules regarding requirements for receptacles and must be consulted for proper applications.

10-9 Control of Lighting and Other Lighting Layout Circuits

Lighting circuits can be controlled either locally or from lighting panels. Where large open areas are involved, the least expensive method of control is to use the circuit breaker in the lighting panelboard as the switching means. In smaller areas, where a more convenient switching requirement must be met, local switches can be used. They must be applied in accordance with the code. These may be single-pole, three-way, or four-way switches and either surface or flush mounted to meet the finish requirements of the area. Normally circuits for receptacles and other small power loads are not switched. Some designers require that a *no-switch* device, which is simply a cover, be installed over the breaker handle of the panelboard circuit breakers so that the circuits are not inadvertantly switched off when lighting circuits are being switched.

10-10 Emergency Lighting

Emergency lighting is required to meet many building codes and requirements of other agencies. Emergency lighting systems should be so designed and installed that the failure of the normal system will not leave any space in total darkness. These systems are required by the NEC to include all required emergency exit lights and all other lights specified as necessary to provide sufficient illumination for egress to the outside of a building. Usually exit lights are provided with at least two lamps so loss of a lamp is not likely to put the sign in darkness.

Many emergency lighting systems utilize selected fixtures in the total fixture layout as the emergency lights. These fixtures are wired to an emergency section of a circuit breaker panel for protection and the entire section is wired to automatic transfer switches that act automatically to transfer the supply from a normal source to an emergency source.

Other systems are designed as two or more separate and complete systems with independent power supplies. Automatic transfer means are provided to energize the lighting system from one system to the other. The NEC sets rules for these systems that must be observed.

Emergency lighting units are also used for emergency service consisting of a rechargeable battery, a battery charging means, provision for one or more lamps mounted on the unit, and a relay device arranged to energize the lamps automatically upon failure of the normal supply to the building. The NEC makes specific requirements for the design and installation of these units.

10-11 Symbols for Lighting Layouts

The American National Standards Institute publishes a pamphlet showing graphic symbols for electrical wiring and layout diagrams used in architecture and building construction Some of the more commonly used symbols used on lighting layout drawings follow:

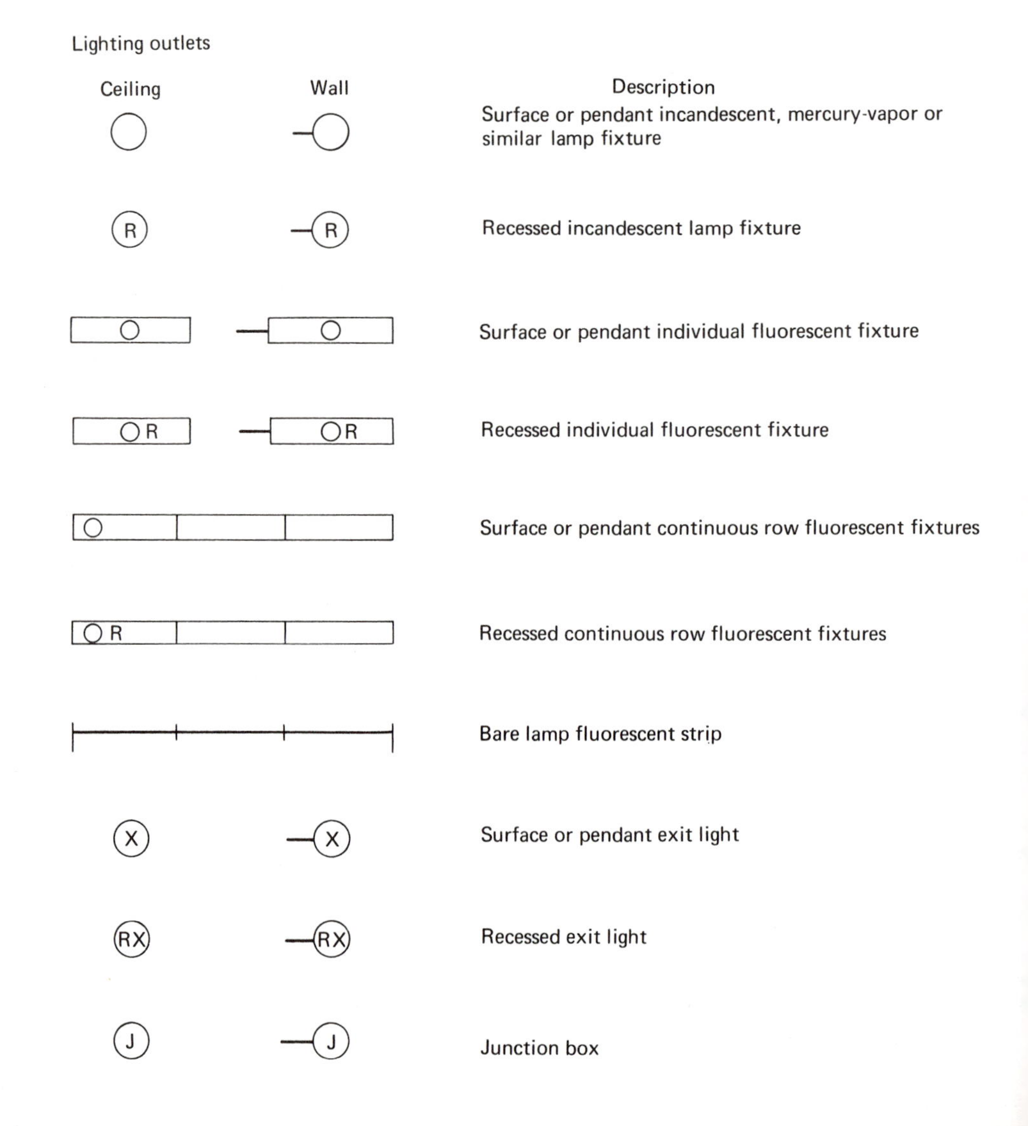

Receptacle outlets

(All symbols shown are for grounded receptacles, ungrounded receptacles require *UNG* note near symbol.)

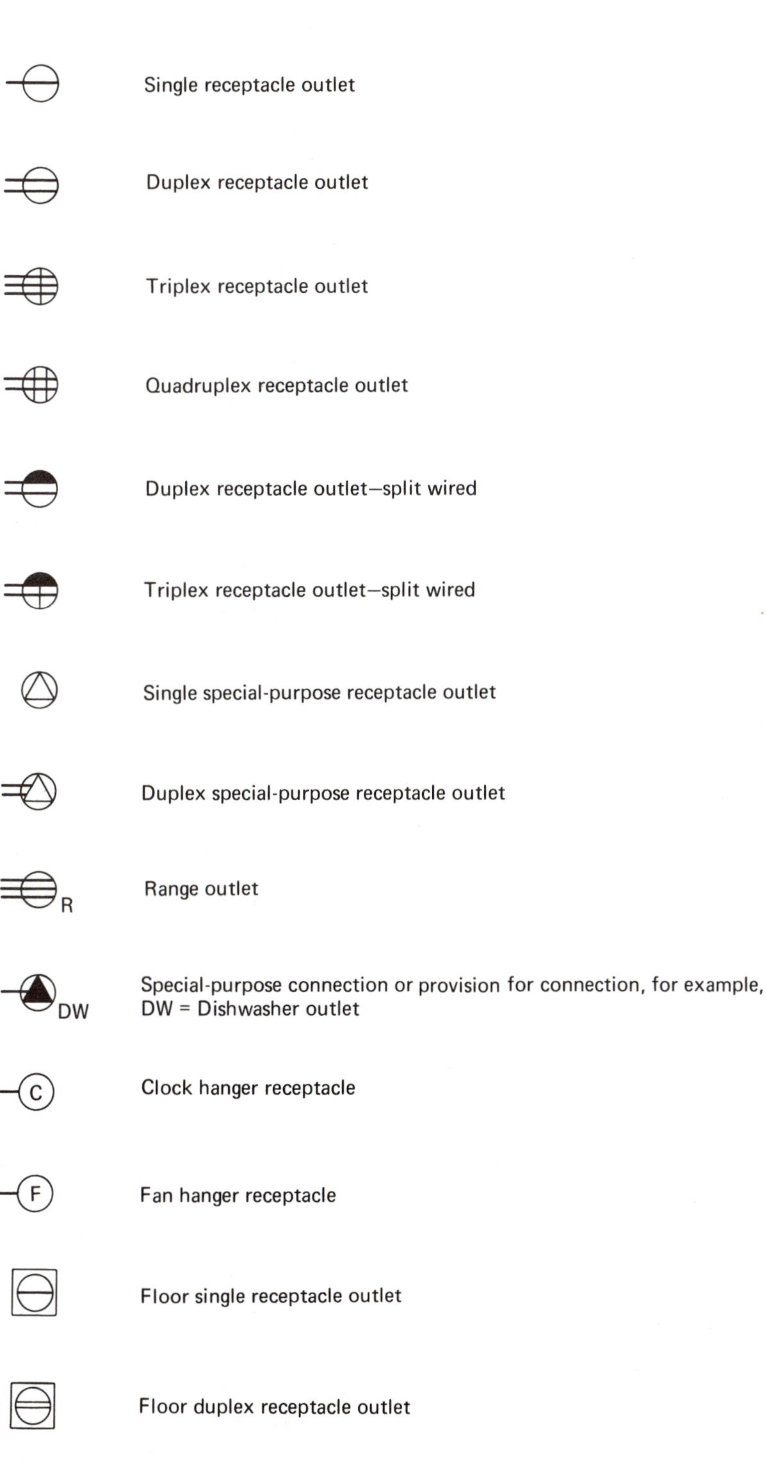

Floor special-purpose outlet

Floor telephone outlet—public

Floor telephone outlet—private

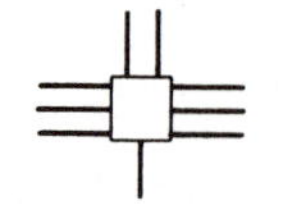

Under-floor duct and junction box—number of lines indicates number of duct cells

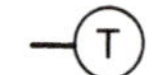

Thermostat

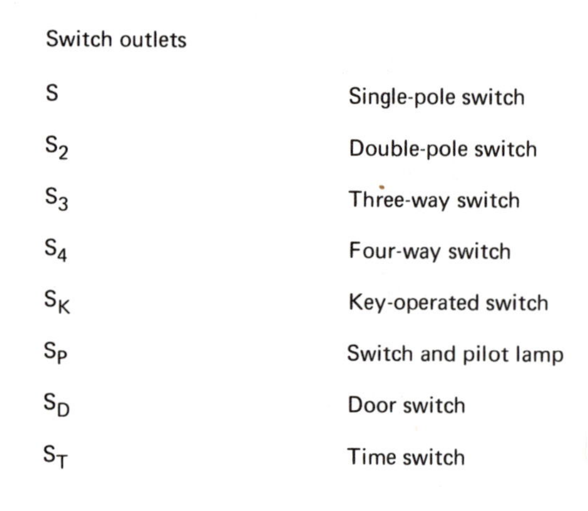

Switch outlets

S	Single-pole switch
S_2	Double-pole switch
S_3	Three-way switch
S_4	Four-way switch
S_K	Key-operated switch
S_P	Switch and pilot lamp
S_D	Door switch
S_T	Time switch

Panel boards

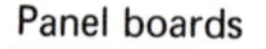

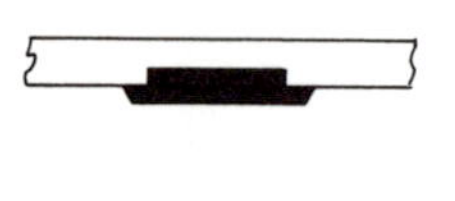

Flush mounted panel board and cabinet

Surface mounted panel board and cabinet

Circuiting

Wiring concealed in wall or ceiling

Wiring concealed in floor

Wiring exposed

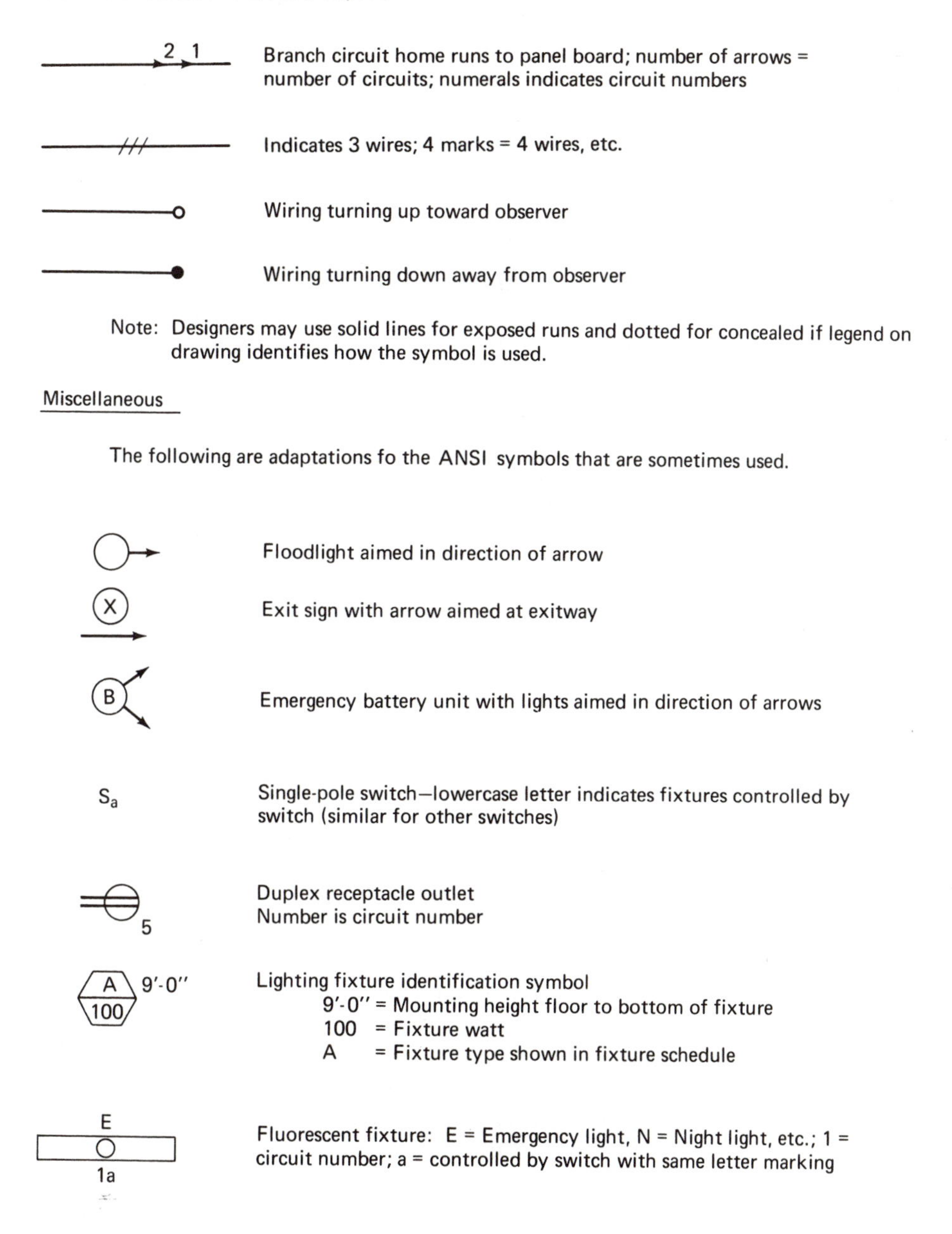

10-12 The Out-of-Function Part of a Layout

An out-of-function framework of physical walls, columns, stairways, etc., must be prepared prior to starting a lighting layout. This requirement is described for raceway layouts in Sec. 9-3. The source of information and the techniques for starting a lighting layout are similar to those for raceway layouts.

Lighting layouts are often drawn at a scale of $\frac{1}{8}'' = 1'\text{-}0''$. Smaller scales are used for large areas where the $\frac{1}{8}''$ scale might force the layout into several drawings. Decisions to use smaller scales should be weighed carefully because they are difficult to draw and hard to read.

The layout should be placed on the sheet similar to that shown in Fig. 9-14.

10-13 The In-Function Part of a Layout

The lighting fixtures, connecting circuiting, receptacles, etc., comprise the in-function part of the lighting layout. The fixture locations should be laid onto the out-of-function to scale because the spacing requirements are important and the layout should portray the balance intended. Many times lighting locations are dimensioned—other times scaled layouts are sufficient. If dimensioned, they should relate to a column center line or to fixed locations such as walls. Receptacles should be located and panelboard locations indicated. Local switches should be indicated and any other device symbols should appear where they are to be installed. After all fixtures, devices, etc., are shown, the circuiting must be planned before circuiting lines are drawn and wiring is determined.

10-14 Circuiting Lighting Layouts

At some point in the design development a decision has probably been made as to the voltage and type of system that will be used for lighting. This data will often appear on the one-line diagram and will usually be one of the following systems:

1. 480/277 V, three phase, four wire
2. 208/120 V, three phase, four wire
3. 120/240 V, single phase, three wire

Large areas to be lighted often use the 480/277-V, three-phase' four-wire system with the fixtures rated for 277-V service. This system is ideal for installations with large lighting loads and motor loads being supplied 480 V, three phase because the one system does the whole job. Of course, 120-V, single phase loads such as receptacles must be supplied from another source.

The 208/120 V, three-phase, four-wire system is used a great deal because it provides lighting service at 120 V, which if properly balanced in design reduces the requirement for additional neutral wiring. Similar to the 480/277-V system, three-branch circuits properly balanced can be run using only one neutral return. This balance is illustrated in Fig. 10-11.

Assume in Fig. 10-11 that all fixtures are 300 W each. The neutral (N) connection will serve to join a combination of two lamps in series across 208 V so there will be no current flow back along the neutral. With the perfectly balanced load, such as illustrated, there is no neutral current; however, any unbalance will be reflected into the neutral so the one neutral path is required.

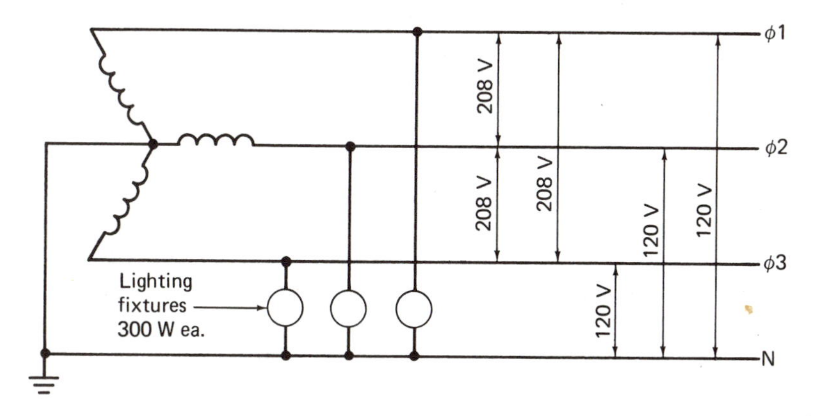

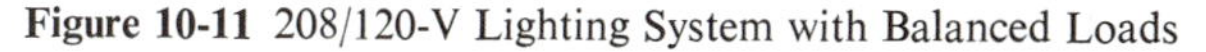
Figure 10-11 208/120-V Lighting System with Balanced Loads

Here then is the equivalent of three circuits and one neutral, which saves copper by eliminating the need for a neutral for each circuit.

Most standard panelboards for lighting service are arranged to stagger the connections to the phase busses so consecutive odd- or even-numbered circuits are fed from different phases. This arrangement is shown in Fig. 10-12.

By careful circuiting and balancing of loads, only one neutral needs to be run from the panel in Fig. 10-12 to serve three-branch circuits. For example,

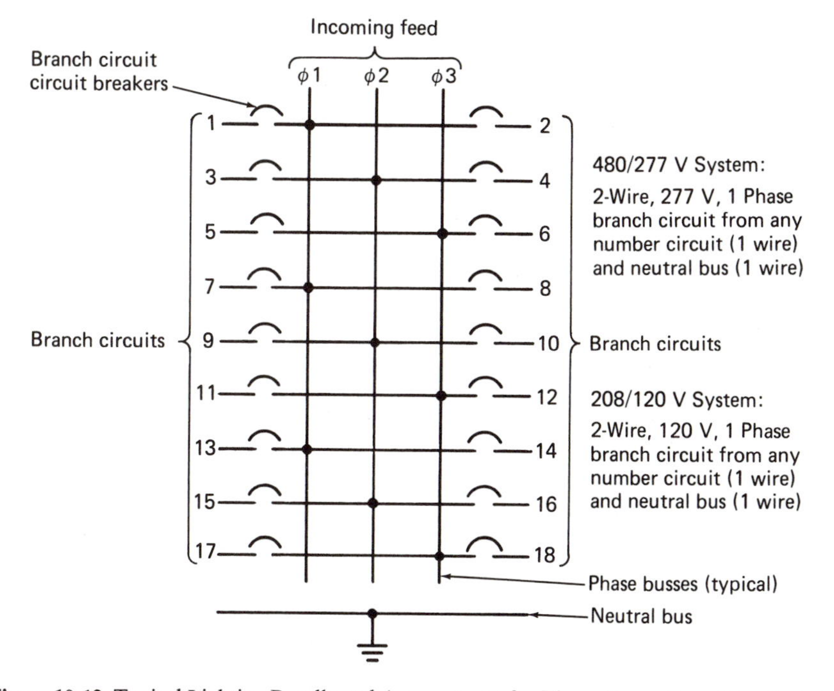

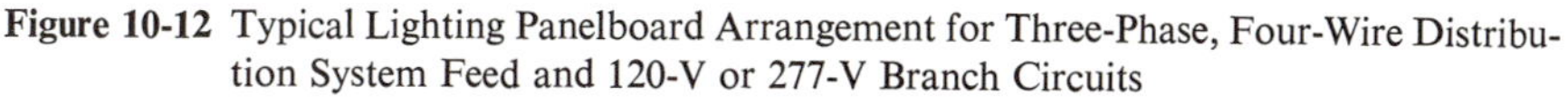
Figure 10-12 Typical Lighting Panelboard Arrangement for Three-Phase, Four-Wire Distribution System Feed and 120-V or 277-V Branch Circuits

circuits 1, 3, and 5 are all from different phases. Another similar combination is 1, 4, and 6, etc. If circuits 1, 7, and 13 were used, a separate neutral would be required for each—a total of six wires instead of four as in the separate phase selection.

The single-phase, 120/240-V system is common in residence wiring and to some degree is used in commercial and industrial design. Two circuits and one neutral can be run from panelboards fed by this system, as shown in Fig. 10-13.

It is fairly common to find panelboards similar to those shown in Figs. 10-12 and 10-13 specified with 20-A branch circuit breakers. Wiring for circuits to make full use of the 20-A rating will generally be a minimum of No. 12 gauge. The NEC limits the load on the branch circuits to 80% of the circuit rating for lighting. Therefore a 20-A circuit cannot be loaded beyond 16 As. At 120 V, this is approximately 1,920 W at unity power factor (16 A × 120 V = 1,920 W).

Taking the system balance criteria and load limit restrictions into account, circuiting can be started on the lighting layout.

As an example, assume the lighting fixture layout shown in Fig. 10-10 is for an industrial pump house. The building will be an isolated structure constructed of a concrete slab on grade, steel columns, and beam framing and a concrete slab roof with built-up roofing. The walls will be concrete block. There will be a large equipment door at one end and personnel doors at both ends. Other than for windows in the personnel doors, there will be no windows.

Switching of the majority of the lights will be from the panelboard; however, certain select fixtures will be controlled from three-way switches near each personnel door to enable a person entering the building to turn on a few lights prior to switching others required at the panelboard. Receptacles will be furnished as required by the NEC. The power supply will be from a dry-type transformer mounted above and feeding power to the panelboard at 120/208 V, three phase, four wire similar to the arrangement shown in Fig. 10-12.

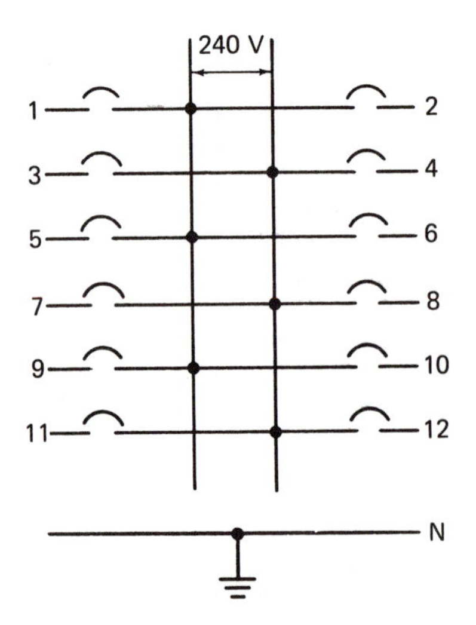

120/240 V System:
2-Wire, 120 V, 1 Phase branch circuit from any number circuit (1 wire) and neutral bus (1 wire)

Figure 10-13 Typical Lighting Panelboard for 120/240-V Service

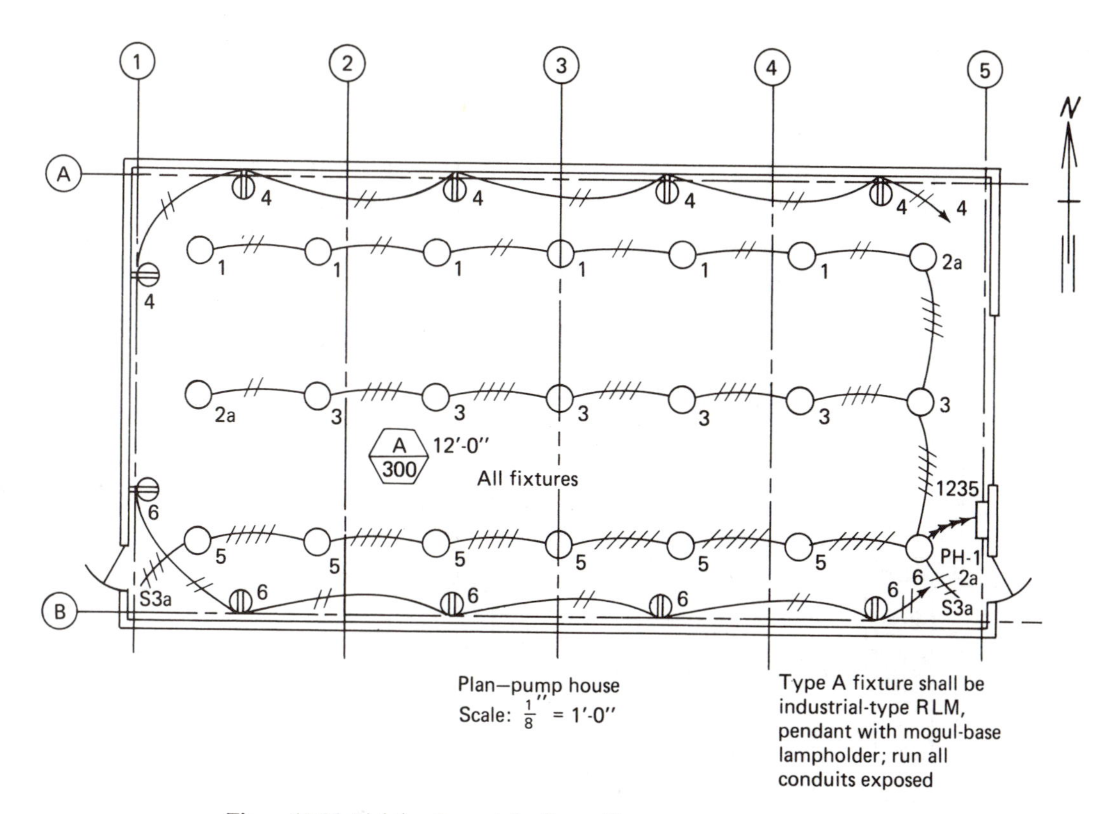

Figure 10-14 Lighting Layout for Pump House

The completed lighting layout for the criteria above is shown in Fig. 10-14. Circuiting lines between fixtures and devices are often made in an arc form rather than as straight lines because the drawing is intended to show how circuiting is to be made without getting into detail of precise locations for runs. These lines are made with a French curve.

10-15 Power Consumption—Fluorescent versus Incandescent

It was noted in Sec. 10-4 that the principal advantage of the fluorescent fixture over the incandescent is the very large increase of lumens per watt. To demonstrate this difference, the lighting requirements in the pump house of Fig. 10-14 could be met by using 16 to 18 two-lamp, 40 W, industrial-type fluorescent fixtures instead of the incandescents used in the example. Comparison of power consumption is approximately as follows:

Incandescent

$$21 \times 300 \text{ Watt each} = 6{,}300 \text{ W}$$

Fluorescent (includes ballast)

$$18 \times 100 \text{ W each} = 1{,}800 \text{ W}$$

No. PH-1 Location PUMP HOUSE

208 V, 3 PH., 4 W 100 A Mains, 100 A Solid neutral; Grd. bus
20 A AIC bkrs. @ 120 V; Surface mtg. X Flush mtg.

Designation	Load watts	No. outlets	CKT. bkr Trip	CKT. bkr No.	Phase	CKT. bkr. No.	CKT. bkr. Trip	No. outlets	Load watts	Designation
NORTH ROW GENERAL LTG	1800	6	20A	1	1	2	20A	3	900	WEST & EAST END ON SWITCH
CENTER ROW GENERAL LTG	1800	6		3	2	4		5	250	NORTH RECEPTACLES
SOUTH ROW GENERAL LTG	1800	6		5	3	6		5	250	SOUTH RECEPTACLES
SPARE				7	1	8				SPARE
				9	2	10				
				11	3	12				
				13	1	14				
				15	2	16				
				17	3	18				

Figure 10-15 Typical Lighting Panelboard Schedule

In this comparison the incandescents use $3\frac{1}{2}$ times more power than the fluorescents require. Usually the fluorescent fixture makes the best choice. Incandescents are usually less expensive fixtures to buy and if power is small (as it might be in an unattended pump house), the initial investment may be an important consideration.

10-16 Lighting Panelboard Schedules

Standard forms can be made with decal-type adhesives that can be stuck onto the back of lighting layout drawings showing the panelboard circuiting. A typical schedule for the pump house panelboard of Fig. 10-14 is shown in Fig. 10-15.

Chapter Review Topics and Exercises

10-1 Light on a surface 1 ft from a light source of 1 candlepower is called what? How is this term used in measuring illumination?

10-2 How is the term *lumen* applied?

10-3 The ratio of lumens on the area to be lighted to the lumens generated by the light source is called what?

10-4 The factors effecting the ratio value of Exercise 10-3 is influenced primarily by four factors. Name them and explain the influence of each.

10-5 What is the maintenance factor? How does it affect lighting levels?

10-6 How can the proper level of illumination be determined for a particular seeing task? What is a recommended level for school drafting rooms?

10-7 What is the principle advantage of a fluorescent lamp over an incandescent lamp of equal wattage?

10-8 Write the basic formula for determining the number of footcandles in a given area and explain why this formula provides an in-service value.

10-9 Why is the zonal cavity method in more common use than the older methods?

10-10 Using the zonal cavity method, calculate the number of recessed lighting fixtures required to light the following room to 75 fcs: room measures 60 ft wide by 60 ft long. Reflectances are ceiling, 50%; walls, 50%; and floor, 30%. Fixtures are as shown in Fig. 10-7 mounted in hung ceiling 11′-6″ above the floor. The structural floor to ceiling height is 14′-0″. Lamps are rated 3,100 lumens each. Maintenance factor is .80.

10-11 What would be a good spacing for the fixtures in Exercise 10-10?

10-12 Why do some row and spacing arrangements make better layouts than others?

10-13 Transfer switches are used in what way in emergency lighting systems?

10-14 What advantage is there to a balanced load on a three-phase, four-wire system?

10-15 Referring to Fig. 10-12, how many neutral conductors would be required if circuits 13, 14, and 15 were run in one conduit?

10-16 Lay out the out-of-function of the pump house of Fig. 10-14 using the data given in Exercise 9-25.The building is to be lighted using industrial fluorescent fixtures lamped with two-40-W cool-white tubes in each fixture (initial lumens from appendix, Table 6, and 3,050 lumens per tube). The fixture data is as shown below.

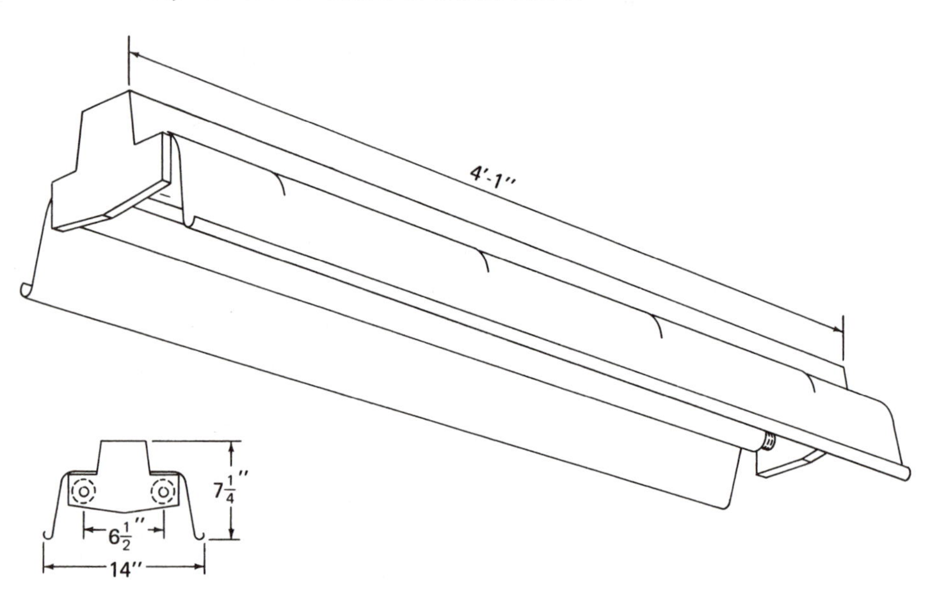

C.U.—Obtain from table below (may be used for any lamping)

Maintenance factor (M.F.): Good 0.75; Medium 0.70; Poor 0.60 (RS and Slimline)
Good 0.70; Medium 0.65; Poor 0.55 (HO)
Good 0.65; Medium 0.60; Poor 0.50 (P-G, SHO, VHO)

Pcc	80			50			10		
Pw	50	30	10	50	30	10	50	30	10
RCR	COEFFICIENTS OF UTILIZATION (X.01)* FOR 20% EFFECTIVE FLOOR CAVITY REFLECTANCE								
1	0.88	0.85	0.82	0.79	0.77	0.74	0.69	0.67	0.66
2	0.78	0.72	0.67	0.70	0.66	0.62	0.61	0.58	0.56
3	0.69	0.62	0.57	0.62	0.57	0.53	0.54	0.51	0.48
4	0.61	0.54	0.48	0.55	0.50	0.45	0.49	0.45	0.41
5	0.54	0.46	0.41	0.49	0.43	0.39	0.43	0.39	0.36
6	0.48	0.41	0.36	0.44	0.38	0.34	0.40	0.35	0.31
7	0.43	0.36	0.31	0.40	0.34	0.29	0.35	0.31	0.27
8	0.39	0.32	0.27	0.36	0.30	0.25	0.32	0.27	0.24
9	0.35	0.28	0.23	0.32	0.26	0.22	0.28	0.24	0.20
10	0.32	0.25	0.20	0.29	0.23	0.19	0.26	0.21	0.18

Pcc = Effective ceiling cavity reflectance
Pw = Wall reflectance
RCR = Room cavity ratio

Using the data above, calculate the number of fixtures required to light the area to 20 fcs (in-service). Mount fixtures 12′-0″ above finish floor. Use short form of zonal cavity calculations. Provide receptacles and any local switches; circuit all runs back to a panelboard located on the east wall between the personnel door and the equipment door. Use a panel similar to that shown in Fig. 10-12.

11

POWER DISTRIBUTION AND SUBSTATIONS

11-1 General

Power is distributed from central power plants and other generating facilities at various voltage levels. Starting at the generator itself, the voltage may be in the area of 18,000 to 24,000 V similar to the machine shown in Fig. 4-1 where the generated voltage is 22,000 V. Higher generating voltage aggravates the insulation problem and lower voltages increase the conductor size and cooling requirements. The generated voltage is stepped up to meet transmission requirements. An old rule of thumb is to allow 1,000 V per mi when judging transmission line levels. A 110-kV line would serve loads 100 mi or more away from the generating point. With sources of water power often times far removed from cities and other major loads, transmission levels have increased dramatically over the past years in order to bring the power from remote stations.

Lower voltages are adequate to the needs of cities and towns. These generally range from 13,800 V down to 2,400 V. Almost all these systems are three phase. Some voltage *ranges* and their common general classifications are as follows:

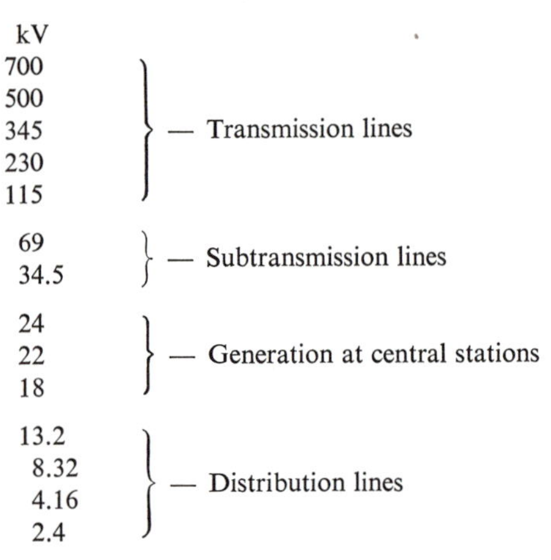

kV	
700 500 345 230 115	— Transmission lines
69 34.5	— Subtransmission lines
24 22 18	— Generation at central stations
13.2 8.32 4.16 2.4	— Distribution lines

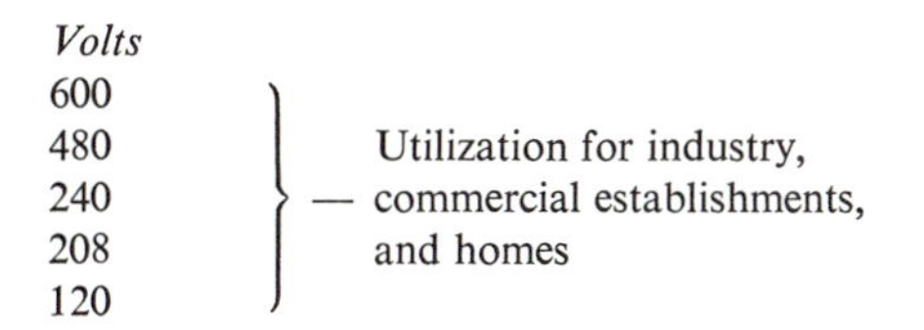

Volts	
600	
480	Utilization for industry,
240	— commercial establishments,
208	and homes
120	

Variations of the list above are common; however, the general bracketing shows the common range levels in use.

11-2 Power Distribution from Generation to Utilization

By taking the liberty of restricting the routing of power distribution to the supply from a single generating plant and a single path of distribution, the flow can be demonstrated in a simple pictorial form as shown in Fig. 11-1. Major items of equipment are identified.

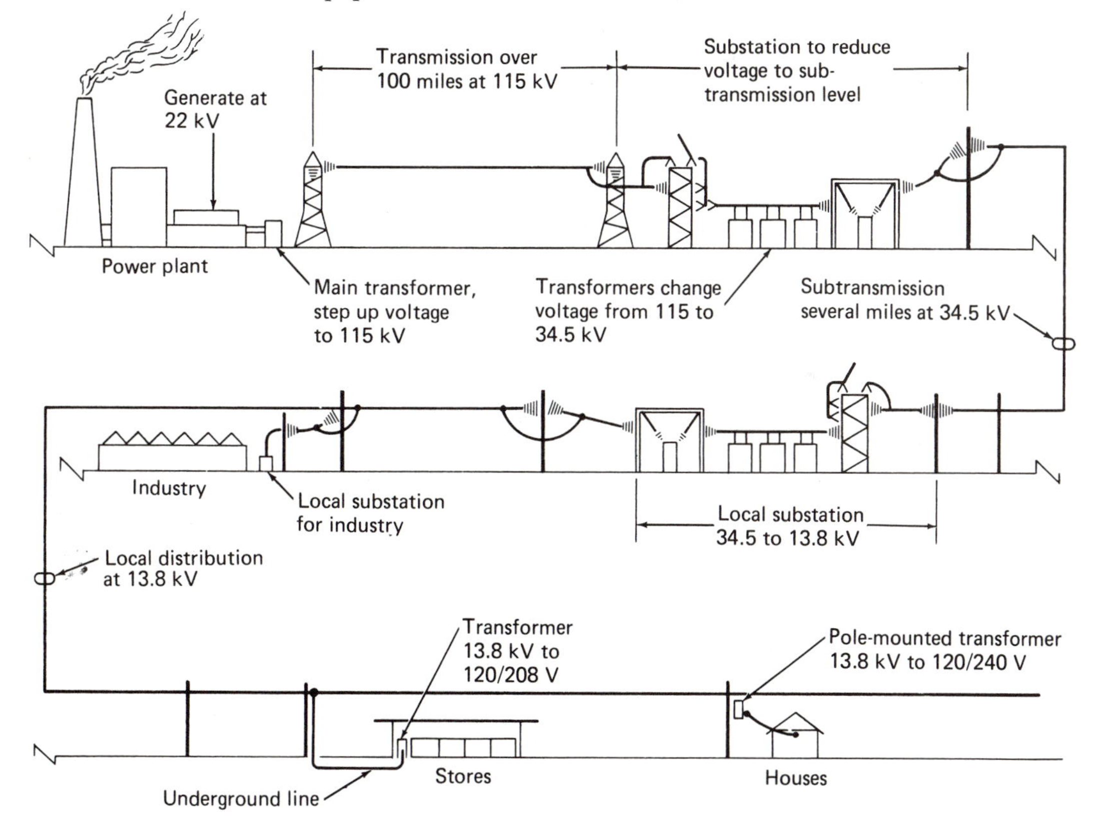

Figure 11-1 Power Distribution from Generation to Utilization (Typical example)

11-3 Transmission Line Design

Transmission line design is a somewhat specialized field of engineering. The towers are mostly a structural design problem with the support of the overhead conductors falling into some rather common arrangements. Spacing of the conductors between each other and away from the grounded structure varies with the voltage level. Good reference material is available in several texts dealing with this subject.

11-4 Distribution Lines

Design of the common distribution systems seen throughout the countryside are generally created from long-standing standard designs that have been developed by the utility companies who generally build these lines. If the system is an overhead one, each pole design has usually been drafted into a standard so the corner pole, the transformer pole, street lighting pole, etc., are part of design standards. Naturally variations must be handled as they occur and unique designs may need to be made based on the standards. Underground design has been developed in much the same way. Manholes, duct banks, and cable arrangements in manholes, transformer vaults, etc., are standardized as much as possible. Texts are available treating this subject in detail.

11-5 Arrangement of Primary Distribution Circuits

Primary distribution circuits generally range from 13,800 V down to 2,400 V. Higher voltages are used and the tendency toward higher levels is on the increase. Primary systems are usually classified as radial or network. The *radial system* is usually supplied from a single source and the lines radiate from a substation out along roads and streets to serve the community loads. A one-line diagram of a simple radial system is shown in Fig. 11-2.

This same radial system design is often formed into a loop arrangement that amounts to two radial lines leaving the substation, making a *loop* through high-density load areas and returning to the substation. Sectionalizing switches

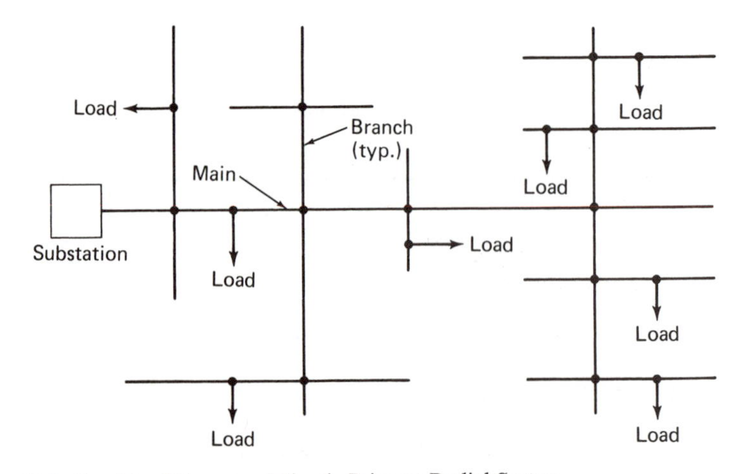

Figure 11-2 One-Line Diagram of Simple Primary Radial System

in the main lines provides the versatility of feeding the line from both ends. A schematic diagram showing such an arrangement is shown in Fig. 11-3.

The power from the substation normally feeds out on lines No. 1 and 2 to an open switch somewhere in the loop. If a line is down due to storm damage or for some other reason, the *section* in trouble can be *sectionalized* out of service through use of the switches and the balance of the system stays on the line. This arrangement is extremely useful for maintaining service as much as possible.

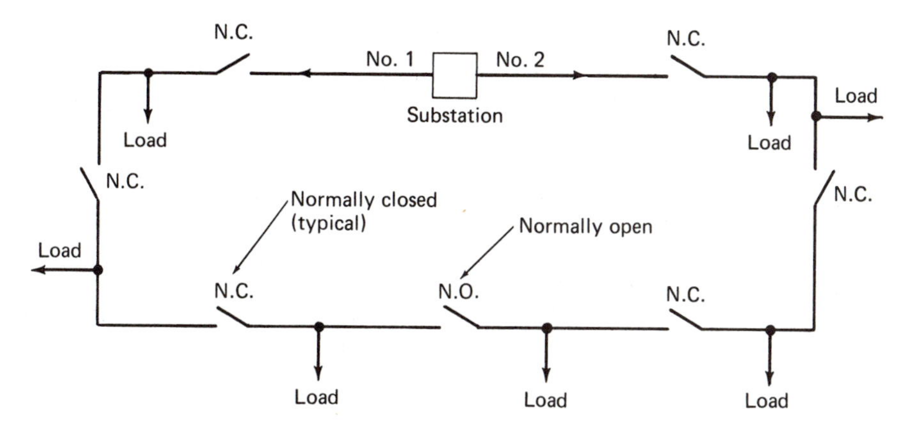

Figure 11-3 Loop Arrangement for Distribution Lines

Primary network systems are used a great deal in metropolitan areas and to some degree in suburban areas. The basic system provides for tying the feeds from several sources together in order to feed selected *load centers* from more than one supply. Failure of any one source normally will not shut down the system. Figure 11-4 shows a one-line arrangement of a typical system.

11-6 Arrangement of Secondary Distribution Circuits

Secondary systems are usually at the utilization levels of voltage from 600 V down to 120 V. Systems may be either single or three phase depending on load requirements. Circuit arrangements are similar to primary systems. Systems for industry and commercial applications are described in detail in Chapter 3. Distribution of secondary voltages for stores, homes, and similar loads are generally made from utility lines running overhead on poles or underground in duct lines. The radial system is used for low-density loads with banking of transformers a fairly common practice. This is simply a case of running secondary lines and connecting all transformers in a parallel arrangement, as shown in Fig. 11-5.

The transformers in Fig. 11-5 might be pole mounted and separated by several hundred feet (depending on load conditions). This *banking* arrangement helps to decrease disturbances on the secondary line and improves reliability. In addition, smaller transformers may be used because load demands are shared to some degree by all the transformers.

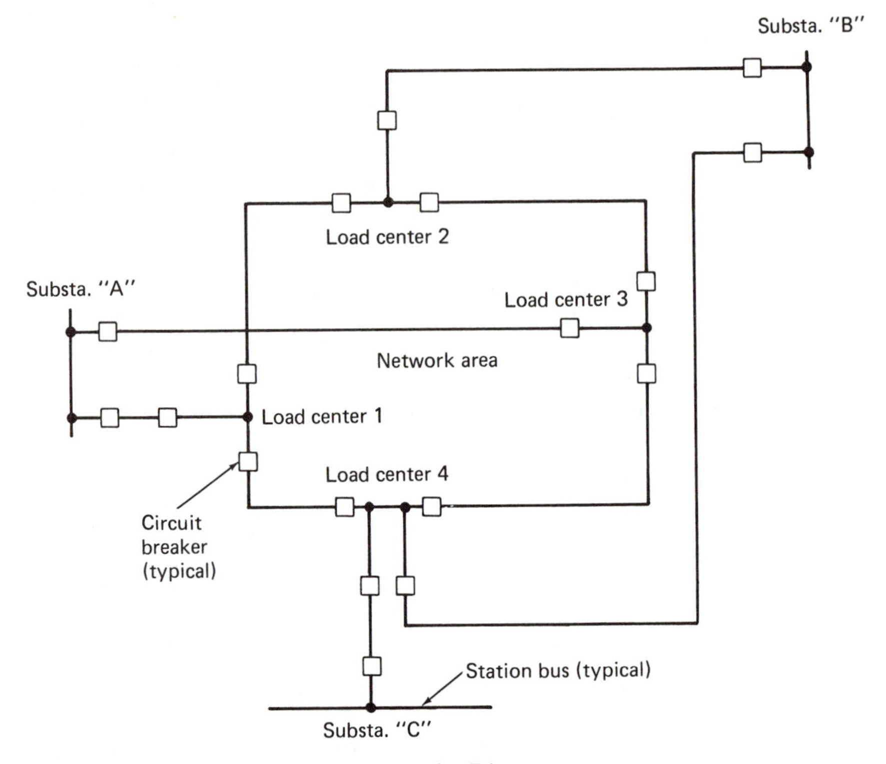

Figure 11-4 Primary Network System One-Line Diagram

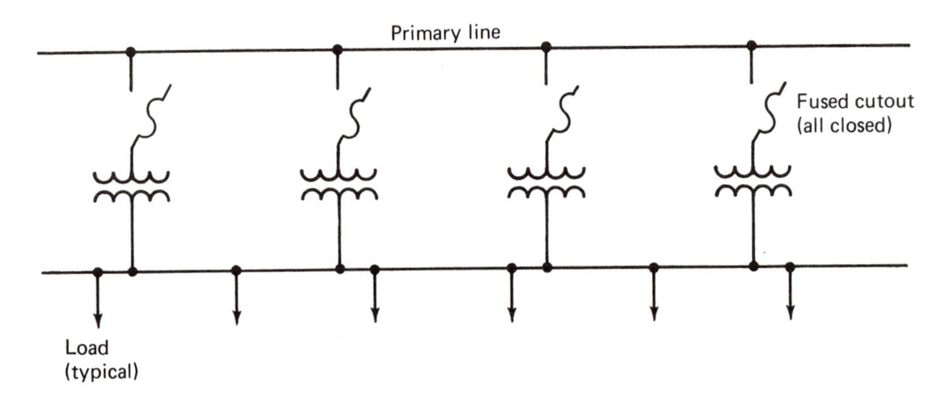

Figure 11-5 Transformer Banking for Secondary Distribution

Secondary network systems are similar in principle to the banked transformer secondary connection except that they are much larger in capacity. In the network arrangement, the secondary mains form a solidly connected secondary mesh or grid. Figure 11-6 shows such a network.

The transformers of Fig. 11-6 are often located in vaults beneath the sidewalk. Special network transformers are used and protective devices are provided in order to remove faulty equipment from service without shutting down the

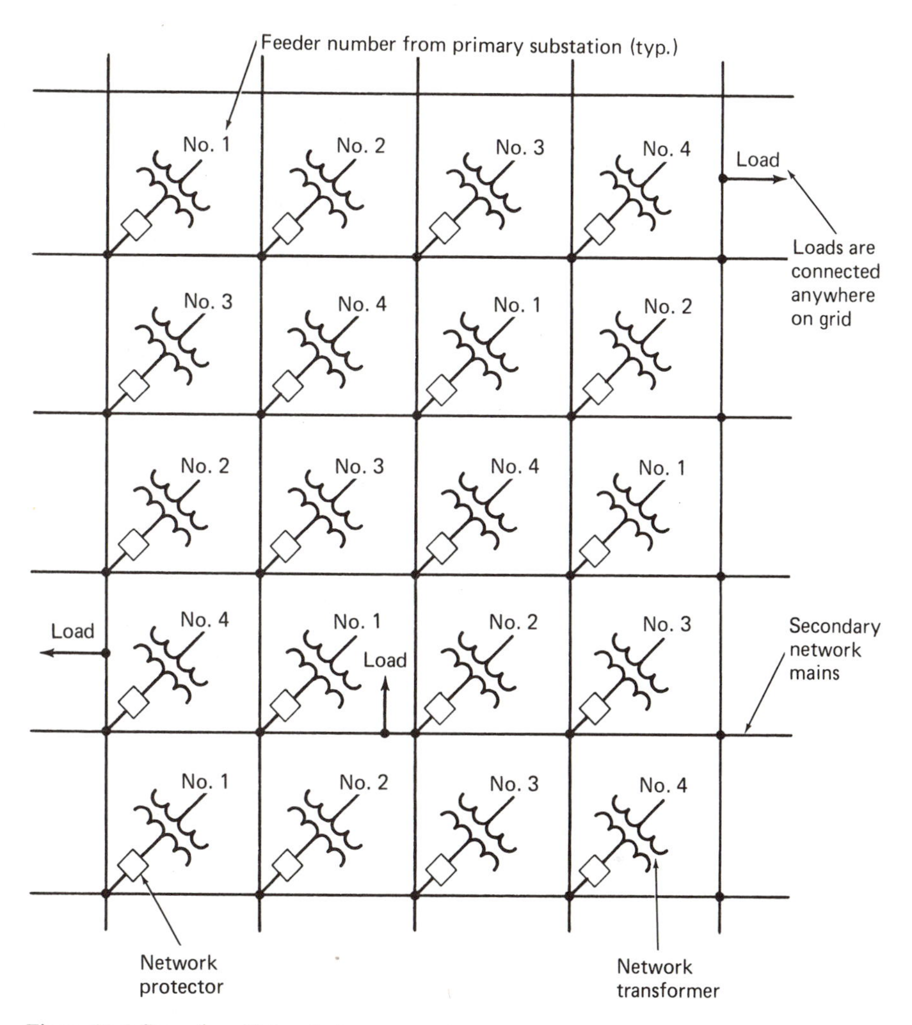

Figure 11-6 Secondary Network Arrangement

entire network. Careful design keeps feeders reliable by taking them off separate busses, substations, etc., so a bus or substation fault will not take several feeders out of service that could cause collapse of the network. Several networks would comprise the total city distribution system.

11-7 Substation Design

Substations are so common in distribution systems as to have become part of a *package*-type installation. A one-line diagram and specification document may be all the technical data necessary to purchase a package of structures, equipment, wiring, etc., to erect a substation on a given site. Built-up substations composed of several items of equipment arranged in a particular way may require some layout and detail design.

Chapter Review Topics and Exercises

11-1 If a hydroelectric power plant was built about 200 mi away from a city where the major load was to be served, what would probably be the transmission line voltage?

11-2 What major piece of electrical equipment is used to step generation voltage up to transmission line levels?

11-3 Describe the radial system of primary distribution.

11-4 Why are loop feeders a desirable form of distribution?

11-5 Suppose a line has been broken through storm action on a loop feed. How can unaffected sections of the line be kept in service?

11-6 What are the obvious advantages of a primary network system?

11-7 Explain transformer banking and its advantages.

11-8 Why would a secondary network system be the logical choice for distribution in a densely populated city area?

11-9 What basic function does a substation serve?

12

SYSTEM PROTECTION REQUIREMENTS AND DEVICES

12-1 General Requirements

Electrical systems are designed to provide a continuous supply of electrical energy to the equipment or devices being served. Reliability of service is becoming increasingly important. Power supply to computers with memory banks, supply to medical experiments requiring uninterrupted service in order to achieve the planned results, and similar loads are completely dependent on a continuous supply of electrical energy. The greatest hazard to this service is in the flow of currents greater than the normal currents that are expected to flow. These currents are generally called *overcurrents*. Overcurrents may develop from a great many causes. They may range in magnitude from a few amperes greater than normal to 100,000 A or more. Very often they tend to divide into two categories: (1) overloads or (2) short circuits.

12-2 Overloads

Overloads are defined as currents that are greater than normal current flow, confine themselves to the normal conducting path, and will cause overheating of the conductor if they are allowed to continue to flow.

Overloads are caused in several ways. For example, in a motor circuit the motor bearings or bearings of the equipment being powered by the motor may require lubrication and therefore run hot. They expand in heating and seize onto the shaft. The motor cannot rotate up to speed and may indeed stop entirely. Excess current is drawn that is *seen* by the overcurrent protective device as an overload. As another example, a branch circuit in a home may be properly sized and protected by overcurrent devices but an additional appliance is plugged in causing excess current over the circuit rating and the fuse blows. This too is an overload condition.

In general, overcurrents that do not exceed five to six times the normal current fall within the overload classification because although they may actually be short circuits, they are *seen* by overcurrent protective devices as overloads.

12-3 Short Circuits

Short circuits are defined as currents that are out of normal current ranges. Actually some short-circuit currents are no larger than load currents, whereas others can be many thousands of times larger than normal currents.

Short circuits can be caused in many ways. Vibration in equipment may cause insulation to wear through and expose the bare conductors to each other or to ground. Insulators carrying high-voltage conductors may become excessively dirty from polluted atmospheres and in the presence of rain or mist a flashover to the grounded structure occurs. Whatever the cause, short circuits are generally the result of insulation breakdown. They can occur whether the insulation is rubber, wood, or merely an air gap when the current no longer stays within the intended current path but *jumps* across and flows to an unintended path.

Short circuits generally have three effects as follows:

Arcing

This is similar to the arc that is seen when a welder is working. A very bright and hot arc occurs that can range in current flow from a few to many thousand amperes. The effect at the *fault* is very dramatic. The arc burns up everything in its path. Copper wires are similar to the welder's electrode and many feet of wire may be vaporized or melted into splatters of copper and scattered across the surrounding areas. Sometimes enclosing conduits are burned through, which fills the surrounding area with showers of sparks and molten steel. Heat at the arc causes vaporization of the organic insulating materials. This tends to liberate explosive gases, which in turn are ignited by the arc.

Heating

When short circuits are of a high-current magnitude, they cause severe heating effects. For instance, even a moderate fault current of 15,000 A will raise the temperature of No. 6 copper wire over 400° F in less than one cycle. This in turn can initiate fire in surrounding materials.

Magnetic stressing

Because a magnetic field is formed around any conductor carrying current, it can be readily seen that when short-circuit currents of thousands of amperes flow, this field is magnified many times and the stress caused by the field becomes significantly larger. As an example, a fault current of 25,000 A in parallel bus bars can produce stresses of over 750 lb per ft of bus. This level of force tends to cause the bus to break away from the supports. If the supports fail, the bus may short against the other busses or fault to ground and arcing will occur.

Short circuits in today's large electrical systems can easily exceed 100,000 A. The results can be damaging in loss of power and possibly in loss of life. *The need to design systems that are safe from the effects of short circuits is an ever-present responsibility for the electrical design discipline.*

12-4 The Definition of Some Terms about Short-Circuit Currents

These are helpful to the designer and draftsman in understanding the application of protective devices. Some of these follow.

Effective Current

Because alternating current varies continuously from 0 to a maximum value first in one direction and then in the other, it is not readily apparent just what the *true* current value is.

The current at any point on a sine wave is called the *instantaneous current.* The current at the top of the wave is called the *peak* or *crest* current. It is possible, of course, to determine the arithmetic *average value* of the alternating current but none of these values correctly compare alternating to direct current. It is desirable to have 1 A of alternating current do the same work as 1 A of direct current. This current is called the *effective current* and 1 A of effective alternating current will do the same heating, for instance, as 1 A of direct current. So *effective current* must be the basis of comparison of the two systems.

RMS Current

Effective current is more commonly called *rms current,* rms is the abbreviation for *root-mean-square.* It is the square root of the average of all the instantaneous currents squared. When the peak current of a sine wave is 100 A, the effective or rms value is 70.7 A. Relating this to dc current, means that the sine wave peak current of 100 A is equivalent to a direct-current flow of 70.7 A in heating effect.

Symmetrical Current

A symmetrical current wave is one that is equal or balanced around the zero current axis. This is shown in Fig. 12-1.

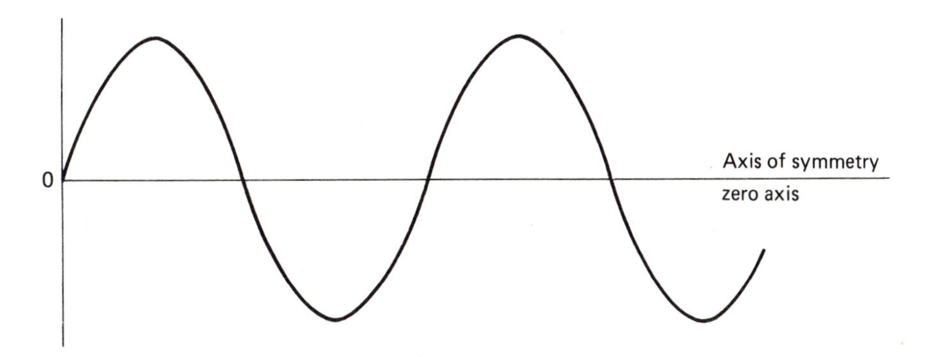

Figure 12-1 Symmetrical Sine Waveform

Asymmetrical Current

An asymmetrical current wave is not symmetrical about the zero axis. The axis of symmetry is offset or displaced from the zero axis. This is shown in Fig. 12-2. This offset condition generally takes place when a short circuit occurs. In three-phase systems the asymmetrical current is greater than the symmetrical current by a factor of 1.25 (commonly used).

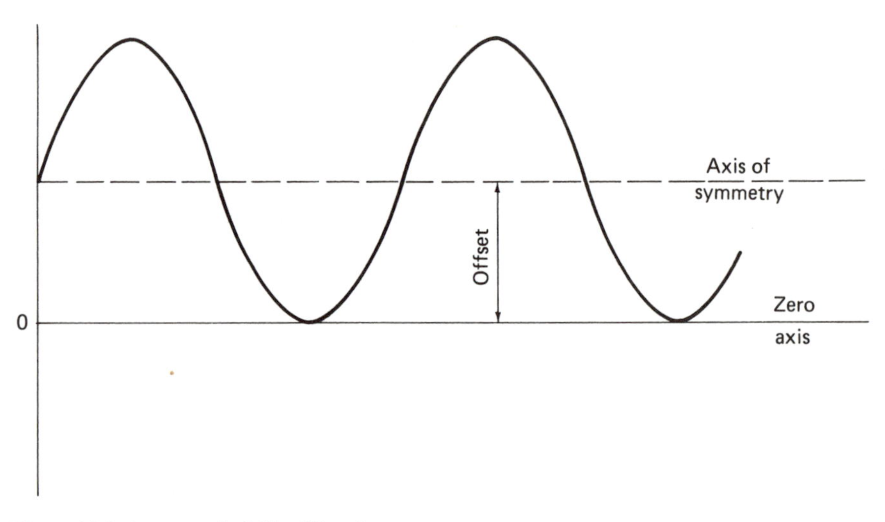

Figure 12-2 Asymmetrical Sine Waveform

12-5 Conditions at the Instant of Short Circuit

Conditions at the instant of short circuits are quite different than under normal conditions. On an alternating-current system, under normal operation, a circuit draws current in proportion to the voltage being applied and the impedance (Z) of the load being served. When a short circuit occurs, the applied voltage no longer *sees* the normal opposition to current flow that the load presented. Instead, the voltage is applied across a load of much lower impedance. This is generally made up of the impedance of the conductors from the source of voltage to the point of short-circuit fault, the impedance of the transformer in the circuit feeding into the fault, and any other impedance due to equipment in the circuit relating to the fault. In determining fault currents, the impedance at the fault itself is assumed to be zero and the fault point itself is commonly called a *bolted fault*, meaning that it is a hard, solid connection that has developed, which is, of course, the worst condition possible. Actually in a real practical sense the bolted fault condition may seldom occur but all calculations must assume that it can and does happen. With zero impedance at the fault, it can be readily seen that the current flow is going to increase tremendously.

12-6 Determining Short-Circuit Currents

Determining short-circuit currents can involve rather extensive study on large complex systems, which in larger engineering firms will be done by the project engineers; however, designers and draftsmen should have a reasonable knowledge of simple short-circuit calculations.

Short-circuit currents do not come from the point of fault. All the *fireworks* occur there but the fault current itself is poured into the fault from several possible sources, as follows:

1. The utility system supplying the power
2. Induction motors
3. Generators other than at the utility
4. Synchronous motors
5. Synchronous condensers

The two principal sources are the utility system and the induction motors because customer-owned generators, synchronous motors, and condensers are not too common in industry. If they are present, of course, they must be considered.

For purposes of this explanation, they will be ignored. Also, for purposes of simplification, only a radial-type distribution will be considered because networks become complicated in analysis and are beyond the scope of this discussion.

Figure 12-3 shows a simple radial system condition with a 750-kVA, three-phase transformer connected to the primary system at 13,800 V. Also indicated are motor loads on the secondary bus.

The amount of short-circuit current from any source is called the *available fault current*. In this example, the available fault current from the utility system

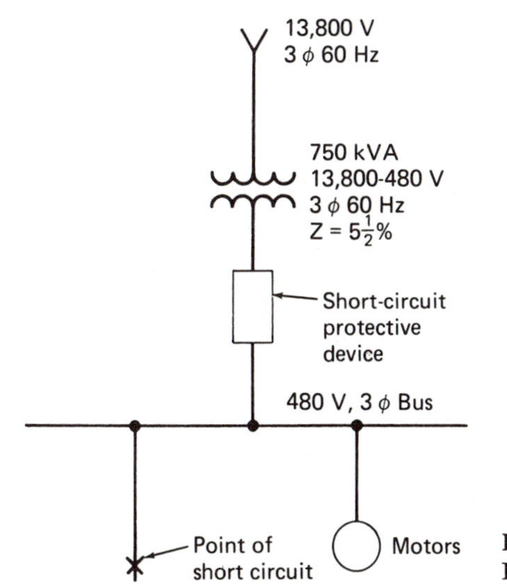

Figure 12-3
Diagram for Calculating Short-Circuit Current

will be considered as *unlimited*, which is the worst condition. (Short-circuit studies may use actual data supplied by the utility company as to available short-circuit currents from their systems in order to refine the calculations.)

To make the simple calculations for the example given, first the continuous full-load current of the transformer is determined.

(1) $$I = \frac{\text{kVA} \times 1{,}000}{E \times \sqrt{3}}$$

(2) $$I = \frac{750 \times 1{,}000}{480 \times 1.73}$$

(3) $$I = 903 \text{ A}$$

The available symmetrical fault current is calculated:

(4) $$\text{Max } I_{sc} = \frac{100\%}{\%Z_T} \times \text{secondary full-load current}$$

(5) $$= \frac{100}{5.5} \times 903$$

(6) $$= 16{,}418 \text{ A symmetrical}$$

Because this is a three-phase transformer, a factor of 1.25 will be used to determine the asymmetrical current.

(7) $16{,}418 \times 1.25 = 20{,}522$ A asymmetrical (many available tables will round this figure to 20,500)

During the first few cycles of a fault, induction motors contribute short-circuit current that cannot be ignored. Where a fault occurs on a system, the induction motors are driven by the inertia of the load they have been driving. Momentarily motors act like generators. The output is of very short duration and lasts only a cycle or two. This contribution to the short circuit is called the *motor contribution.*

The short-circuit current put out by induction motors varies widely from motor to motor. For this reason, it is seldom practical to determine the motor contribution precisely. Very often the exact number of motors and their characteristics are unknown. Also motors are frequently changed and added to systems. Because of these complications, approximations are used based on the assumption that connected motor loads form some percentage of or perhaps all of the connected load. Commonly used assumptions are as follows:

System voltage	*Percent motor load*	*Motor contribution factor times normal current*
208 V	50	2.5
240 V	100	5
480 V	100	5
600 V	100	5

(Where the circuit voltage is less than 600, 480, or 240 V, the motor contribution current should be multiplied by the ratio of the voltage used to the voltage shown above. For example, at 550 V, the multiplying factor would be $\frac{600}{550} =$ 1.09 times the current determined at 600 V.)

In the example (Fig. 12-3), the motor load is assumed to be 100% with a factor of 5 times the transformer full-load current as the total short-circuit current from the motors.

$$903 \text{ A} \times 5 = 4{,}515 \text{ A short circuit}$$

The combined total asymmetrical short-circuit current is therefore 20,500 (rounded from 20,522) + 4,500 (rounded from 4,515) = 25,000 A.

This is the approximate value of the current available at the secondary terminals of the transformer if a short circuit occurs. Short-circuit currents for various-sized transformers at different secondary voltages are given in Table 7 in the appendix. Examination of this table will show that at lower system voltages the available short-circuit currents are higher and a 1,500-kVA transformer with a $5\frac{1}{2}\%$ impedance and with a three-phase secondary voltage of 240 V will allow a 100,000-A fault to develop from the combined total of the 100% motor contribution and an unlimited primary source.

12-7 Reduction of Short-Circuit Currents

Reduction of short-circuit currents from those calculated to be present at the transformer and those that are likely to occur at some distance from the transformer are a function of the characteristics of the circuit conductors and their lengths. For example, at the end of a 100-ft length of No. 4 copper cable from the 750-kVA, 480-V transformer with 100% motor load and unlimited primary source used as an example in the previous explanation, the asymmetrical fault current of approximately 25,000 A has been reduced to approximately 9,400 A (from Table 9, 7,500 × 1.25 for asymmetrical value) asymmetrical. The short circuit has been reduced in magnitude because the conductors present an impedance to the flow of current and thereby reduce it.

Data showing the approximate available fault currents for various sizes of transformers, sizes of wires and cables, and distances from the transformer to the fault are published in curves or in table form. Representative tables are included in the appendix as Tables 8 and 9. *Special note should always be taken of the table heading and footnotes to be sure of the characteristics of the data shown.*

To demonstrate how these tables can be used, the one-line diagram of Fig 12-3 is expanded upon in Fig. 12-4 to show what could be the distribution system for a small industrial building. The size of circuit breakers, feeder cables, and their lengths are noted on the one-line diagram. Note that all pertinent data are recorded. This is always a good practice because at a later date the infor-

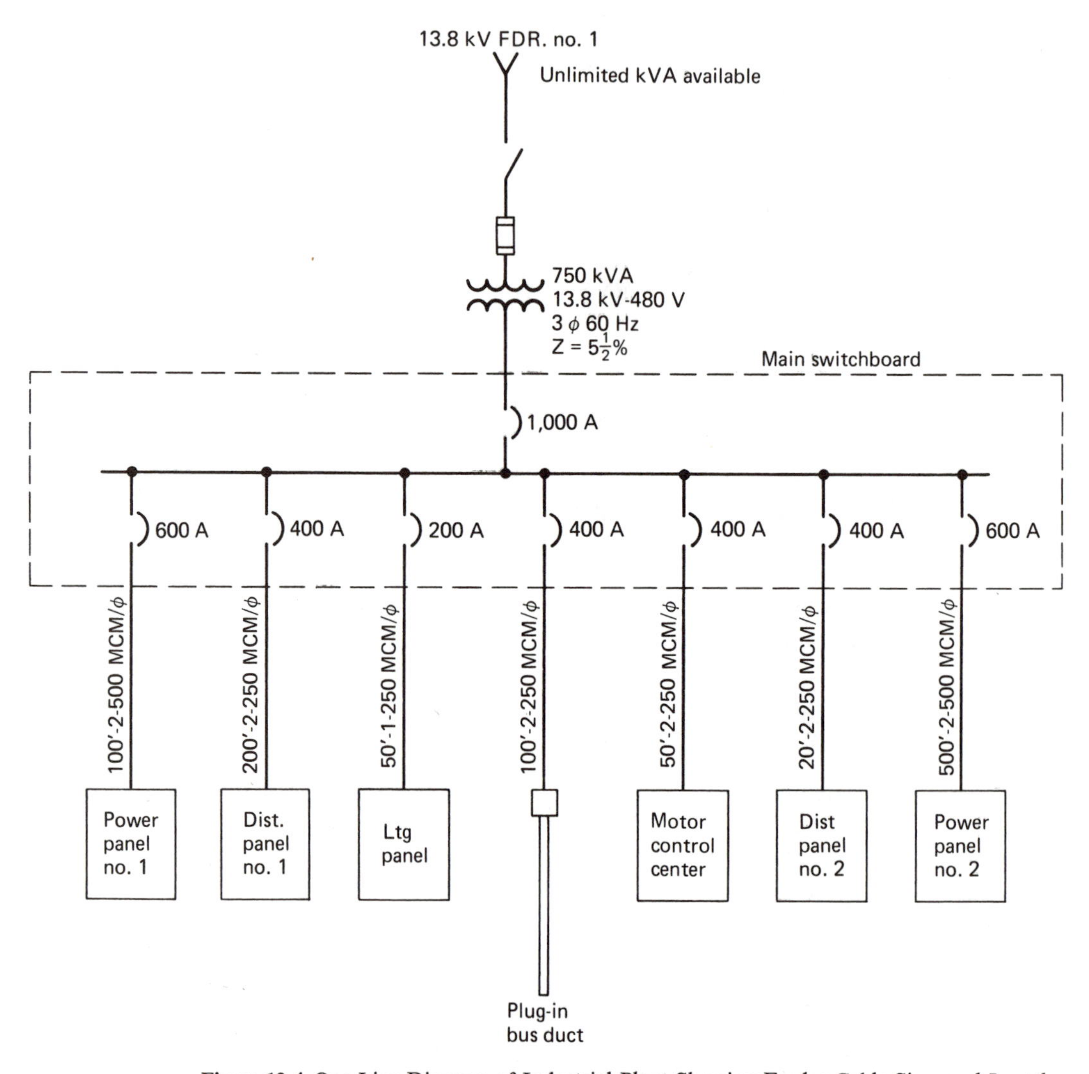

Figure 12-4 One-Line Diagram of Industrial Plant Showing Feeder Cable Sizes and Lengths to Determine Available Fault Current at Distribution Points

mation may need to be reviewed and if it is down in black and white, it can be easily checked.

From the previous example, it was calculated that the available fault current at the secondary terminals of the transformer is approximately 25,000 A, asymmetrical. The main secondary circuit breaker will be exposed to this current if it is located close to the transformer as is often the case. Therefore, for all practical purposes the main switchboard bus will also *see* a possible 25,000-A fault. But use of Table 9 shows the symmetrical rms fault currents at the points of distribution are as follows:

	Symmetrical (rms amperes)
Power panel No. 1	18,200
Distribution panel No. 1	15,000
Lighting panel	17,500
Plug-in bus duct	17,500
Motor control center	19,000
Distribution panel No. 2	19,700
Power panel No. 2	12,000

12-8 Overcurrent Protective Devices

Overcurrent protective devices are the elements that have been devised to protect the electrical system from damage by overload and short-circuit currents. For this reason, it is obvious that these devices perform an extremely important function. The National Electrical Code states their purpose as follows: "Overcurrent protection for conductors and equipment is provided for the purpose of opening the electric circuit if the current reaches a value which will cause an excessive or dangerous temperature in the conductor or conductor insulation."

Almost all electrical circuits must have overcurrent protection in some form. In some rare instances, circuits are designed without overcurrent devices because the risk of having a loss of power through a tripped circuit breaker or blown fuse is greater than the overcurrent effects on the circuit. Such a case is in the trip circuit for power circuit breakers. The source of power is usually from a battery and the assurance of tripping power when needed is mandatory (see Chapter 5)

Overcurrent protective devices must fulfill the following general requirements:

1. Be completely automatic
2. Carry normal currents without interruption
3. Interrupt overcurrents immediately
4. Be replaceable or easily reset
5. Be safe under both normal and overcurrent conditions

To meet the requirements for short-circuit protection fully and completely, an overcurrent protective device should fulfill the following basic specifications:

1. It should be capable of being safely closed in on any load current or short-circuit current within the momentary rating of the device.
2. It should safely open any current that may flow through it up to the interrupting rating of the device.

3. It should automatically interrupt the flow of abnormal current up to the interrupting rating of the device.

There are two fundamental devices that are commonly used for or have short-circuit protection as one of their functions. These are

1. Circuit breakers
2. Fuses

12-9 Circuit Breakers for Systems above 600 V

Circuit breakers for systems above 600 V divide into four basic groups:

1. Air breakers
2. Vacuum breakers
3. Oil breakers
4. Gas-filled breakers

All these breakers operate in conjunction with protective relays (see Sec. 4-10) in order to fulfill the requirement for automatic operation. They are not generally supplied with any overcurrent sensing device built into the breaker itself but rely upon other devices in the circuit to initiate opening of the breaker under abnormal conditions.

Air Breakers

Air breakers tend to be favored for indoor installations and as their name indicates, they break contact in air only. These breakers are also being used extensively in outdoor installations where the switchgear mechanisms, controls, etc., are enclosed in a weather-tight housing. These breakers are built by major manufacturers to operate generally at voltages from 2,400 to 34,500 V. Manufacturers have developed many innovations in their designs in order to make these breakers perform reliably and without harm to the equipment or danger to personnel. Of course, the entire assembly is totally enclosed in a metal housing so that all live parts are inaccessible to operators. They have designed some equipment so that the operating mechanism is retractable from its normal operating position and can be moved on wheels away from the enclosure for periodic testing and servicing. A cross-section view of a typical, air blast, air circuit breaker mechanism and description of its operation are shown in Fig. 12-5.

The illustration in Fig. 12-5 shows one pole of the breaker. Air from a compressed air source is used either to close or to open the moving contact blade into or away from the fixed contact. This function is electrically controlled through the magnetic air valves. When the contact blade opens under electrical load, an arc will start to form between the moving blade and the fixed contact.

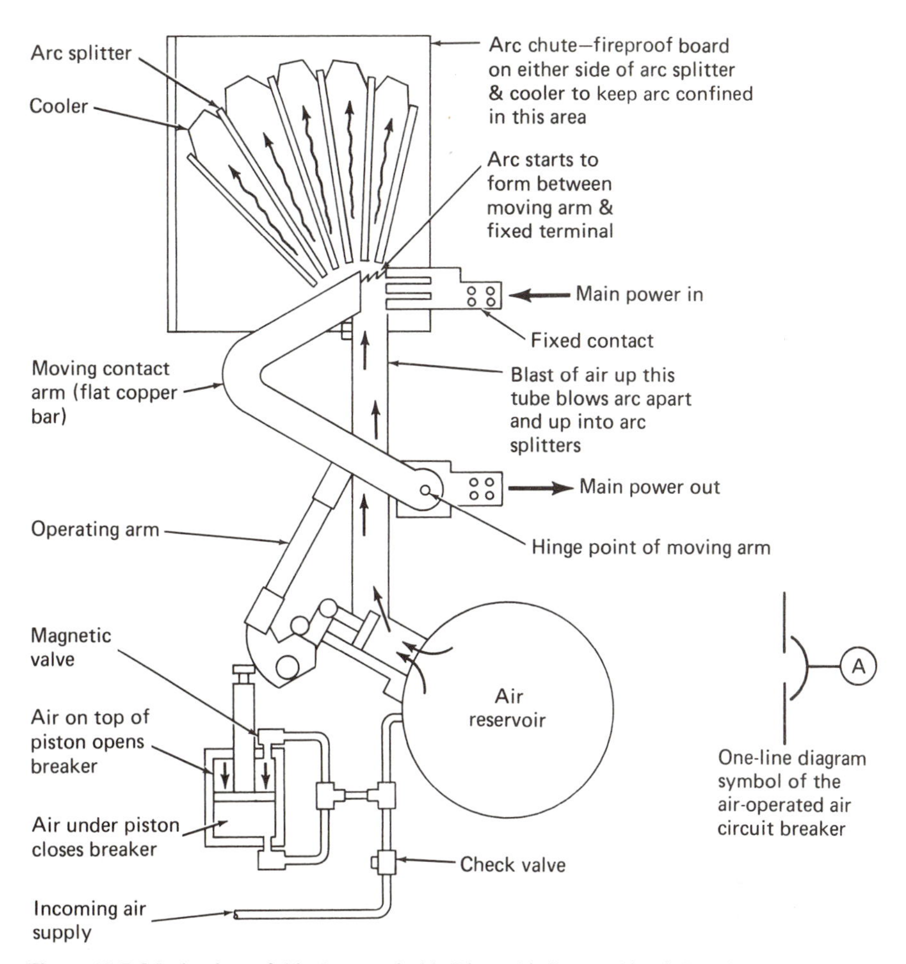

Figure 12-5 Mechanism of Air-Operated, Air Blast, Air Power Circuit Breaker

To prevent damage from burning, a blast of air is released at precisely the right moment to blow the arc upward and force it to break by cutting it into pieces through the splitters and then cooling the arc and gases on the metal surfaces of the coolers. The arc chutes contain the flame and gases from spreading to surrounding surfaces.

Vacuum Circuit Breakers

The best conductors of electricity are those materials that offer the most free electrons and, conversely, the best insulators or dielectrics offer the least free electrons. Because a vacuum constitutes an absence of any substance and therefore the absence of electrons, in theory it represents the best dielectric.

Based on this theory there would be great advantages to be realized if mechanically operated electric contacts were opened in a vacuum chamber.

Major manufacturers have been able to build such devices for use on high-voltage service. Among the advantages are faster arc extinguishing, less noise and impact from the breaker operation, longer contact life, and the elimination of potentially explosive gases or liquids. Maintenance on these breakers is reduced and because the mechanism is little affected by ambient temperature or other atmospheric conditions, they can be applied in most locations.

The breaker element, shown in Fig. 12-6, is simple in construction. A pair of disk contacts is mounted inside a cylindrical insulator housing. The chamber is evacuated to provide a vacuum. One contact is fixed and the other is arranged to move against the fixed contact, or away from it, by a steel bellows assembly that is controlled from the outside. The distance between contacts can be as little as approximately $\frac{3}{4}$ in. because of the absence of electrons in the vacuum. The contacts are made from metal material that is as gas-free as possible; however, there is some minor arcing so collector rings are provided to accumulate the metallic arc by-products.

Oil Circuit Breakers

Oil circuit breakers can also be used indoors but they have been largely superseded by the air breakers for indoor installations. Oil breakers are used extensively for systems above 14,400 V in outdoor installations. These breakers are basically an oil reservoir into which the contact making and breaking mechanism is installed so that the arc that forms is extinguished under oil. There are several different ingenious designs that have been developed by the manufac-

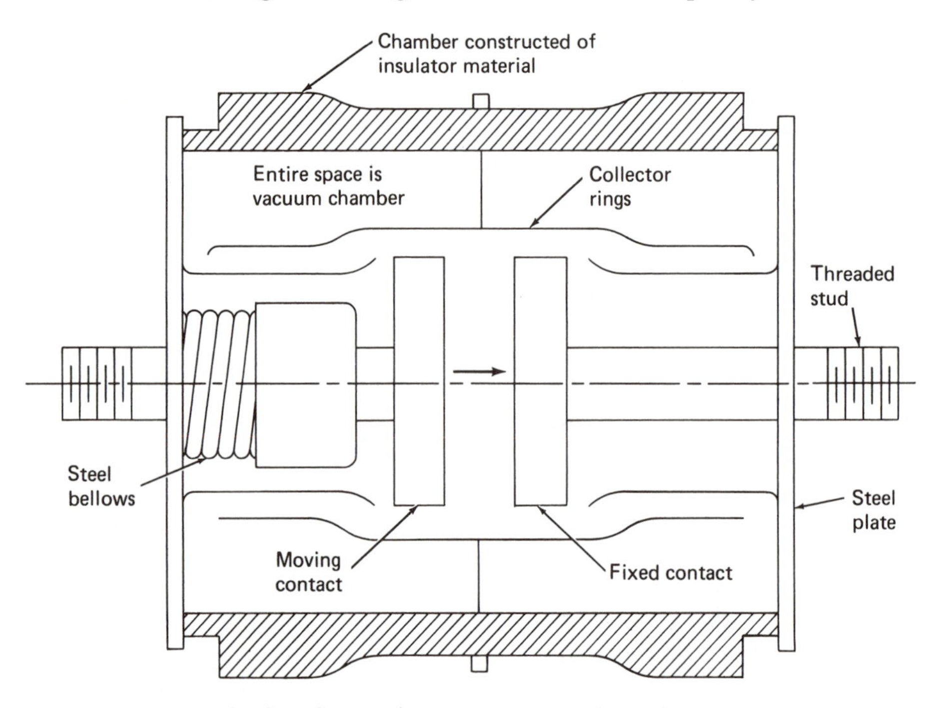

Figure 12-6 Cross Section of One Pole of a Vacuum Circuit Breaker

turers to minimize the harmful burning effects of the arc. Descriptive data on these breakers are published telling how the arc is controlled. Breakers of this type range from 2,400 to 362,000 V. At voltages up to 69 kV, all three poles of a three-phase breaker may be enclosed in the same tank, although a separate tank for each pole may be preferred and is available. At the higher voltages, separate tanks are almost always used for each pole. A cross section through a typical oil breaker is shown in Fig. 12-7.

The oil circuit breaker shown in Fig. 12-7 indicates in a simplified way the general design idea behind oil circuit breakers. The moving contact will vary in shape as will the nature of the fixed contacts depending on the manufacturer and, of course, on the voltage class, etc. In all cases, however, a moving contact makes and breaks the circuit connection while immersed in oil, which aids in minimizing damage to the contacts.

Gas-Filled Circuit Breakers

Gas-filled circuit breakers are also used on the higher voltages and particularly on the extra high voltages (EHV) up to 765,000 V. They use an inert gas in the *interrupter modules*, which are the gas-filled chambers where the making

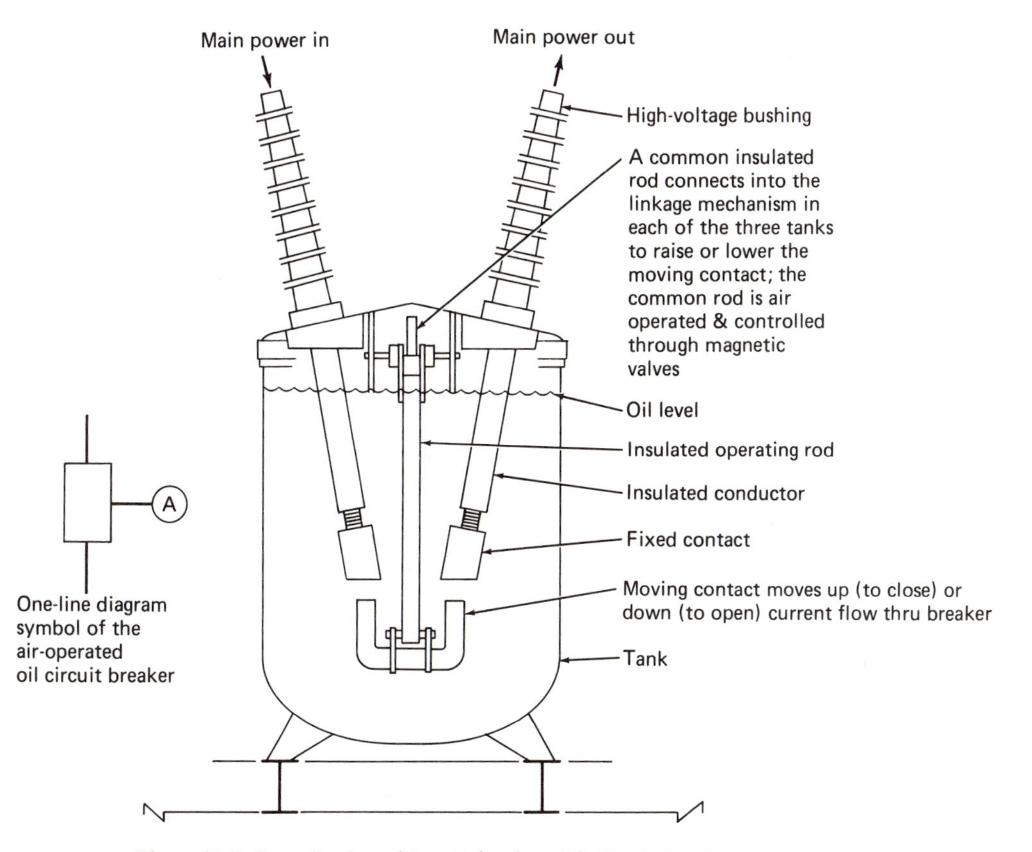

Figure 12-7 Cross Section of One Pole of an Oil Circuit Breaker

and breaking of the contacts takes place. Two or three modules (three on 765 kV) are connected in series in each pole of the breaker so that the circuit is opened simultaneously at more than one point. The particular gas used is chosen for its inert and insulating qualities. These circuit breakers are enormous in size with overall heights from the ground ranging between 20 and 30 ft. At 765 kV, the space between breaker poles may be as much as 35 ft! Major manufacturers can provide more descriptive data on this class of breaker which has limited use because EHV transmission is small compared to transmission at lower voltages which use the air and oil-circuit breakers extensively.

12-10 Fuses for Systems Above 600 V

There are many types of power fuses available for circuits rated 2,400 V and above. They divide into four major categories as follows:

1. Current-limiting power fuse
2. Noncurrent-limiting power fuse
3. Oil fuse cutout
4. Distribution-type fuse for use in fuse cutouts

Current-Limiting Power Fuse

Current-limiting power fuses are designed to blow before the short-circuit current has time to reach its peak value. They therefore *limit the current* to safe levels for equipment and devices being protected. Briefly, this is accomplished by using a slender silver fuse link that connects at either end of the fuse into heavy ferrule pieces. The slender silver link can carry normal load currents through its small diameter because the heat that is generated is immediately absorbed by the heavy ferrules that are actually *heat sinks.* Abnormal currents, however, instantaneously melt the silver link. The link is enclosed in an insulating tube similar to most cartridge fuses; however, for the current-limiting fuse the tube is filled with quartz sand. When the silver link melts, the heat also melts some of the surrounding sand, which becomes glass and therefore provides insulation inside the tube. The current-limiting effect of this action is shown diagrammatically in Fig. 12-8.

Examination of the curves of Fig. 12-8 shows that melting of the fuse link as the heavy short-circuit current starts to build is very rapid and before the full peak is reached, the current flow has been interrupted. These fuses are not necessarily a cure-all for the short-circuit problem. They have limitations and can only be used in accordance with their capabilities. Manufacturers provide application data that must be followed for proper circuit design.

Noncurrent-Limiting Power Fuses

Noncurrent-limiting power fuses are similar to the commonly used cartridge fuse employed extensively on systems under 600 V. Basically they are constructed of an insulating tube with ferrule ends and a fuse link assembly

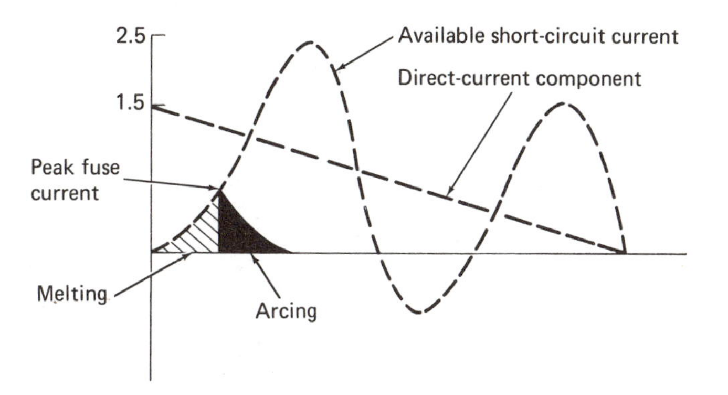

Figure 12-8 Current-Limiting Characteristic of Current-Limiting Fuses

stretched between the two ferrule contacts to form the conducting path. Some are oil filled; others are boric acid filled. Manufacturers have developed various types for different voltages, atmospheric conditions, etc. Some are made as expulsion type, which means that they expel hot gases when they operate. These latter fuses and others of a similar type are not suitable for indoor applications due to the hazard of the hot gases.

Oil Fuse Cutouts

Oil fuse cutouts are a combination of a mechanical switching device called a *cutout* and a fuse immersed in a container of oil. As such, they can be used as a switching device and a protective device because of the fuse element. The container housing the fuse holds a relatively small amount of oil so this type is often used without fireproof enclosures in buildings as protection for dry transformers and transformers using nonflammable liquid coolants. A cross-section view of a typical oil fuse cutout is shown in Fig. 12-9.

It can be seen from the figure that the spring contacts on the insulated shaft make or break contact with the contact studs while immersed in oil. If the fuse link melts due to an overload or short circuit, any arcing takes place under oil. Pipe-type chambers (not shown) provide for the expansion of gases that develop. The entire assembly is relatively small and compact. These units are ganged together on a common support and the operating rods are linked to make the complete assembled mechanism a three-pole cutout. Sizes and application data is published by manufacturers of these devices that will provide more detailed information.

Distribution-type Fuses

Distribution-type fuses for use in fuse cutouts on open-wire, outdoor distribution systems are used extensively by utility companies. They have rather limited use in industrial plant systems because they are not metal enclosed and therefore cannot be readily used indoors.

These fuse cutouts are made in several forms. Most employ a porcelain insulator housing or support for the fuse element. All these cutout units are

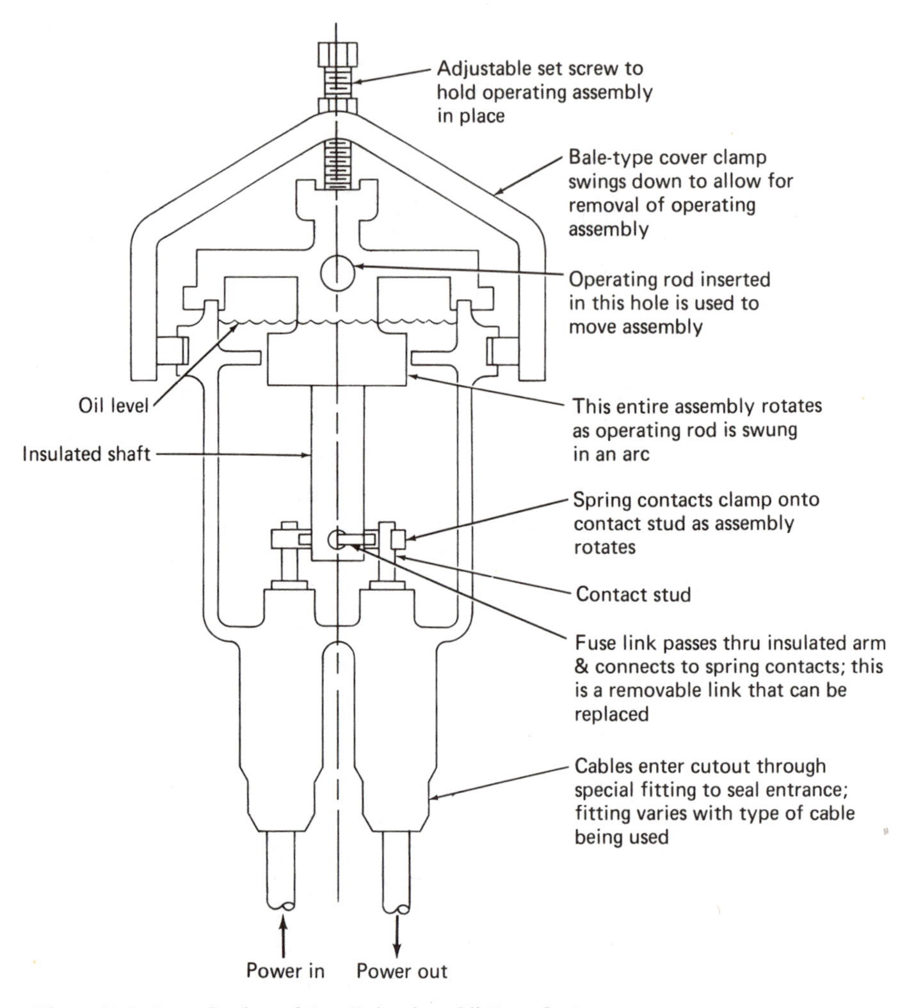

Figure 12-9 Cross Section of One Pole of an Oil Fuse Cutout

designed for replacement of the fuse link that is generally contained inside an insulated tube. (Some distribution cutouts are made with the fuse link connected in the opening between spring-loaded terminals; no tube is used at all and a blown fuse can be readily spotted from the ground.) The fuse link is often made with a button-like top on one end and a stranded wire *tail* on the other end of the fuse link. Some holders are made with spring-loaded base terminals so that if the fuse blows, the spring tension is released and the fuse is moved out of its normal position or the spring action releases a flag device to indicate to the lineman on the ground which fuse has blown. There are others so ingeniously made that the blowing of one fuse with its spring-loaded holder provides enough energy to *throw* a standby fuse into position and if that blows still another fuse may be closed in through the spring action. These are particularly good on rural lines where a momentary fault occurs because of some storm effects. The

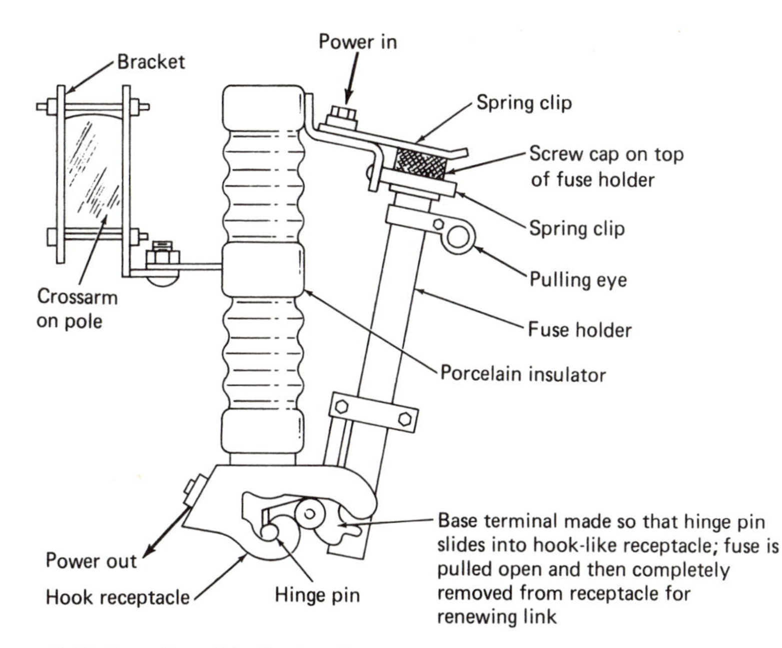

Figure 12-10 Open-Type Distribution Cutout

automatic reclosing feature will save sending a line crew out and may very well maintain service that otherwise would be lost for an hour or more. One type of standard distribution cutout is shown in Fig. 12-10.

12-11 Low-Voltage Systems (600 V and Less)

For low-voltage systems there are three basic short-circuit devices:

1. Circuit breakers
2. Fuses
3. Fuse–circuit breaker combinations

Circuit breakers divide into two basic groups:

1. Air circuit breakers
2. Molded-case circuit breakers

Air Circuit Breakers

Air circuit breakers are generally used as protection for main power feeders. These breakers usually consist of an operating mechanism, contacts, arc interrupters, and a built-in overcurrent tripping device that is connected in series with the load-side breaker conductor. These breakers are characterized by their sturdy construction and are available in high-current-carrying and interrupting ratings. The built-in tripping devices are designed to give accurate tripping protection because they are precisely made and carefully tested for quality. They can be furnished with various characteristics: long time delay, short time delay, and instantaneous tripping so that the protection provided will function properly with the other elements in the system. A diagrammatic sketch of this type of breaker is shown in Fig. 12-11.

The operating mechanism may take several forms. If the breaker is manually operated only, then the handle is used to open or close the breaker-moving contact. This is usually accomplished with a spring-operated device so that the opening and closing action is fast. If the breaker is electrically operated, the mechanism may consist of a magnetic coil with a shaft *plunger* set to move through the center of the coil so that when the coil is energized, the plunger is pulled into the center of the coil and, in so doing, closes the breaker through the linkage arrangement. This device is called a *solenoid* operator. Other operating mechanisms utilize a *stored energy* principle wherein an electric motor winds a spring during both the closing and tripping cycles. The energy is thereby *stored* in the spring and when it is released, it either closes or trips the breaker. (When the breaker trips, the motor is energized to wind the spring for the next closing and, conversely, when the breaker closes, the motor is energized to wind the spring for the next tripping; the spring, not the motor, provides the energy for each closing and tripping.)

The series overcurrent device trips the breaker by direct mechanical action in response to the magnetic force created by the current in the circuit. Almost all of these devices are adjustable by means of movable settings on scaled markings on the face of the tripping unit. Therefore, if currents in excess of the settings flow through the series coil, the magnetic forces that develop will cause the operating linkage between the series trip device and the operating mechanism to move and thereby trip the breaker open.

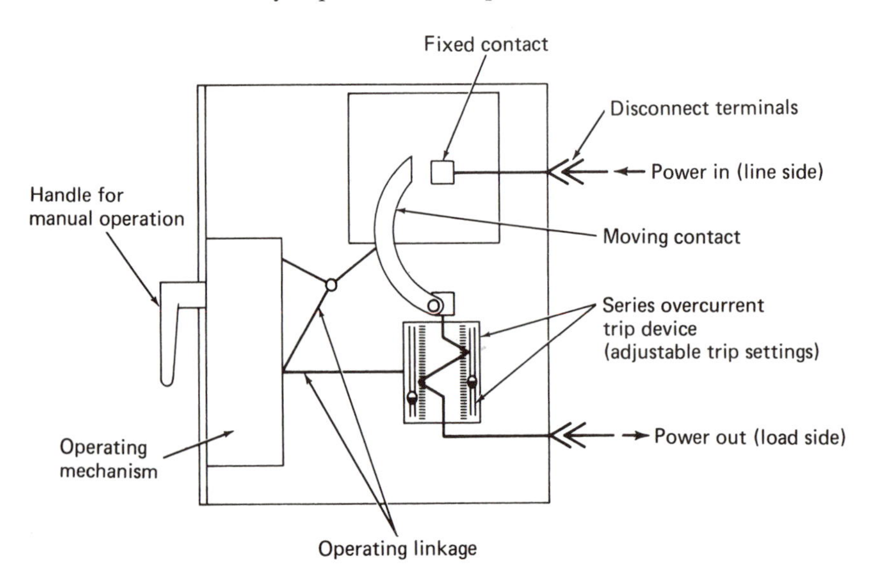

Figure 12-11 Air Circuit Breaker Arrangement

This class of breaker is normally furnished as the drawout type, which means that when the breaker is open (no power flow), it can be drawn out from the housing that supports it much like a suitcase in a luggage locker. This allows for servicing the breaker in a safe way. The disconnect terminal shown in Fig. 12-11 provides for removal and for plugging the breaker back in place.

Breakers of this type generally have a range of continuous current ratings from 15 to 4,000 A and interrupting ratings from 15,000 to 150,000 A, asymmetrical.

Molded-Case Circuit Breakers

Molded-case circuit breakers are often used as protection for secondary feeders and branch circuits. As their name indicates, the breaker mechanism is enclosed in a molded insulated encasement. Being essentially a high interrupting capacity switch with resettable elements that provides repetitive operation, molded-case breakers are comprised of three main functional components. These are trip elements, operating mechanisms, and arc extinguishers. All these elements are generally sealed into the case. The trip settings are often factory set and therefore cannot be changed although some breakers do have adjustable settings that can be changed after they are installed. The trip elements vary in their characteristics but their function is tripping the operating mechanism in the event of a prolonged overload or short-circuit current. Multipole breakers are constructed so that tripping action on one pole will trip all poles.

Thermal trip action is accomplished through the use of a bimetallic strip that is heated by the current flow. The strip is made of two different metals bonded together. The lengths of these strips will increase with a rise in temperature caused by excess current flow. Because the two metals do not increase in length equally, the strip bends more and more until it causes the operating mechanism to work and open the breaker contacts. Because the bimetallic element is responsive to heat emitted by the current flow, it allows a long time delay before tripping on light overloads; yet it has a fast response on heavier overloads. Figure 12-12 shows the principle of operation of this breaker type.

Magnetic trip action is achieved through the use of an electromagnet in series with the load current. This provides an instantaneous tripping action

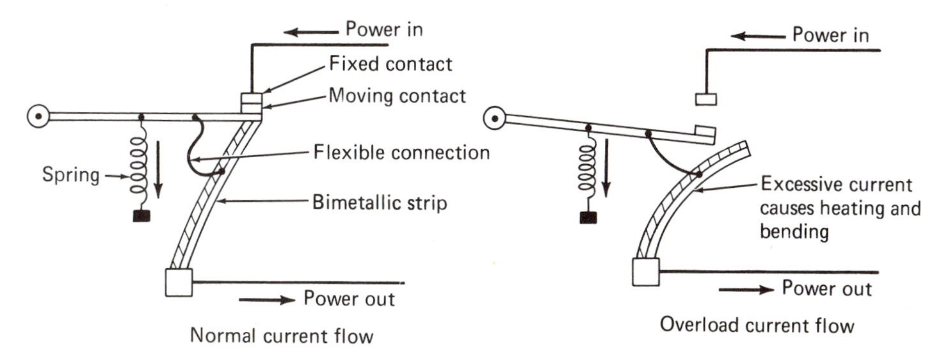

Figure 12-12 Action of Circuit Breaker with Thermal Trip

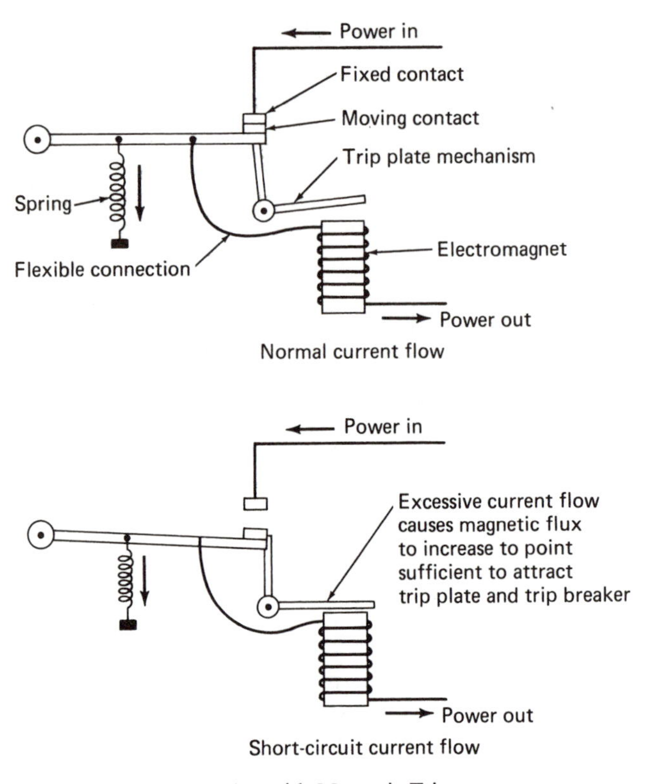

Figure 12-13 Action of Circuit Breaker with Magnetic Trip

when the current reaches a predetermined value. Figure 12-13 shows the principle of operation for this breaker type. Note that the current flow is through the coil of the magnet. The current rating of the breaker is determined by the number of turns and wire size of the coil.

These magnets are often hermetically sealed and have a nonmagnetic tube containing a movable iron core immersed in a fluid. Under normal load the current does not develop enough flux in the magnet to move the iron core up through the fluid and thereby reduce that air gap to attract the trip plate. If short-circuit currents occur, however, the magnet reacts instantaneously to trip the breaker. This breaker type with only the magnetic element is applied where only short-circuit protection is required.

Thermal–magnetic trip actions are general-purpose devices that are suitable for the majority of breaker applications and therefore are often thought of as the standard-type breaker. Combining thermal and magnetic trip elements, they provide accurate overload and short-circuit protection for conductors and connected apparatus. A thermal–magnetic breaker will react to normal overloads in much the same manner as the thermal breaker through action of the bimetallic strips. On heavy overloads or especially on short circuits the magnetic action will *beat* the thermal action and trip the breaker instantaneously. The combination element is constructed of a thermal bimetallic strip to which a magnetic plate has been added. Below heavy short-circuit currents, the magnetic forces are so slight that they are not a factor in the operation of the release linkage. At the lower overload current levels, the thermal element only operates the release.

At heavy short-circuit current levels, the magnetic plate element produces so much force that the release operates immediately. Figure 12-14 illustrates the elementary operation principles.

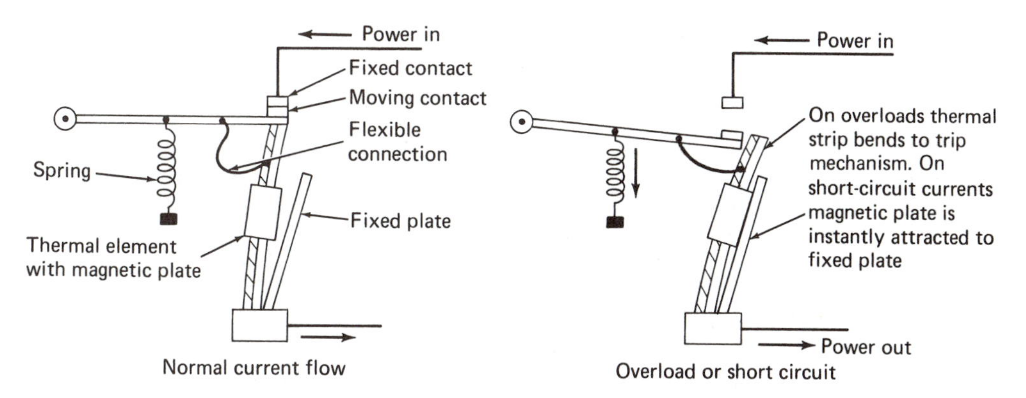

Figure 12-14 Action of Circuit Breaker with Thermal–Magnetic Trip

Molded-case circuit breakers are rated in rms amperes (at a specific ambient temperature), voltage, frequency (usually 60 Hz), and interrupting capacity (in rms symmetrical and asymmetrical amperes). They are generally rated from 15 to 2,500 A; continuous and interrupting ratings range from approximately 5,000 rms amperes symmetrical at 120/240 V acto 65,000 A at 240 V. In addition to their electrical rating, molded-case circuit breakers are also classified by *frame size,* which categorizes the physical dimensions into groups. Breakers of the same manufacture and frame size are physically interchangeable. Naturally, different makes, models, etc., vary in rating. Manufacturers publish specific data on all breaker types.

Fuses

A fuse is an overcurrent protective device with a circuit-opening fusible member that is directly heated and severed by the passage of an overcurrent through it (A.N.S.I. definition).

Fuses are simple in construction, rugged, and relatively inexpensive. All modern fuses are made in such a way that the fuse element itself is totally enclosed. These enclosures take different shapes depending on the size and rating of the fuse. Figure 12-15 illustrates some types. Plug fuses are made in sizes up to and including 30 A for use on circuits not exceeding 125 V. The ferrule contact cartridge fuse is made in sizes from 35 A up through 60 A. The length and diameter of the enclosure varies with the voltage class. Knife blade contact cartridge fuses are made in sizes from 70 to 6,000 A. These too vary in physical dimensions depending on the voltage rating.

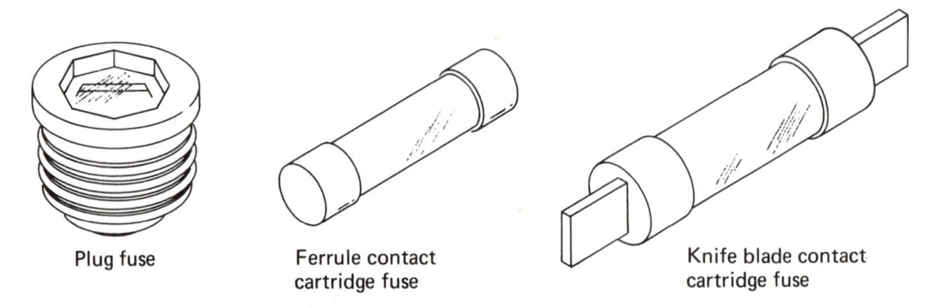

Figure 12-15 Types of Enclosures for Low-Voltage Fuses

Available ratings are published by manufacturers. Standard ratings for both 250- and 600-V fuses are 15, 20, 25, 30, 35, 40, 45, 50, 60, 70, 80, 90, 100, 110, 125, 150, 175, 200, 225, 250, 300, 350, 400, 450, 500, 600, 700, 800, 1,000, 1,200, 1,600, 2,000, 2,500, 3,000, 4,000, 5,000 and 6,000 A.

The cartridge-type fuses are made either as *one-time* fuses (they are thrown away if they blow) or as a renewable link type. The ferrules on the renewable link type are made to unscrew from the insulating tube so that the burned out link can be removed and a new link put in its place. These are secured by various means depending on the make of fuse. Figure 12-16 shows a cutaway section of a renewable link fuse.

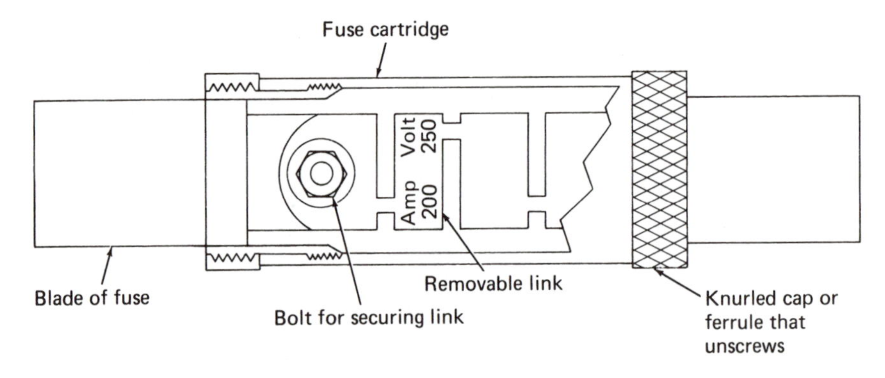

Figure 12-16 Renewable Link-Type Fuse

Time delay in blowing is a desirable condition for some fuse applications. All fuses have some inherent time delay but specific makes are designed to provide delay in their operation. These fuses are designed to blow instantaneously on short-circuit currents but on overloads such as motor starting in-rush currents they will ride through without blowing in order to prevent unnecessary outages. These fuses have compound fusible members. One part provides fast operation in the range of short-circuit currents and a second part provides the time delay in the range of most overcurrents. Construction of the fuse elements for time delay fuses are shown in Fig. 12-17.

On short circuits this fuse type blows just as in the case of other types. On

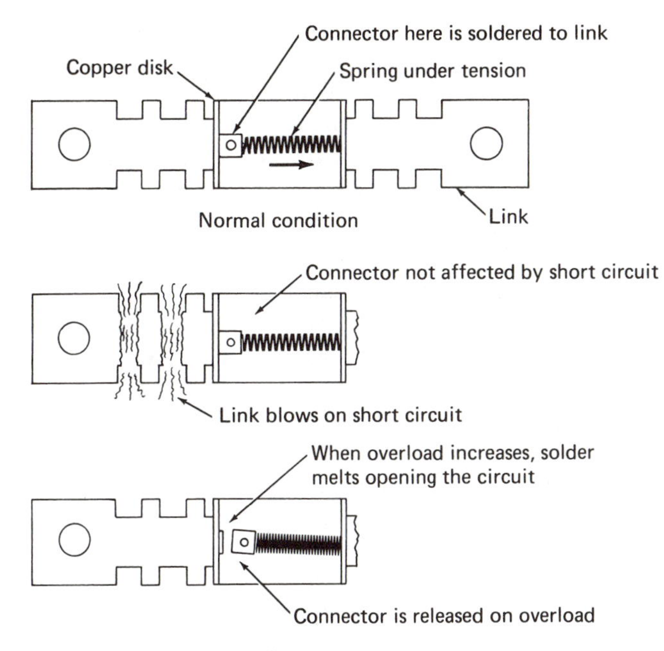

Figure 12-17 Time Delay Fuse Operation

an overload, the fuse link remains inactive. The heat generated by an overload is fed to the center mass of the copper disk on which is mounted the connector and the short spring. The connector is held in place by low-melting-point solder. The connector and spring connect the two link pieces together electrically. When the temperature is increased by the overload to such an extent as to melt the solder (approximately 280°F), this connector is pulled out of place by the spring, thereby opening the circuit. A comparison of time delay fuse operation versus ordinary fuses is graphically illustrated in Fig. 12-18.

It has been previously noted that one of the greatest hazards in electrical systems is the heavy overcurrent at the level of short circuits. This is particularly

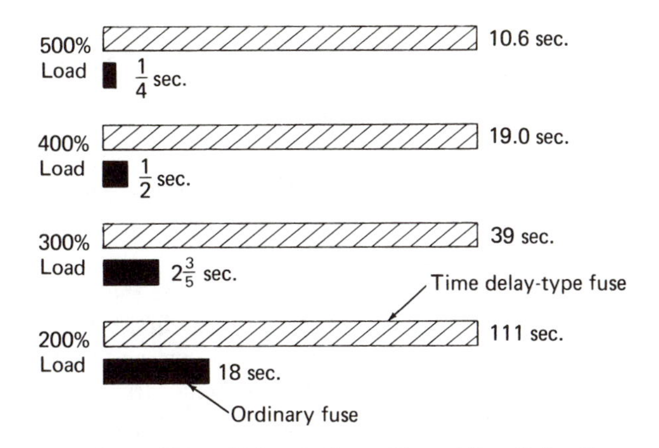

Figure 12-18 Comparison of Time Delay–Ordinary Versus Time Delay Fuses

true in the voltages below 600 V, either alternating or direct current. Utility companies have been increasing their generating capacity year after year at a very high rate. The available fault currents from these systems has therefore increased to very high levels. It is not unusual to have short circuits involving 50,000 kVA of three-phase energy. At 15,000 V, the current per phase would be approximately 2,000 A; at 460 V, the current would be 63,000 A per phase; at 208 V, it would be 140,000 A per phase. It is these high currents that cause the terrific burning at a fault and it is also these amperes that generate the high magnetic flux around conductors resulting in the tremendous forces that tear electrical equipment apart.

Ordinary one-time fuses are not capable of withstanding the heavy currents mentioned above. These fuses have a tested interrupting capacity rating of 10,000 ac, rms amperes symmetrical. Some newer designs are rated up to 50,000 A rms. The dual element, time delay, one-time cartridge fuse may be rated as 100,000 A rms, ac interrupting capacity. Because of these limitations and in order to make it possible for engineers and designers to use fuses in these high fault current areas of circuits, it became necessary to develop a fuse that could withstand the higher currents.

Manufacturers developed the *current-limiting fuse* for this service. These fuses have high interrupting capacities up to 200,000 A rms, ac. The current-limiting fuse should be distinguished from the so-called high interrupting-capacity fuse. The current-limiting fuse does more than the high interrupting-capacity fuse. The current limiter does what its name says for fuses with a relatively low ratio of available current to rated current of the fuse because it *chops* the short-circuit current off in its rise and the current that is let through before the fuse action is complete is limited to a value below the available current. This was shown in Fig. 12-8, which described this same type of fuse for use on high-voltage systems. A high interrupting-capacity fuse is one that may be safely used in circuits of high available current. In operation it is generally slow compared to the current-limiting fuse and it may allow several half cycles of current through before actually interrupting the circuit. The current-limiting fuse tends to minimize damage to *downstream* devices in the system due to mechanical and thermal stresses from high fault currents. The high interrupting-capacity fuse does not provide this added protection; it simply provides a fuse for use on high-current systems.

The development of the current-limiting fuses is interesting. It was found that European manufacturers of high-voltage oil circuit breakers used contacts made of spheres of solid copper. The actual point of contact of the spheres was limited to the small area where the two spheres touched, as shown in Fig. 12-19.

The startling observation was made that the small contact area could carry 600 A continuously in the oil-filled breaker tank without undue temperature rise. How could it do that? It was learned that the copper spheres were functioning as *heat sinks* and continually drawing the heat away from the contact point. The heat in the spheres was conducted away to the oil that in turn was being cooled at the surface of the tank. This showed that the heat absorbtion

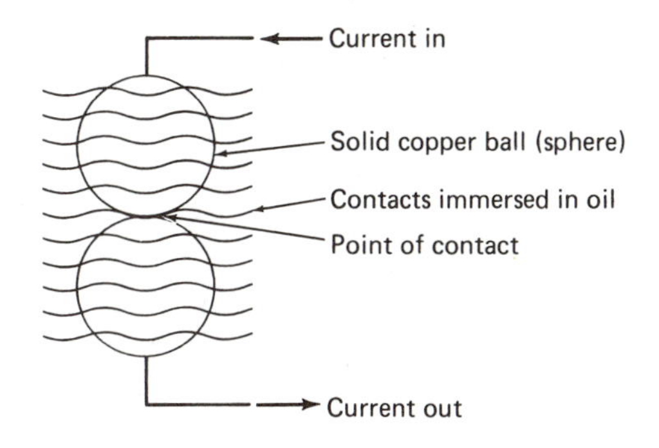

Figure 12-19 Contacts of Oil Breaker Using Copper Spheres in Oil

of a mass of solid copper is very high. To develop this idea, the spheres were separated by about $\frac{1}{32}''$ and connected together with a piece of silver wire having a diameter of 0.030″ (30/1,000″).

This was tested and it was learned that the tiny silver wire would conduct 600 A without undue heating! Why? Because of the same principle as explained for the breaker.

It was further learned that if 20,000 A were suddenly thrust on the minutely small silver wire, it disappeared instantly! This then proved to be a current limiter. A device that would carry normal load currents safely and interrupt abnormal short-circuit currents instantly. This is where a current limiter departs from standard fuse designs. The link in most standard fuses is made of zinc and must have enough mass to carry the current over a long link, as in 250- or 600-V fuses. This large mass cannot be instantaneously volatilized like the minute element of the current limiter can.

Speed is therefore the significant factor between standard fuses and current-limiting fuses. In the latter, the fault currents are interrupted before they can build up to a dangerous level. In the standard fuse, the additional time required causes additional heating, build up of internal pressures, and possible explosive results.

The current-limiting fuse is made in several forms for different voltage and current ratings. The general idea in its construction is shown in Fig. 12-20.

Fuses have some disadvantages. Those commonly listed are

1. Single phasing on the three-phase circuit with only one fuse blown and possible damage to electrical equipment through overheating.

2. Fuses of the same type and rating are not always used as replacement for blown fuses and therefore adequate protection is lost.

3. Surrounding (ambient) air temperatures tend to affect fuses more than circuit breakers. The fuse is a thermal device; therefore operation tends to be subject to changes in the ambient temperature.

4. Fuses cannot be reset and readily restore the same protection like a circuit breaker can.

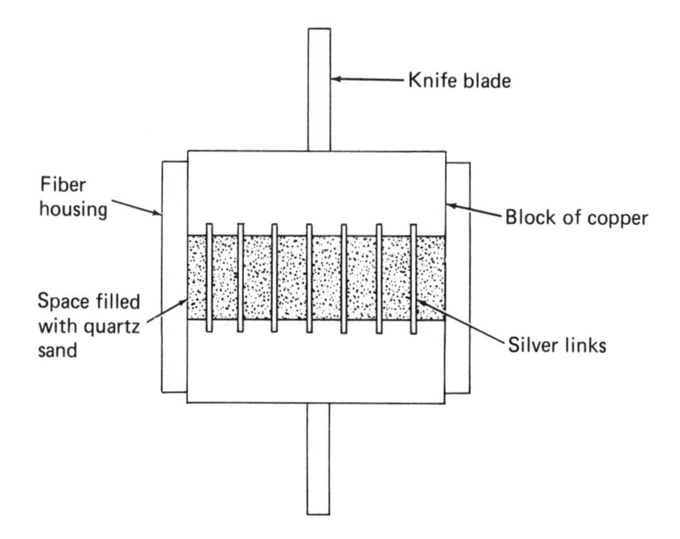

Figure 12-20 Section through Current-Limiting Fuse

5. Fuses do not have adjustable time-current characteristics and therefore are not as easily coordinated in a total system design.

The two outstanding advantages of fuses are their first cost and their speed of operation on high overcurrents. Cost is very often the deciding factor in choosing fuses rather than breakers. This is a judgment decision that will vary with the conditions and the individual's preference.

Because there are some disadvantages associated with fuses and because molded-case circuit breakers may not have sufficient interrupting capacity for some applications, the combination circuit breaker and current-limiting fuse device has been developed.

Combination Circuit Breaker and Current-Limiting Fuses

By combining current-limiting fuses with molded-case circuit breakers, the interrupting rating of the circuit breaker has been increased over the breaker rating alone. The circuit breaker still performs all its functions within its rating and the fuse detects and interrupts short-circuit currents that are above the rating of the breaker. Of course, the two devices must be properly matched to produce the desired results.

This device is useful for the following reasons:

1. It is lower in cost than a fully rated large air circuit breaker.
2. It can be used on systems where the available fault current may be as high as 200,000 A but where continuous current rating requirements are low.
3. It is a fully coordinated device. The current-limiting fuse element will not blow on overloads or low-level fault currents but will operate when the current reaches higher levels.

4. The problem of single phasing is eliminated because breaker action trips all poles and fuse action is also made to trip the breaker.

This type of protective device looks very much like a molded-case breaker except for the added extension of the space for the current-limiting fuses that are usually located on the bottom of the unit.

The current limiters are designed with a pop-up plunger that acts to trip the breaker if the limiter operates.

12-12 Solid State Devices for System Protection

The molded-case circuit breaker has been used as the basic breaker to which a solid state tripping device has been added to provide a relatively new line of breakers for system protection. The solid state circuitry allows for field changes of the tripping characteristic to an almost infinite degree. This versatility allows for changes in the continuous current rating as well as in the time delay characteristics. Literature is available from manufacturers that outlines in detail the many possible applications for these breakers. These breakers have high interrupting ratings (up to 200,000 A at 460 V). They are not current limiting in the way that the fuse-circuit breaker combination is.

Chapter Review Topics and Exercises

12-1 Overcurrents are currents greater than what?

12-2 How do short circuits differ from overloads?

12-3 Describe the possible damages a short circuit can do.

12-4 Why is careful design to limit the possibility of short circuits so important in modern design?

12-5 How does peak current compare to rms current?

12-6 Sketch an illustration of a symmetrical current.

12-7 Sketch an asymmetrical sine wave and indicate the amount of offset.

12-8 Why are short-circuit currents reduced if they occur further away (on the circuit) from the transformer?

12-9 What are overcurrent protective devices used for?

12-10 What does the air blast do in a high-voltage air circuit breaker?

12-11 How does an oil circuit breaker work?

12-12 How does a high-voltage current-limiting fuse operate?

12-13 What is an oil fuse cutout and how does it operate?

12-14 What piece of electrical equipment would a distribution-type fuse cutout probably protect?

12-15 Describe the function and operation of the series trip coil in an air circuit breaker.

12-16 Explain thermal tripping.

12-17 How does a magnetic trip device work?

12-18 Where are magnetic trip breakers used primarily?

12-19 The thermal–magnetic trip device finds common usage on many circuits. Why?

12-20 Why are 600-V class fuses longer than 250-V class fuses for the same current rating?

12-21 What is the advantage of a renewable link fuse? What could be a decided disadvantage?

12-22 Describe the construction of a time delay fuse and how it operates.

12-23 Describe a current-limiting fuse and why large currents can be conducted by a slender silver wire.

12-24 Why was the combination circuit breaker and current-limiting fuse device developed? State its principle advantage.

13

APPLICATION OF SYSTEM PROTECTIVE DEVICES

13-1 General

The detection of faulty conditions and isolation of the trouble as quickly as possible is the job of protective devices. Their proper application depends on many factors involving study and experience.

At the power generation, high-voltage distribution and major equipment protection level, protective relays are applied extensively. These relays are made in many different types for a variety of applications. A brief description and some applications were explained in Chapter 4. Protective relay applications are beyond the scope of this text; however, several publications are available from major electrical manufacturers that outline common applications for those who wish to pursue this further.

On systems of 600 V and less, designers and draftsmen are often required to size protective devices; therefore, this chapter relates to that need. *It is important to remember that the National Electrical Code rules govern most overcurrent protective device applications and any examples used here are hypothetical and intended as a guide only. Direct reference to the code should be made for actual application problems.*

13-2 Basic Objectives in Applying Protective Devices

The ideal electrical power system is a *selective* system. To be selective, the protective devices must be sized and coordinated with each other in such a way as to have only the protective device nearest the fault, on the power source side, operate first. If for any reason it fails to operate, the next device upstream should open the circuit and so forth up to the power source protective device. This operating requirement is illustrated in Fig. 13-1.

To provide selective operation, the designer must be careful to chose devices with the proper interrupting ratings and time–current characteristics to do the job. Fuses have fixed characteristics; i.e., they are adjustable only through changing size or type. Air circuit breakers have different type overcurrent trip

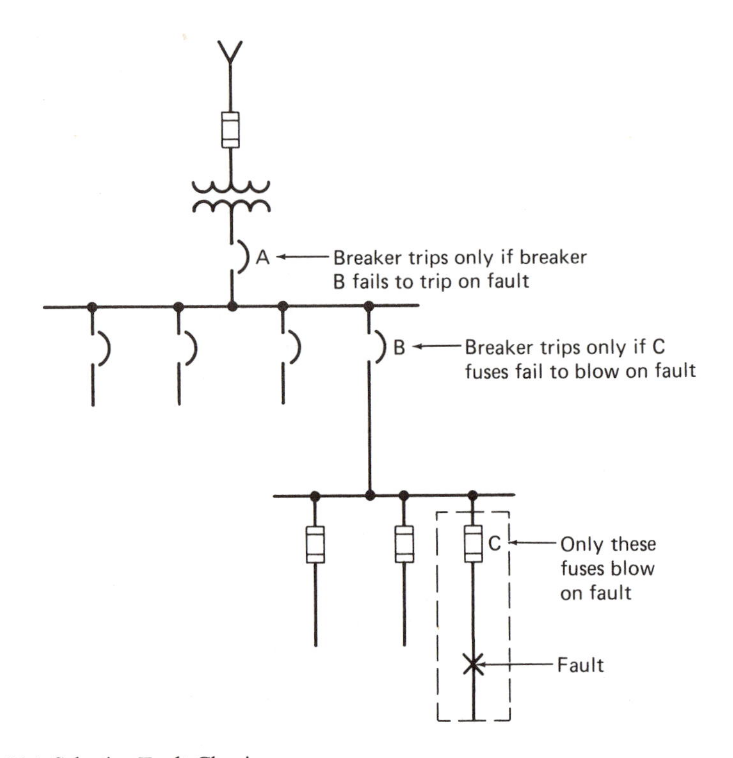

Figure 13-1 Selective Fault Clearing

devices, which are usually adjustable. Molded-case circuit breakers generally have fixed characteristics and are almost impossible to coordinate properly. To prove that devices with varying characteristics will be selective with each other, a coordination study must be done.

13-3 Coordination of Protective Devices

Graphic analysis is the common method used to prove selectivity. This procedure involves plotting the characteristic curves of the overcurrent devices in series with each other to see if any of the curves overlap that would indicate they were not selective.

As an example, the average melting time curves of the three fuses in series in the one line of Fig. 13-2 are plotted on standard log–log time–current characteristic graph paper demonstrating that if a 1000-A fault occurred at the motor terminals, the 80-A fuse in the combination fused disconnect and motor starter would blow in approximately −0.55 sec and thus clear the fault before either the 225 or 600-A fuses blew. This removes only the faulty motor from service and therefore the fault has been cleared selectively.

To plot coordination curves, the curves for each device must be obtanied from manufacturers. Data are usually readily available and the technique that follows for their use allows them to be used over and over again. The most

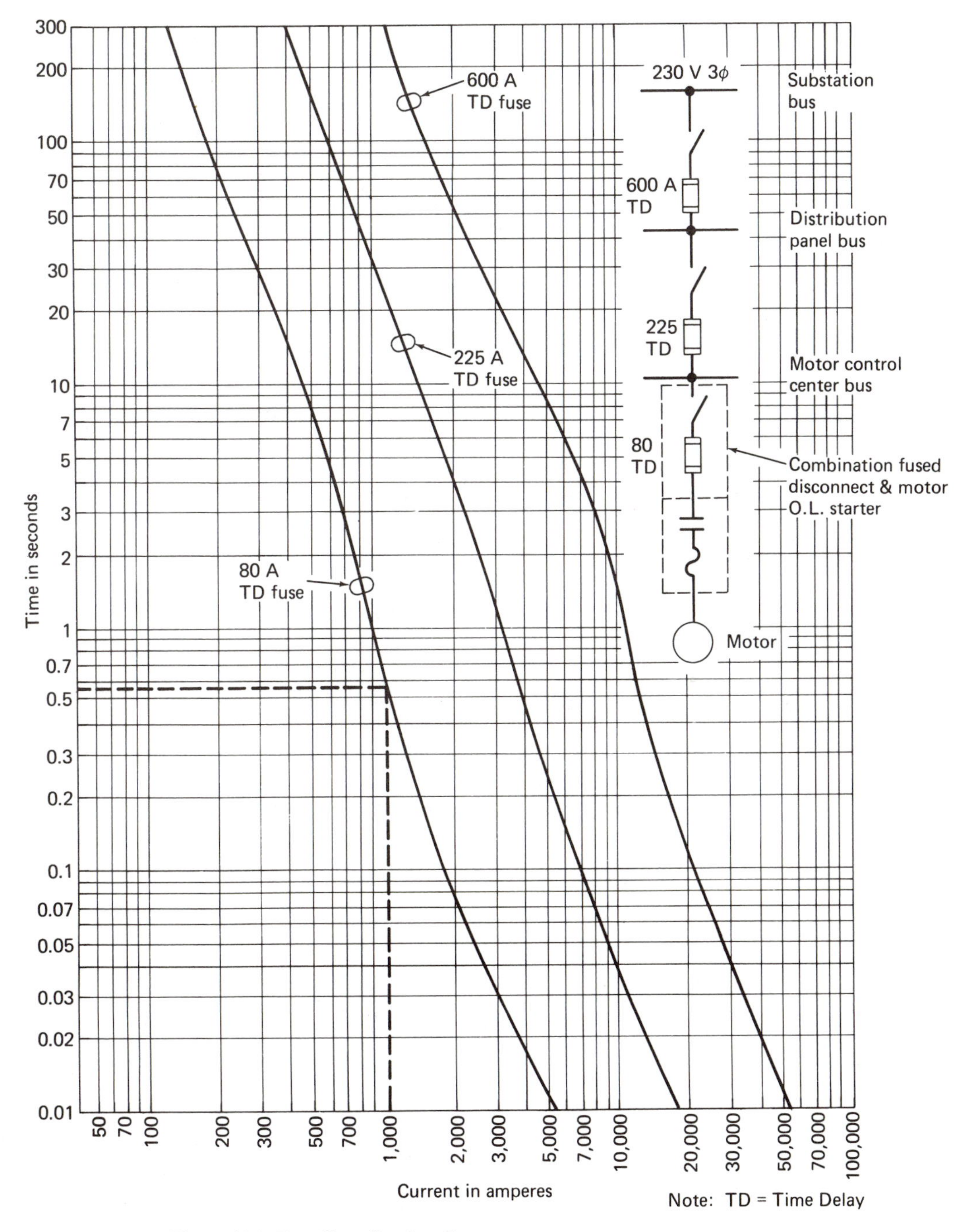

Figure 13-2 Fuse Coordination Curves

convenient way to plot is by tracing the prepared curves onto a blank sheet of standard log–log paper. A lighted table or window is a big help in tracing steps. The curves must be properly indexed to one another based on the current each device will *see*. Particular attention must be given to plotting devices that are applied at different voltages. This can be easily accomplished by establishing different current scales on the same graph sheet. For example, if a transformer in the one line being coordinated were rated 2,400-V primary and 480-V secondary, the voltage ratio is 5 to 1 (2,400 ÷ 480 = 5). If the current scale is plotted for 480 V (in order to deal with the majority of devices), a current of 2,000 A at 480 V would have a corresponding current of 400 A at 2,400-V scale lining up the 400-A point on the 2,400-V scale and maintaining the same ratio the corresponding scale will work out as shown in Fig. 13-3.

Once the tracing paper is prepared, with additional current scales if necessary, and the curves to be used are assembled, the curve of the device farthest removed from the source should be plotted first. Then the curve for the next device is plotted to determine if any overlap occurs. If not, the next device curve is added, etc., until the study is complete. A typical example is shown in Fig. 13-4 including the curve for the utility company protective relays. Of course, during the tracing operation, the time scale on current scales on the curves being traced must line up with the time and current scales on the tracing sheet. Circuit breaker curves are usually plotted against scales marked *percent of continuous rated current* rather than in amperes. If the breaker to be used were to have a continuous rating of, say, 800 A, then the 100% line on the breaker curve scale would be made to line up with the 800-A line on the tracing scale before tracing the curve.

If during the coordination study, the curves overlap or there is not enough space between curves to assure selectivity, then different devices or different device settings must be tried. There is no easy way to solve the problem and only patience and perserverance in trying different combinations will result in a properly designed selective system.

Experience in making coordination studies and careful attention to restriction manufacturers may recommend in their literature on sizing of devices to

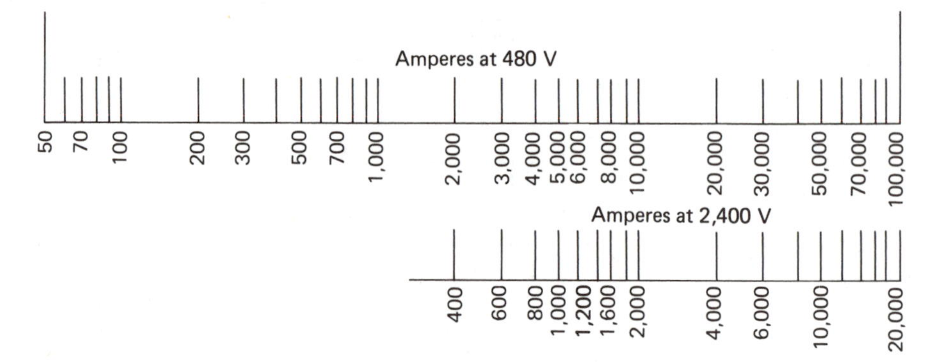

Figure 13-3 Plotting 2,400-V Current Scale Corresponding to 480-V Current Scale

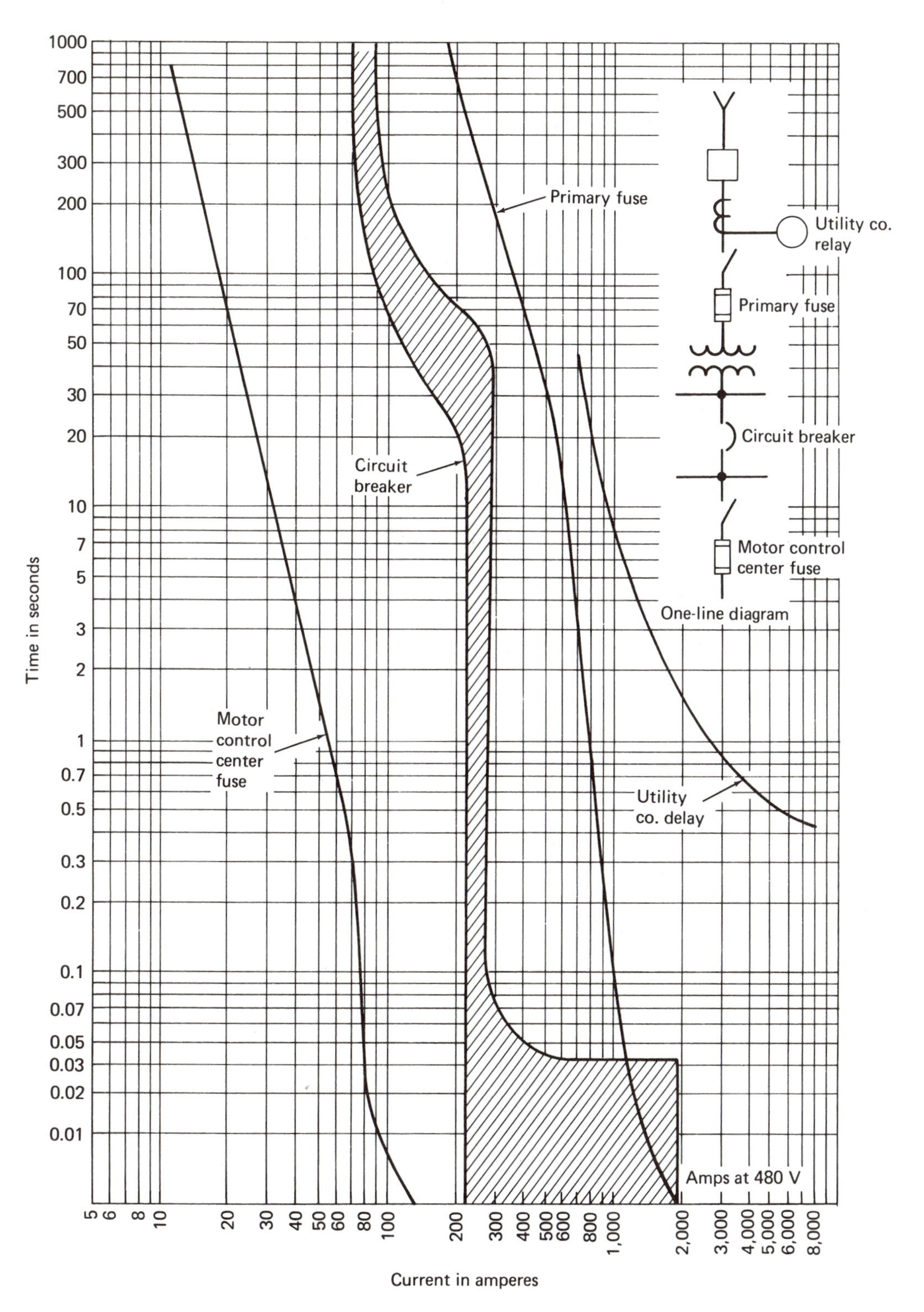

Figure 13-4 Typical Coordination Chart for Selective System

be coordinated is the only means by which designers can develop expertise in this field because the possible device combinations in systems are so varied.

13-4 Overcurrent Protection for Transformers Over 600 V

The NEC requires overcurrent protection for transformers as outlined below. As used in this explanation, the word *transformer* means a transformer or bank of two or three single-phase transformers operating as a polyphase unit.

Primary Protection

Where fuses are used, they shall be rated at not more than 150% of the rated, full-load, primary current of the transformer. The code permits the use of the next larger size standard fuse if the calculated rating at 150% does not correspond with a standard fuse size.

Where circuit breakers are used, they shall not be set at more than 300% of the rated, full-load, primary current. If the calculated rating at 300% does not correspond with a standard rating, the next *lower* rating must be used.

(The code permits some exceptions to these rules and therefore must be consulted for specific applications.)

Primary and Secondary Protection

To understand the code rules governing primary and secondary protection for transformers, the explanation of some terms and phrases is necessary. As used in the rules, the following definitions apply:

The *primary feeder overcurrent device* is the device located at the source of the supply to the transformer such as fuses or a circuit breaker connected to a bus in switchgear.

Individual overcurrent devices in the primary connections are those devices generally located near the transformer itself.

Coordinated thermal overload protection by the manufacturer is a built-in device made as part of the transformer that is arranged to open the lines to the transformer coils under overload conditions. This device is common in distribution-type transformers that are the type generally seen mounted on poles.

The code rules state that transformers with overcurrent protection on the secondary side rated or set to open at not more than the values noted in Table 450-3(a)(2), or transformers with a coordinated thermal overload protection by the manufacturer, shall not be required to have an individual overcurrent device in the primary connection provided the primary feeder overcurrent device is rated or set to open at not more than the values noted in Table 450-3(a)(2).

The application of the rules described above are illustrated in Fig. 13-5.

As some examples of the rules summarized in Fig. 13-5, the following calculations show proper overcurrent ratings in compliance with the NEC:

Example 1: Primary Fusing

a. Transformer rated 1,500 kVA, 13.8 kV-4,160 V, 3 phase, 60 Hz; impedance, 5.5%

b. $$\text{Primary current} = \frac{1{,}500 \times 1{,}000}{13{,}800 \times 1.73} = 63\ \text{A}$$

c. Size primary fuse not larger than 150%:

$$63 \times 1.5 = 95\ \text{A}$$

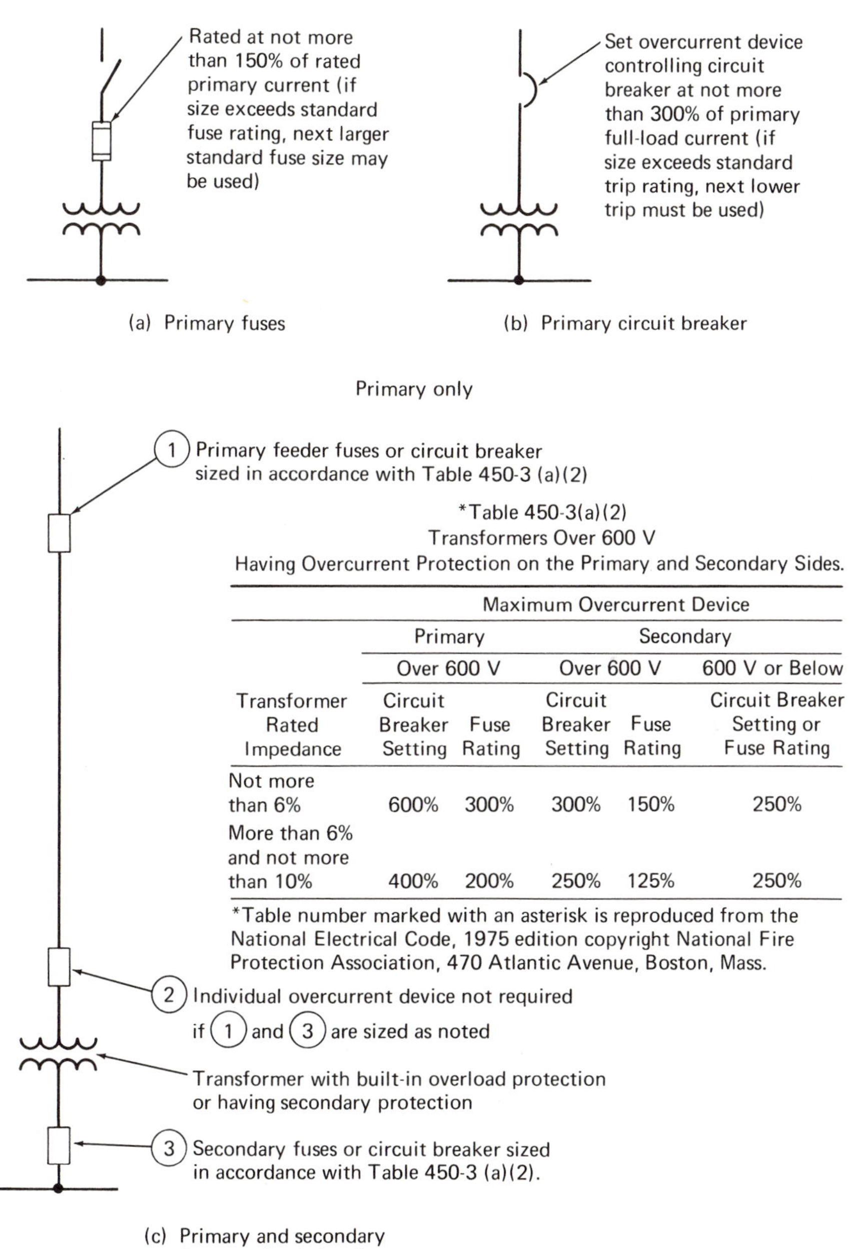

*Table 450-3(a)(2)
Transformers Over 600 V
Having Overcurrent Protection on the Primary and Secondary Sides.

	Maximum Overcurrent Device				
	Primary		Secondary		
	Over 600 V		Over 600 V		600 V or Below
Transformer Rated Impedance	Circuit Breaker Setting	Fuse Rating	Circuit Breaker Setting	Fuse Rating	Circuit Breaker Setting or Fuse Rating
Not more than 6%	600%	300%	300%	150%	250%
More than 6% and not more than 10%	400%	200%	250%	125%	250%

*Table number marked with an asterisk is reproduced from the National Electrical Code, 1975 edition copyright National Fire Protection Association, 470 Atlantic Avenue, Boston, Mass.

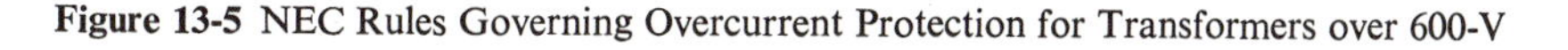
Figure 13-5 NEC Rules Governing Overcurrent Protection for Transformers over 600-V

Use standard fuse rated 100 A.
d. Check for selectivity with other devices.

Example 2: Primary Circuit Breaker
a. Same transformer as in Example 1.
b. Full-load current = 63 A.
c. Size primary circuit breaker at not more than 300%.

$$63 \times 3.0 = 189 \text{ A}$$

Set overcurrent device controlling breaker tripping at not more than 189 A or if fixed rating, use breaker rated 175 A.
d. Check for selectivity with other devices.

Example 3: Primary and Secondary Overcurrent Device
a. Same transformer as in Example 1.
b. Assume that overcurrent protection is to be fuses in both the primary and secondary circuits and that the transformer is not furnished with a built-in thermal overload device.
c. Primary full-load current is 63 A. Referring to Table 450-3(a)(2), the fuse rating on the primary side must not exceed 300% for a transformer with rated impedance of not more then 6%.

$$63 \times 3.0 = 189 \text{ A} \quad \text{(use maximum 200-A fuse)}$$

d. Calculate full-load secondary current:

(1) $$I = \frac{\text{kVA} \times 1{,}000}{E \times 1.73}$$

(2) $$I = \frac{1{,}500 \times 1{,}000}{4{,}160 \times 1.73}$$

(3) $$I = 208 \text{ A}$$

e. Table 450-3(a)(2) requires that secondary fuse not exceed 150% of the full-load current for a transformer with an impedance of not more than 6%.

$$208 \times 1.5 = 312 \text{ A} \quad \text{(use 300-A fuse)}$$

f. Check coordination of fuses selected to prove selectivity.

13-5 Overcurrent Protection for Transformers 600 V or Less

The NEC requires overcurrent protection for transformers rated 600 V or less to be rated as follows:

Primary Protection

Overcurrent devices on the primary side shall be rated or set at not more than 125% of the rated full-load primary current of the transformer. This basic rule is modified in the code by an exception that is as follows:

1. If primary current is 9 A or more, the overcurrent device cannot be rated or set at more than 125% of the rated primary current unless the calculated rating does not correspond to the standard rating of protective devices in which case the next higher rating, as set by the code, may be used.

2. If primary current is less than 9 A, the transformer may be protected by overcurrent devices rated or set at not more than 167% of rated primary current.

3. If primary current is less than 2 A, the overcurrent device may be rated or set at not more than 300% of the primary current.

(The code permits other exceptions and therefore must be consulted for specific applications.)

Primary and Secondary Protection

Transformers having an overcurrent device in the secondary side rated or set at not more than 125% of the rated full-load secondary current of the transformer are not required to have individual overcurrent devices on the primary side if the primary overcurrent device is rated or set at a current value not more than 250% of the rated full-load primary current (see Sec. 13-4 for explanation of some of the terms used above).

If transformers are equipped with coordinated thermal overload protection by the manufacturer that is arranged to interrupt the primary current, the circuit is not required to have individual overcurrent devices on the primary side if the primary feeder overcurrent device is rated or set at a current value not more than six times the rated current of the transformer having not more than 6% impedance and not more than four times the rated current of the transformer having more than 6% but not more than 10% impedance. (The code makes exceptions to these rules and must therefore be consulted for specific applications.)

Application of the rules described above are illustrated in Fig. 13-6.

Some examples of the rules summarized in Fig. 13-6 are shown in the following calculations:

Example 1: Primary Overcurrent Protection

a. Transformer rated 50 kVA, 480-120/208 V, 3 phase, 4 wire, 60 Hz. Impedance is less than 5%.

b.
$$\text{Primary current} = \frac{50 \times 1{,}000}{480 \times 1.73} = 60 \text{ A}$$

c. Size primary overcurrent protection not larger than 125% of full-load primary current.

$$60 \times 1.25 = 75 \text{ A}$$

Overcurrent protection can be sized at standard size fuse or circuit breaker rated either 70 or 80 A (smaller rating is preferable.)

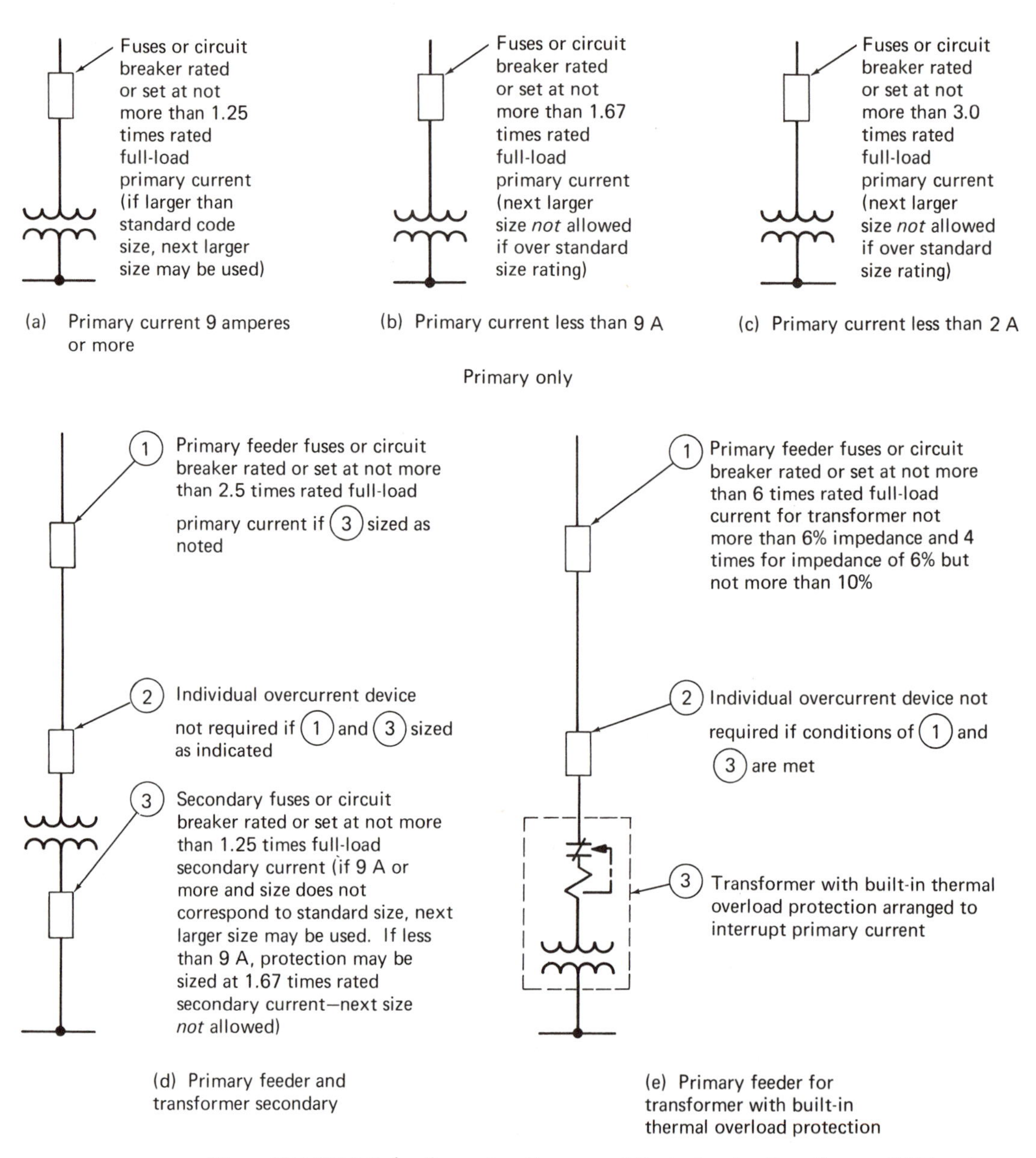

Figure 13-6 NEC Rules Governing Overcurrent Protection for Transformer 600-V or Less

Example 2: Primary and Secondary Overcurrent Protection

a. Same transformer as in Example 1 above.

b. Primary protection set at maximum of 250% of full-load current:

$$60 \times 2.5 = 150 \text{ A}$$

c. $$\text{Secondary full-load current} = \frac{50 \times 1{,}000}{208 \times 1.73} = 139 \text{ A}$$

d. Rate or set secondary device at not more than 125% of full-load current:

$$139 \times 1.25 = 173.75 \text{ A}$$

Use standard size overcurrent device rated 175 A.
e. Check for coordination with other devices.

13-6 Combination of Fuses and Circuit Breakers for Transformer Protection

A very common transformer protection arrangement is to provide high-voltage fuses in the primary circuit and an air circuit breaker as the main over-current device in the secondary circuit. Coordination studies are needed to check selectivity after the overcurrent device sizes have been determined as required by the NEC.

13-7 Rating or Setting of Overcurrent Devices for Individual Motor Branch-Circuit, Short-Circuit, and Ground Fault Protection

The NEC requires that the branch-circuit overcurrent device shall be capable of carrying the starting current of the motor. Short-circuit and ground fault protection is considered to be in compliance with the code when the overcurrent device has a rating or setting not exceeding the values given in Table 430-152 on p. 312. Exceptions and other rulings are given in the code relating to special conditions so the code should be consulted in all cases.

Table 430-152 shows different percentage values for the type of overcurrent devices and the type of motor. The overcurrent devices types are self-explanatory but *Type of Motor* column requires further explanation.

Code letters are an industry standard required to be indicated on the motor nameplate to show the motor input kVA if the rotor of the motor is locked so that it cannot turn. This kVA value is called the *locked rotor kVA*. Because this can vary from motor to motor depending on size and design, a set of code letters has been established to characterize each motor. Note that a classification of *no code letter* is also given in Table 430-152 and is allowed by industry standards. Code letters (where applicable) are indicated on nameplates of all motors but very seldom does the electrical designer see the motor and this data is not generally included in catalogs listing motors, so there is no way of knowing the code letter during the design period! For this reason designers often use the value for *no code letter*.

Manufacturers of fuses and other overcurrent devices will recommend, in their published data, ratings for their devices used with a wide range of motor horsepowers at different system voltages. Data are published in table form and therefore it is very handy for specifying overcurrent device ratings. Typical tables (13, 14, and 15) are shown in the appendix. (The ratings in these tables may vary slightly from those determined by hand calculation. The code some-

Table 430-152 MAXIMUM RATING OR SETTING OF MOTOR BRANCH-CIRCUIT PROTECTIVE DEVICES*

	Percent of Full-Load Current			
Type of motor	*Nontime delay fuse*	*Dual-element (Time-delay) fuse*	*Instan-taneous trip breaker*	*Inverse time breaker*
Single-phase, all types				
No code letter	300	175	700	250
All AC single-phase and polyphase squirrel-cage and synchronous motors with full-voltage, resistor or reactor starting:				
No code letter	300	175	700	250
Code letter F to V	300	175	700	250
Code letter B to E	250	175	700	200
Code letter A	150	150	700	150
All AC squirrel-cage and synchronous motors with autotransformer starting:				
Not more than 30 amps				
No code letter	250	175	700	200
More than 30 amps				
No code letter	200	175	700	200
Code letter F to V	250	175	700	200
Code letter B to E	200	175	700	200
Code letter A	150	150	700	150
High-reactance squirrel-cage				
Not more than 30 amps				
No code letter	250	175	700	250
More than 30 amps				
No code letter	200	175	700	200
Wound-rotor—No code letter	150	150	700	150
Direct-current (constant voltage)				
No more than 50 hp				
No code letter	150	150	250	150
More than 50 hp				
No code letter	150	150	175	150

For explanation of Code Letter Marking, see Table 430-7(b) in National Electrical Code.

*Reproduced from the National Electrical Code 1975 Edition, copyright National Fire Protection Association, 470 Atlantic Avenue, Boston, Mass.

times permits using the next largest fuse size and sometimes manufacturers will indicate that rating. Very often, their time delay fuses do not need to be rated so high as the maximum allowed and so their recommended ratings are smaller.)

The one-line diagram of Fig. 13-7 will be used to give examples for sizing overcurrent branch-circuit protection in accordance with the maximum values shown in Table 430-152. Tables 10 through 12 in the appendix give full-load motor currents.

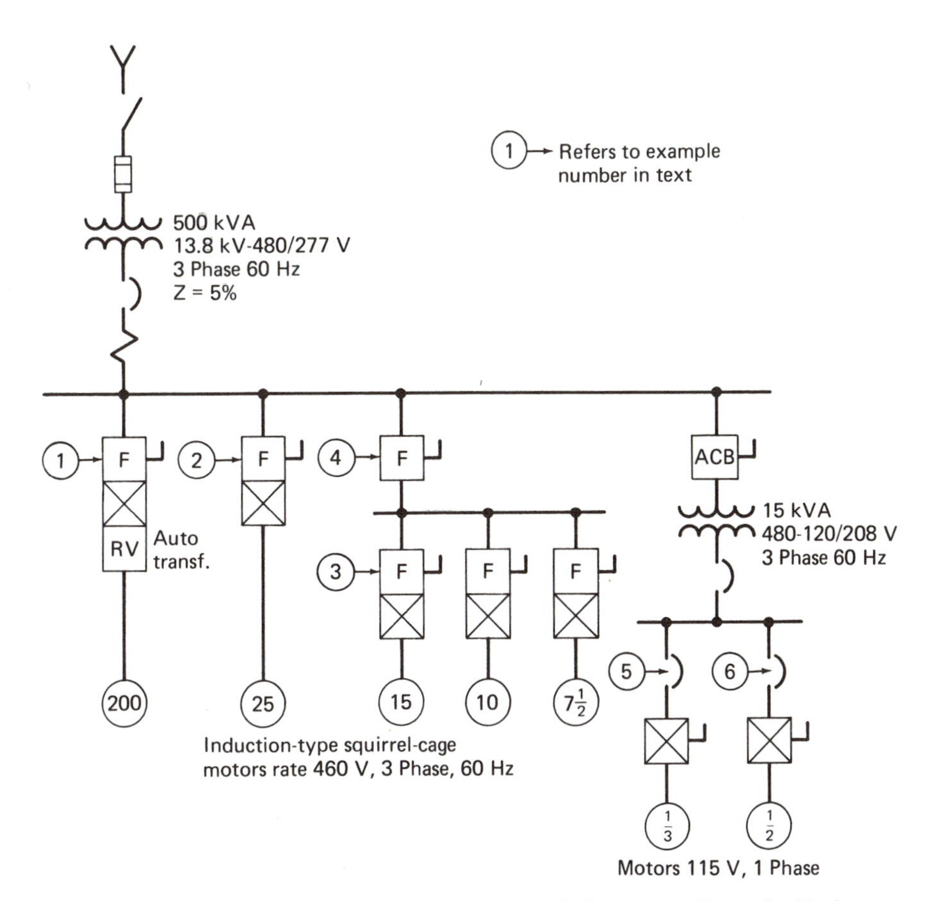

Figure 13-7 One-Line Diagram for Sizing Branch-Circuit Overcurrent Protective Devices

Example 1: Size Fuses To Protect 200-hp Motor Branch Circuit

The starter for this motor is of the reduced voltage type using auto transformer to reduce the starting voltage. By using such a starter the voltage drop on the system, caused by motor starting, in-rush currents will be kept to acceptable limits.

From Table 12 in the appendix the full-load current is 240 A. From Table 430-152, the maximum rating of the overcurrent device is 200% for nontime delay fuses and 175% for dual element (time delay) fuses.

Maximum fuse sizing:

$$240 \times 2 = 480 \text{ A} \quad \text{(use 500-A nontime delay fuses)}$$

$$240 \times 1.75 = 420 \text{ A} \quad \text{(use 400 A dual element fuses)}$$

Example 2: Size Fuses To Protect 25 hp Motor Branch Circuit

This circuit is equipped with a full-voltage starter commonly called an *across-the-line-starter.*

From Table 12 in the appendix the full-load current is 34 A. From Table 430-152, the maximum rating of the overcurrent device is 300% for nontime delay fuses, and 175% for dual element (time delay) fuses.

Maximum fuse sizing:

$$34 \times 3 = 102 \text{ A} \quad (\textit{use } 100\text{-}A \textit{ nontime delay fuses})$$

$$34 \times 1.75 = 59.5 \text{ A} \quad (\textit{use } 60\text{-}A \textit{ dual element fuses})$$

(*Note*: The typical tables in the appendix indicating manufacturers recommended rating for this same motor, show 110-A nontime delay and 50-A dual element ratings.)

Example 3: Size Fuses To Protect Each Branch Circuit for a 15-, 10-, and 7½-hp Motor.

The procedure is the same for each motor as described for the 25-hp motor in Example 2. When several motors are being hand calculated, preparing a simple table may make the job easier.

		Nontime Delay		*Dual Element Time Delay*	
HP	*Table 12 (in appendix) Full-load current*	*Times 300% equal*	*Use fuse*	*Times 175% equal*	*Use fuse*
15	21	63	60 (70)	36.75	35 (30)
10	14	42	40 (45)	24.5	25 (20)
7½	11	33	30 (35)	19.25	20 (20)

Noted in parentheses are the ratings from the manufacturers' tables.

It is worthwhile to consider in the case of the 15-hp motor whether to use 60- or 70-A nontime delay fuses because fuse holders change size and type at 60-A. If the 70-A fuses are used, a 100-A switch will be required as opposed to a 60-A switch with the 60-A fuses.

Example 4: Size Fuses for Branch-Circuit Protecting Group of Motors in Example 3

The code rules as follows for the overcurrent device required to provide branch-circuit protection for a circuit supplying power to a group of motors:

A feeder that supplies a specific fixed motor load and that consists of conductor sizes based on the proper section of the code shall be provided with overcurrent protection and shall not be greater than the largest rating or setting of the branch-circuit protective device for any motor of the group (based on Table 430-152) plus the sum of the full-load currents of the other motors of the group.

Where two or more motors of equal horsepower rating are the largest in the group, one of these motors shall be considered as the largest for the calculations above.

Where two or more motors of a group must be started simultaneously, it may be necessary to install large feeder conductors and correspondingly larger rating or setting of feeder overcurrent protection.

For large-capacity installations, where heavy-capacity feeders are installed to provide for future additions or changes, the feeder overcurrent protection may be based on the rated current-carrying capacity (ampacity) of the feeder conductors.

Applying the code ruling above to the fixed motor group of Example 3, the calculations for feeder overcurrent protection are as follows:

	Nontime delay	*Time delay*
1. The 15-hp motor has	60 (70)	35 (30)
2. Full-load current of 10-hp motor	14 (14)	14 (14)
3. Full-load current of $7\frac{1}{2}$-hp motor	11 (11)	11 (11)
Totals	85 (95)	60 (55)
Proper fuse size is	80 (90)	60 (50)

Ratings in accordance with the manufacturers' typical table ratings are shown in parentheses.

The code does not allow sizing fuses up to the next larger size fuse in these cases.

Example 5: Size Branch-Circuit Molded-Case Circuit Breaker To Protect a Single-Phase, 115-V, $\frac{1}{3}$-hp Motor Being Controlled by an Across-the-Line Manual Starter

From Table 11 in the appendix the full-load current is 7.2 A. From Table 430-152, the maximum rating of the overcurrent device for a time delay breaker is 250%.

$$\text{Maximum breaker size} = 7.2 \times 2.5 = 18 \text{ A} \quad (\textit{use } 20\text{-}A \textit{ breaker})$$

Example 6: Same as Example 5 Except Motor Is $\frac{1}{2}$ hp Full-load current is 9.8 A.

$$\text{Maximum breaker rating} = 9.8 \times 2.5 = 24.5 \text{ A} \quad (\textit{use } 25\text{-}A \textit{ breaker})$$

From the examples above, it is clear that the dual element, time delay fuses are always of a smaller rating than one-time nontime delay fuses. The time delay fuse has been designed for motor service and is generally used for motor protection rather than the nontime delay type. Either can be used if properly sized.

Overload protection for motors is provided in the motor starter. These devices are described in Chapter 5 and an illustration of a typical overload device (called a *heater*) is shown in Fig. 5-7.

Coordination studies may be required to make certain that overcurrent devices protecting motor circuits will be selective with other protective devices in the total circuit.

13-8 Overcurrent Protection for Conductors

Overcurrent protection is provided for conductors for the purpose of opening the electric circuit if the current reaches a value that will cause an excessive or dangerous temperature in the conductor or conductor insulation. Overcurrent devices must be selected or set at the current-carrying capacity of the conductors as given typically in Tables 16 and 17 in the appendix. If the current-carrying capacity of the conductors is such that a standard nonadjustable overcurrent device rating does not match the conductor rating, the next higher standard rating may be used but only where the rating is 800 A or less. The NEC lists other exceptions to the basic requirement for fixture wires and cords, motor circuits (see Chapter 14), control circuits, transformer secondary conductors, and tap conductors.

The governing rules stated above might lend the impression that conductors are sized first and then devices are chosen to protect them. Actually, the overcurrent devices for feeders, branch circuits, etc., are usually sized to suit the design load or future load and then the correct-sized conductors are specified to suit the capability of the device. (The exceptions mentioned above have a bearing on those particular applications, of course.)

As an example, a large 480-V feeder might have current-limiting fuses protecting it, rated at 400 A, which was sized to handle the connected load plus future growth. Conductors to carry 400 A would be needed. Table 16 shows the allowable ampacity (current-carrying capacity) for not more than three copper conductors in a raceway or cable or direct burial. If the insulation were to be type RHW, then 500,000 circular mil (MCM) conductors, which are

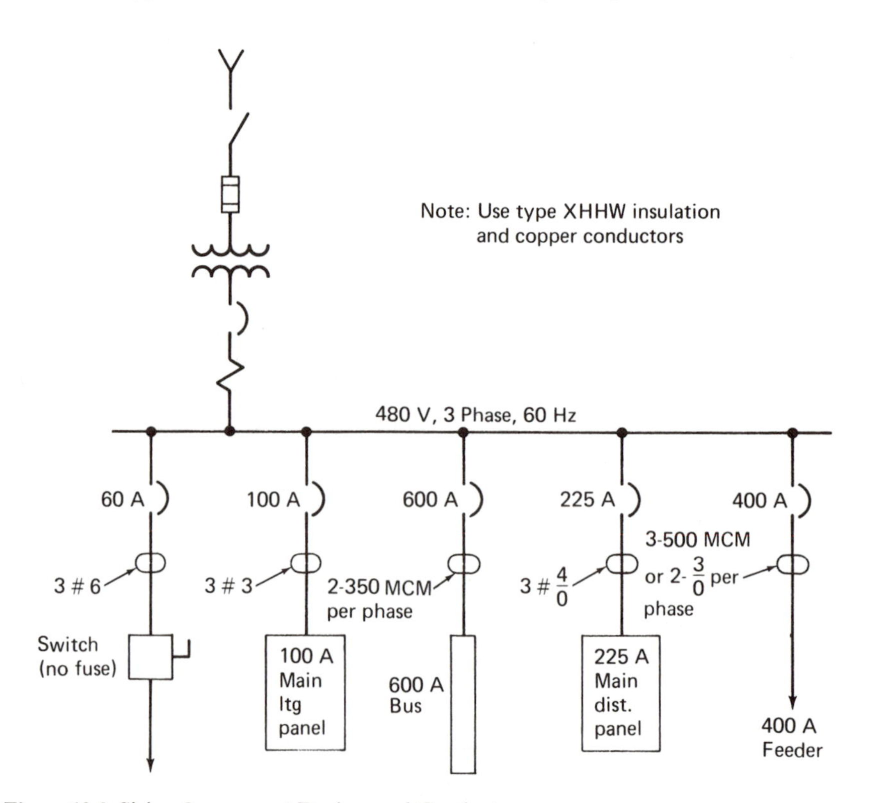

Figure 13-8 Sizing Overcurrent Devices and Conductors

rated at 380 A would meet the code requirement; i.e., the conductors can carry 380 A and the next larger standard fuse is a 400-A rating. In this particular case, a better design might be to put two smaller conductors in parallel, for each phase, rather than to use the big 500 MCM cable, which would not quite meet the fuse rating. Two 3/0 cables, each rated 200 A for a total of 400 A per phase, would exactly meet the code requirement. Judgment and experience must be used in these cases.

The one-line diagram of Fig. 13-8 shows overcurrent devices properly rated to protect the conductors being used.

Chapter Review Topics and Exercises

13-1 What kind of protective devices are widely used to protect major electrical equipment at high-voltage levels?

13-2 How does the National Electrical Code relate to application of protective devices?

13-3 Explain how protective devices in a selective system are planned to operate.

13-4 How are fuses adjusted to make them selective in operation?

13-5 Referring to Fig. 13-2, suppose that for some reason the 80 A failed to blow. How long would the fault of 1,000 A hold on until the 225-A fuse blew and cleared it?

13-6 Describe the procedure for making a coordination study.

13-7 Calculate the maximum fuse permitted as primary protection for a transformer rated 300 kVA, 2,400-480/277 V, three phase, four wire, 60 Hz, 5% impedance.

13-8 With the same transformer used in Exercise 13-7, what are the maximum fuse sizes permitted in primary and secondary protection?

13-9 Referring to Fig. 3-38, assume all motor branch-circuit fuses in motor control centers LM-1 and LM-2 are to be time delay type. Size all fuses.

13-10 Suppose the current-carrying capacity of the conductors does not match the standard sizes of overcurrent protective devices. What can be done?

14

SIZING WIRES, CABLES, AND CONDUITS

14-1 General

Wires and cables are manufactured in a great many different types and forms of construction. The various types and forms have developed over the years to meet the demand for installation under all manner of circumstances. Cables that run in ducts under the street must be able to function when completely under water because many times the duct lines are flooded. Wires that serve controls and other devices needed for operation of ovens and furnaces must be able to withstand high temperatures. Cables supplying power to electric mining equipment must have tough jackets and insulations to take the abuse from abrasion against rocks as they are dragged across the ground. The proper choice of a wire or cable may depend on consideration of many factors to make sure it will perform safely.

14-2 Metals Commonly Used as Conductors

Annealed copper is the material generally used as a conductor in insulated wires and cables. When covered with rubber insulation, the copper is coated with tin or a lead alloy for protection of both insulation and the copper because each contains substances that attack each other.

Aluminum is also used as a material for conducting electricity. The conductivity of aluminum is 61% of copper. To carry the same current as the copper conductors, the aluminum conductors must be considerably larger. If aluminum conductors are large enough in circumference to have the same conductivity as a given copper conductor, their weight per unit length will be about half that of the copper conductor.

14-3 Wire Gauges

The American Wire Gauge (AWG) is used exclusively in the United States for copper, aluminum, and bronze wire. Wires and cables are sized according to AWG from small sizes such a #18, #16, #14 up through #0000 (commonly written 4/0 and called *four aught*). From 4/0 up through all the larger sizes, cables are rated for size in circular mils. This is basically an expression in terms of total cross-sectional area of the conductor. The area of a circle, in circular mils, is equal to the square of its diameter in mils. For example, a wire with a 10-mil diameter would be rated for wire size as 100-circular mil wire (10 mils $\times$ 10 mils $=$ 100 mils).

Standard wire sizes listed in tables in the NEC in AWG and circular mils are given in the table of Fig. 14-1. The abbreviation MCM is used for circular mil cables with the first letter M signifying 1000. As an example, 250 MCM has cross-sectional area of 250,000 circular mils.

14-4 Conductors for General Wiring

The National Electrical Code lists the many types of conductors that are used for general wiring. Each type has been given a letter designation and these wires and cables are often called *code grade*. The code provides a table for conductor applications and a table for conductor insulations. A majority of wire and cable installations use the code grade conductors.

14-5 Allowable Ampacities of Code Grade Wires and Cables

The code grade wires and cables described in Sec. 14-4 have been rated in the NEC for their current-carrying capacity under several conditions of installation and operating temperature. Tables 16 and 17 (included in the appendix) are representative. Because the NEC is periodically revised, the latest edition must be consulted for actual applications.

14-6 Allowable Ampacities for Wires and Cables Other than Code Grade

Where wires and cables may be required that are not listed as code grade, catalog data from manufacturers is usually available for determining the allowable ampacities.

14-7 Selection of Cable Size

Conductor size is generally based on three considerations:

1. Current-carrying capacity
2. Voltage drop
3. Short-circuit requirements

The National Electrical Code tables will provide the *current-carrying capacity* for many design problems where the code has jurisdiction (some states publish rules and conductor capacities that differ from the NEC and these govern the

Table 8 PROPERTIES OF CONDUCTORS*

		Concentric lay stranded conductors		Bare conductors		DC resistance ohms/m ft. at 25°C. 75°F.		
						Copper		Alumi-num
Size AWG MCM	Area cir. mils	No. wires	Diam. each wire inches	Diam. inches	†Area sq. inches	Bare cond.	Tin'd. cond.	
18	1620	Solid	.0403	.0403	.0013	6.51	6.79	10.7
16	2580	Solid	.0508	.0508	.0020	4.10	4.26	6.72
14	4110	Solid	.0641	.0641	.0032	2.57	2.68	4.22
12	6530	Solid	.0808	.0808	.0051	1.62	1.68	2.66
10	10380	Solid	.1019	.1019	.0081	1.018	1.06	1.67
8	16510	Solid	.1285	.1285	.0130	.6404	.659	1.05
6	26240	7	.0612	.184	.027	.410	.427	.674
4	41740	7	.0772	.232	.042	.259	.269	.424
3	52620	7	.0867	.260	.053	.205	.213	.336
2	66360	7	.0974	.292	.067	.162	.169	.266
1	83690	19	.0664	.332	.087	.129	.134	.211
0	105600	19	.0745	.372	.109	.102	.106	.168
00	133100	19	.0837	.418	.137	.0811	.0843	.133
000	167800	19	.0940	.470	.173	.0642	.0668	.105
0000	211600	19	.1055	.528	.219	.0509	.0525	.0836
250	250000	37	.0822	.575	.260	.0431	.0449	.0708
300	300000	37	.0900	.630	.312	.0360	.0374	.0590
350	350000	37	.0973	.681	.364	.0308	.0320	.0505
400	400000	37	.1040	.728	.416	.0270	.0278	.0442
500	500000	37	.1162	.813	.519	.0216	.0222	.0354
600	600000	61	.0992	.893	.626	.0180	.0187	.0295
700	700000	61	.1071	.964	.730	.0154	.0159	.0253
750	750000	61	.1109	.998	.782	.0144	.0148	.0236
800	800000	61	.1145	1.030	.833	.0135	.0139	.0221
900	900000	61	.1215	1.090	.933	.0120	.0123	.0197
1000	1000000	61	.1280	1.150	1.039	.0108	.0111	.0177
1250	1250000	91	.1172	1.289	1.305	.00863	.00888	.0142
1500	1500000	91	.1284	1.410	1.561	.00719	.00740	.0118
1750	1750000	127	.1174	1.526	1.829	.00616	.00634	.0101
2000	2000000	127	.1255	1.630	2.087	.00539	.00555	.00885

*Reproduced from the National Electrical Code, 1975 Edition, copyright National Fire Protection Association, 470 Atlantic Avenue, Boston, Mass.

†Area given is that of a circle having a diameter equal to the overall diameter of a stranded conductor.

The values given in the Table are those given in Handbook 100 of the National Bureau of Standards except that those shown in the 8th column are those given in Specification B33 of the American Society for Testing and Materials, and those shown in the 9th column are those given in Standard No. S-19-81 of the Insulated Power Cable Engineers Association and Standard No. WC3-1964 of the National Electrical Manufacturers Association.

Figure 14-1 Properties of Conductors

minimum requirements in their area). *Careful thought should be given to load growth.* Experience in modernization of older installation has revealed feeder capacity as being inadequate to meet modernization requirements. The incremental cost of larger conductors over those that meet the known load demands plus some minor growth may very well pay dividends in the future.

Grouping of cables must be watched carefully because large groupings reduce the current-carrying capacity due to mutual heating effects. The NEC requires the application of derating factors when several conductors are grouped.

Parallel conductors should be considered if conductor size is larger than 500 MCM. The material costs of the two smaller cables per phase is generally less than for one large conductor; however, this may be offset by higher installation costs. Therefore, cost-wise there may be no significant difference but the two smaller cables are generally easier to work with for connections, etc.

Voltage drop must be carefully considered in sizing conductors. When wire or cable runs start to exceed 50 ft in length and expected peak current demands are near to the allowable current rating of the conductor sized by load current considerations only, then the drop in voltage over the circuit length may be excessive. Recommended limits of voltage drop are shown in Fig. 14-2.

If voltage drop is calculated to exceed the limits recommended using the conductors that are sized for the connected load, then the wire or cable size will have to be increased to keep the voltage at the utilization point at the correct level.

To understand the reasons for voltage drop, it is necessary to review some fundamentals. In Chapter 2, it was noted that resistance in a conductor varies directly as its length and inversely as its area. The greater the length, the greater the resistance; the greater the cross-sectional area, the smaller the resistance.

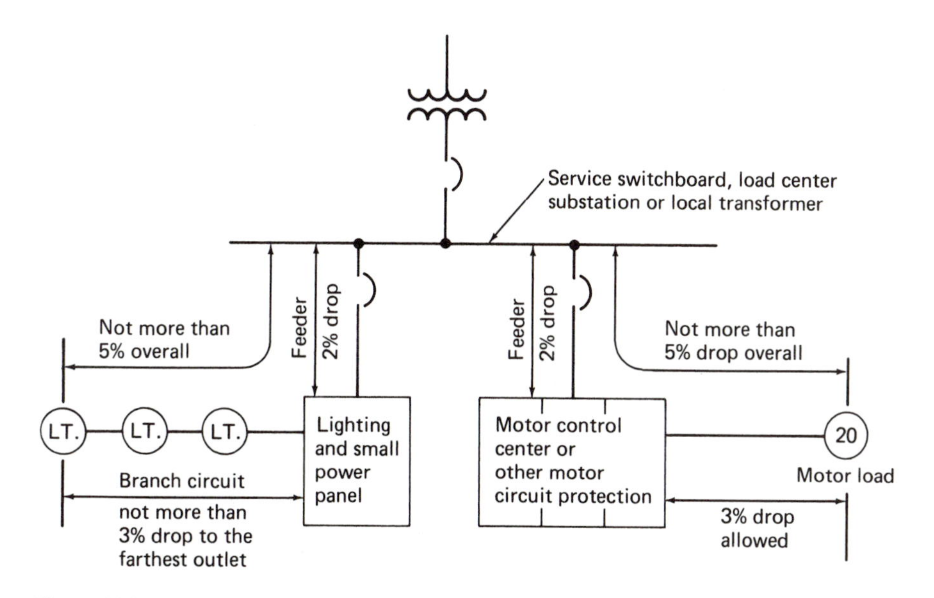

Figure 14-2 Recommended Limits of Voltage Drop

In accordance with Ohm's law, the resistance of the conductor will cause a loss of voltage for a given current flow. This is often called the *IR* drop because, by Ohm's law, $E = IR$, where E is the voltage drop, I is the connected load current that flows along the conductor, and R is the resistance of the total length of the conductor, i.e., from source to load and return to source. If conductors are selected having sufficient cross-sectional area for a given length to limit their resistance, then the voltage (*IR*) drop can be kept to acceptable levels. The value of permissible conductor resistance can be easily calculated. The proper conductor size can be determined by referring to the table of Fig. 14-1. For example, assume that a load of 100 A must be supplied 200 ft from its 230-V, single-phase source; the voltage drop must be limited to 2%. The solution is as follows:

1. Allowable drop $= 230 \times 0.02 = 4.6$ V
2. $R = \frac{E}{I} = \frac{4.6}{100} = 0.046\ \Omega$ for 400 ft of conductor
3. Because Fig. 14-1 lists resistance per 1,000 ft, a conversion to the 1,000-ft basis must be made. R of 1,000 feet $= 0.046 \times \frac{1{,}000}{400} = 0.115$
4. From Fig. 14-1 the nearest wire size listed under tinned copper conductors for the 0.115 Ω is No. 1/0 at 0.106 Ω per 1,000 ft.

The voltage drop using the conductor chosen on the basis of the load current can be determined as follows:

1. Using RHW copper conductors from Table 16 in the appendix, a #3 AWG cable would be required to carry the 100-A load current.
2. From Fig. 14-1, No. 3 tinned copper has a resistance of 0.213 Ω per 1,000 ft. Converting to the resistance of 400 ft of this cable,

$$R \text{ of 400 ft} = 0.213 \times \frac{400}{1{,}000} = 0.0852$$

3. Voltage drop $E = IR = 100 \times 0.0852 = 8.52$ V
4. Percent of line voltage $= \frac{8.52}{230} = 0.037 \times 100 = 3.7\%$

From the foregoing examples, it is clear that if the conductor were chosen for load current requirements only, then the voltage drop would be excessive.

Voltage drop curves are prepared in many forms, which make these calculations simple. These curves are usually available from wire and cable manufacturers. Voltage drop, slide rule-type calculators have been made up by manufacturers and these are extremely handy for voltage drop problems. Figure 14-6 is a typical voltage drop curve.

To complete the selection of cable size, exposure to *short-circuit current affects* must be considered. When a short-circuit current flows, it may amount to several thousand amperes, which causes a rapid rise in the temperature of the conductor. When the fault is cleared, the conductor cools off slowly because of the surrounding insulation, sheath, etc. Wires and cables are designed to withstand heating to a degree but if the limits are exceeded, permanent damage may be done to the insulation. Still another hazard is present if the insulation materials are organic and disintegration takes place accompanied by smoke and combustible vapors that might touch off a fire or explosion. Sometimes when conditions such as these develop, the sheath of the cable may expand, producing voids. This can be a serious problem with cables operating in the 5- to 15-kV range.

To check for short-circuit capability, a series of steps must be taken starting with the amount of symmetrical short circuit at the source of the cable circuit and progressing through prepared tables relating to temperature rise and the clearing time of protective devices to determine the smallest conductor that can be safely used as far as short-circuit exposure is concerned. The procedures, tables, and other pertinent data are generally available in manufacturers catalogs.

14-8 Sizing Branch-Circuit Conductors

The NEC requires the following conditions for branch-circuit wiring:

1. The current-carrying capacity of branch-circuit conductors must not be less than the maximum load to be served.
2. The current-carrying capacity of branch-circuit conductors must not be less than the rating of the overcurrent device rating protecting the circuit.
3. The total load must not exceed the branch-circuit rating and must not exceed 80% of the rating if the load is classified as being *continuous*, such as store and office lighting.

Other code rulings may apply to a particular circuit condition; therefore reference to the code for actual applications is important.

As an example, Fig. 14-3 shows a typical lighting branch circuit for office lighting (continuous load).

In Fig. 14-3, there are eight 200-W fixtures for a total of 1,600 W. The load is continuous; therefore the circuit must not be loaded over 80% of its rating, or 1,920 W ($20 \times 120 \times 0.80 = 1{,}920$ W). This circuit design complies with that ruling. Because the overcurrent protection is rated 20 A in order to meet

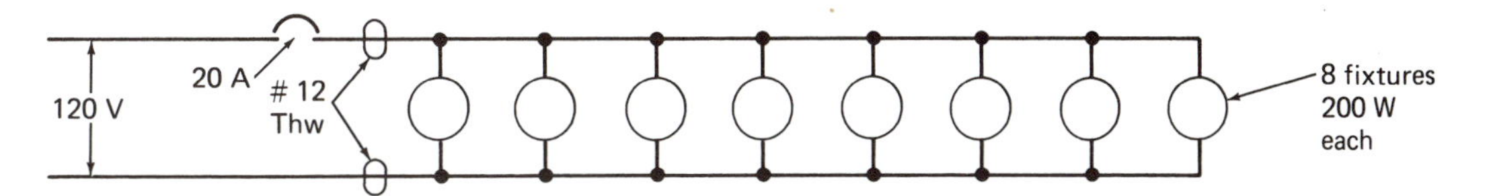

Figure 14-3 A 20-A Branch Lighting Circuit

the load demand, the branch-circuit conductors must be capable of carrying 20 A. Number 12 AWG copper with THW insulation has this ampacity (Table 16, appendix).

14-9 Sizing Feeder Conductors

Conductors for feeders must be sized to carry the current allowed to flow by the overcurrent device protecting the circuit. The code permits sizing an overcurrent device up to the next higher standard rating, provided the rating is 800 A or less. Where standard ampere ratings of fuses and nonadjustable circuit breakers do not correspond with the allowable ampacities of the conductors, conductor ampacities can be less than the overcurrent device rating. Other code ruling exceptions are allowed and must be reviewed for actual applications. As an example of conductor sizing for various ratings of an overcurrent device, the one-line diagram in Fig. 13-8 shows the sizes of conductors required.

14-10 Sizing Conductors for Single Motor Load

The NEC requires that the branch-circuit conductors supplying a single motor must have an ampacity not less than 125% of the full-load motor rating. In the case of single multispeed motors, the highest full-load current must be used. Code exceptions are made and must be investigated. The one-line diagram of Fig. 14-4 is used to provide an example for conductor sizing for several motors. Full-load motor currents are taken from the appendix, Tables 11 and 12, and conductor sizes (Table 16) for copper conductors.

Preparation of a simple calculation table is often handy when several similar problems must be solved.

Motor HP	*Voltage and phase*	*Full-load current*	× 1.25 =	*Wire size reqd.**
5	460–3PH	7.6	9.5	14
10	″	14	17.5	12
15	″	21	26.25	10
$7\frac{1}{2}$	″	11	13.75	14
50	″	65	81.25	4
30	″	40	50	6
3	208–3PH	10.56	13.2	14
$\frac{3}{4}$	115–1PH	13.8	17.25	12
$1\frac{1}{2}$	208–3PH	5.72	7.15	14

*Based on Code Grade RHW, Copper, Minimum Size to be used—#14.

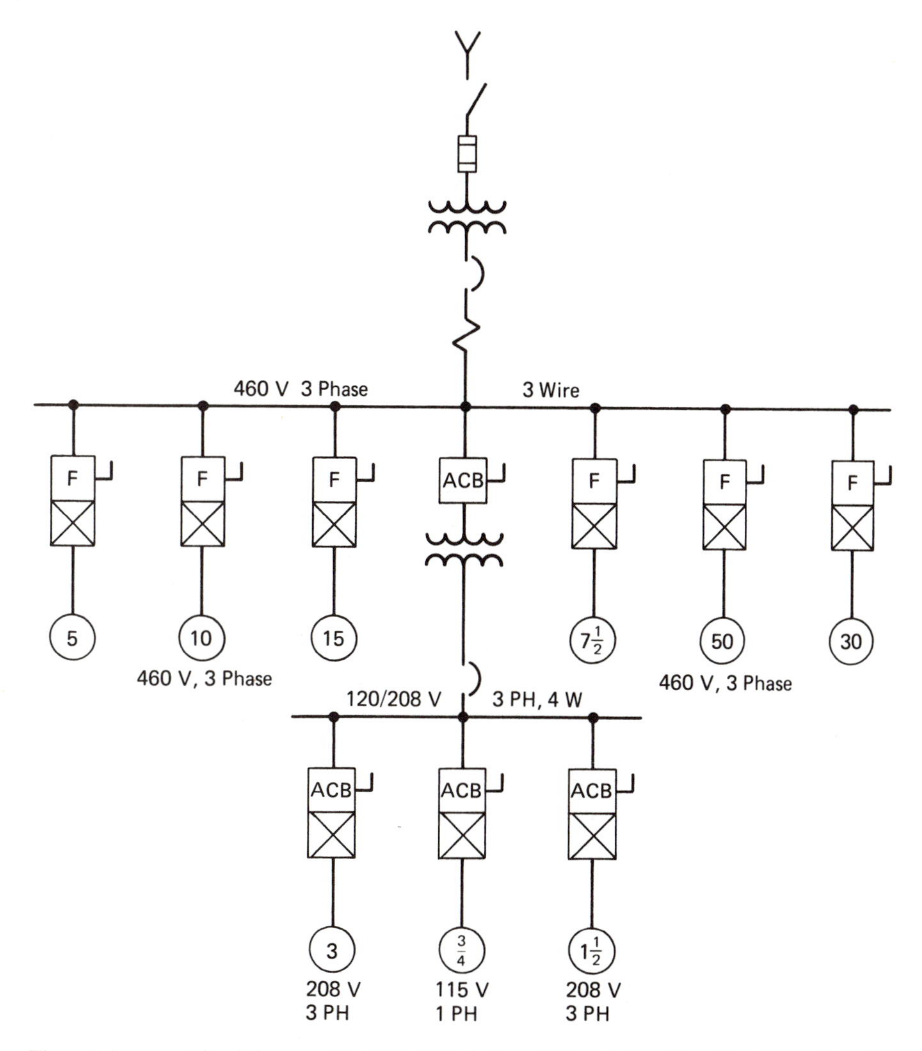

Figure 14-4 One-Line Diagram Used for Calculation of Conductor Sizes for Motor Circuits (See text examples)

14-11 Sizing Conductors for Conductors Supplying Several Motors

The NEC requires that conductors supplying several motors must have ampacity equal to the sum of the full-load current rating of all the motors plus 25% of the highest rated motor in the group. As an example, if all the 460-V motors (excluding the motors served at 115 or 208 V) in Fig. 14-4 were being supplied through one set of conductors from one overcurrent device, the conductor size would be determined by adding the full-load current of all the motors plus 1.25% of the full-load current of the 50-HP motor.

This calculation would total up to 174.85 A. From Table 16 in the appendix it can be determined that #2/0, RHW, copper conductors rated 175 A just meet the requirement.

14-12 Sizing Conductors for Transformer Circuits

Conductors in both primary and secondary circuits for transformers must be capable of carrying the current flow that will be allowed by the overcurrent device protecting the circuit. The NEC makes exceptions to this rule in some cases and therefore must be consulted for actual applications.

14-13 Sizing Conduits and Other Raceways

The NEC requires that in general the percentage of the total interior cross-sectional area of conduits and other raceways occupied by the conductors must not be more than will permit a ready installation or withdrawal of the conductors and allow for dissipation of the heat generated without injury to the insulation of the conductor. The NEC makes specific rules for each type of raceway.

Conduits form a large percentage of the raceway system in most design projects; therefore, this text will relate the general requirements for conduit systems only and other raceway requirements should be researched from the latest edition of the code.

The number of conductors permitted in a single conduit must not exceed the percentage fill limitations specified in the NEC. In the majority of cases this limit is 40% of the cross-sectional area of the inside of the conduit for either new wiring or rewiring of existing conduits. Tables 18, 19 and 20 in the appendix indicate the maximum number of nonlead-covered conductors permitted in trade sizes of conduit and electrical metallic tubing to meet the 40% limitation. These tables cover the code grade wires and cables.

When different size conductors are included in the same conduit, the conduit size must be determined by calculation using the actual cross-sectional area of the particular conductors being installed. As an example, the conduit in Fig. 14-5

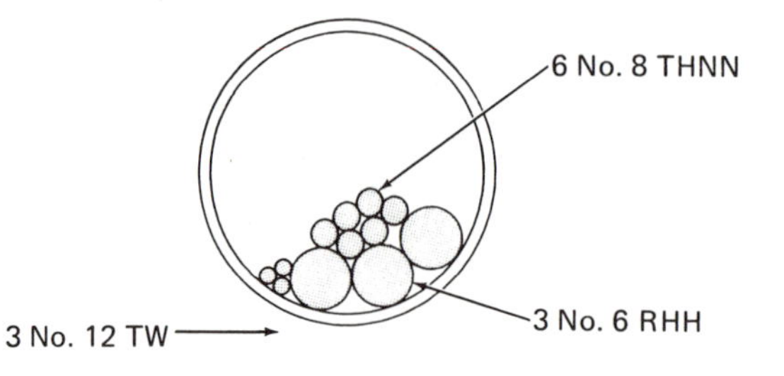

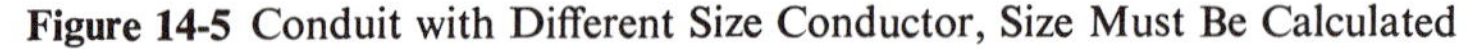

Figure 14-5 Conduit with Different Size Conductor, Size Must Be Calculated

has six No. 8 THHN conductors, three No. 6 RHH conductors, and three No. 12 TW conductors and the conduit size needs to be determined.

From Table 22 in the appendix:

No. 12 TW has cross-section area of 0.0172 sq in.
No. 8 THHN has cross-section area of 0.0373 sq in.
No. 6 RHH has cross-section area of 0.1238 sq in.

Total area occupied by conductors:

3 No. 6 RHH	$= 3 \times 0.1238 =$	0.3714 sq in.
6 No. 8 THNN	$= 6 \times 0.0373 =$	0.2238 sq in.
3 No. 12 TW	$= 3 \times 0.0172 =$	0.0516 sq in.
Total Area		0.6468 sq in.

From Table 21 in the appendix: The fifth column from the left that is headed by "Not Lead Covered—Over 2 Conds—40%" gives the square inch area that is 40% of the cross section of the conduit sizes listed in the first column at the left. The calculated area of 0.6468 sq in. determined above exceeds the 0.60-sq in. area for $1\frac{1}{4}$-in. conduit; therefore, a $1\frac{1}{2}$-in. conduit will be required, having a 40% fill limit of 0.82 sq in.

When more than three conductors are included in any one conduit, the current-carrying capacity of the conductors must be derated in accordance with code requirements.

Chapter Review Topics and Exercises

14-1 Why can copper conductors be smaller in cross-sectional area than aluminum conductors and carry the same current?

14-2 Is a No. 12 AWG wire smaller or larger in diameter than a No. 6?

14-3 How does the designer provide for load growth?

14-4 What causes voltage drop?

14-5 Assume a lighting circuit is operating at 120-V single phase and current demand is 12 A. The load is 100 ft away from the panelboard. What size conductor should be used? (Power factor is 1.0.)

14-6 What size conductor should be used for a 20 hp motor, operating at 460 V, three phase, that is 300 ft from the transformer feeding the circuit?

14-7 Why should cables be checked for short-circuit capability?

14-8 Referring to Fig. 3-38, size all wires and conduits for motor circuits fed from motor control centers LM-1 and LM-2 using code grade RHW conductors. Assume that a green grounding conductor of the same size as the phase conductors will be included in the conduit (See Chapter 8 for reference).

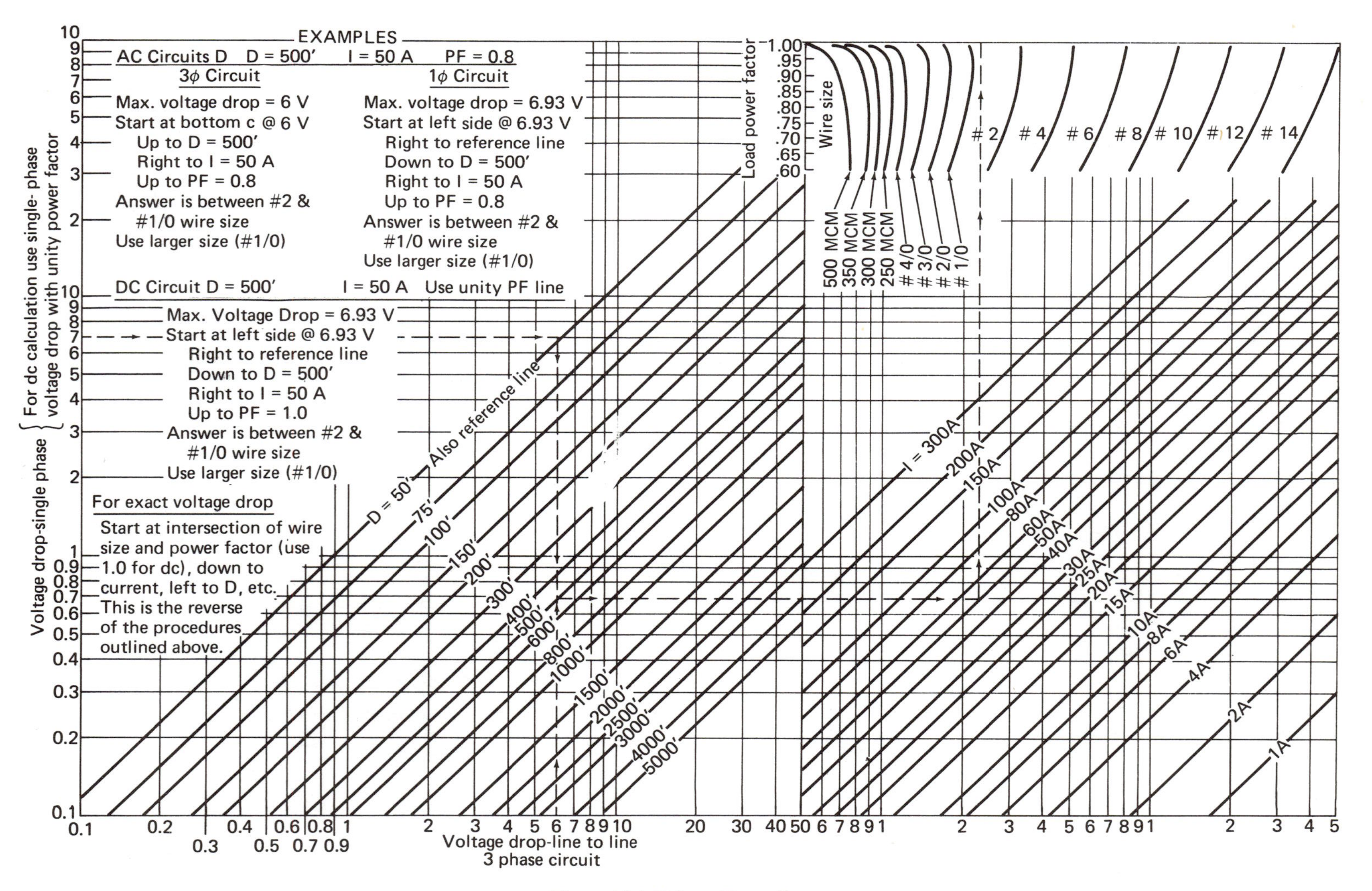

Figure 14-6 Voltage Drop Curve

14-9 If the feeders in Fig. 3-38 were to be code grade type RHW, what size phase conductors would be required?

14-10 A conduit run includes the following conductors: No. 4 RHW, No. 10 THNN, 6 No. 8 TW, 12 No. 12 THWN. What size conduit must be used?

APPENDIX

One-line diagram
Typical electrical symbols

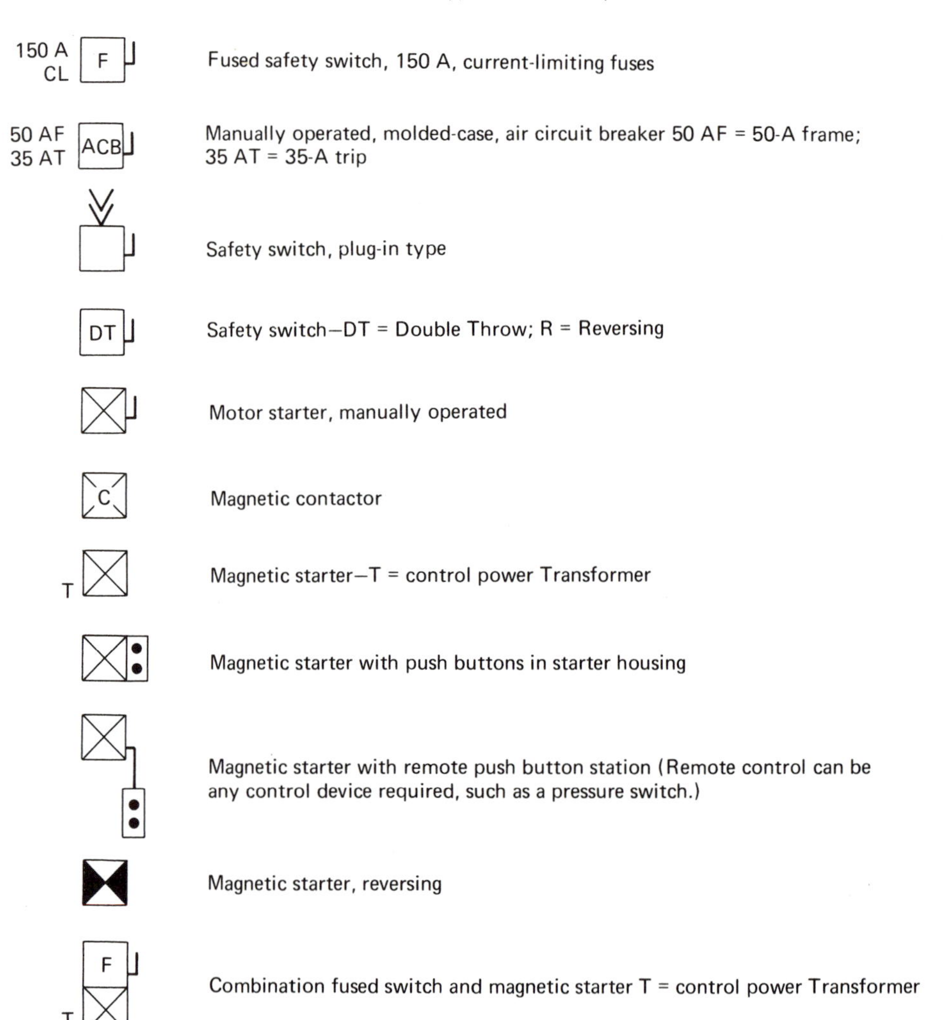

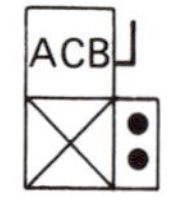

Combination molded-case air circuit breaker and magnetic starter with push buttons in housing

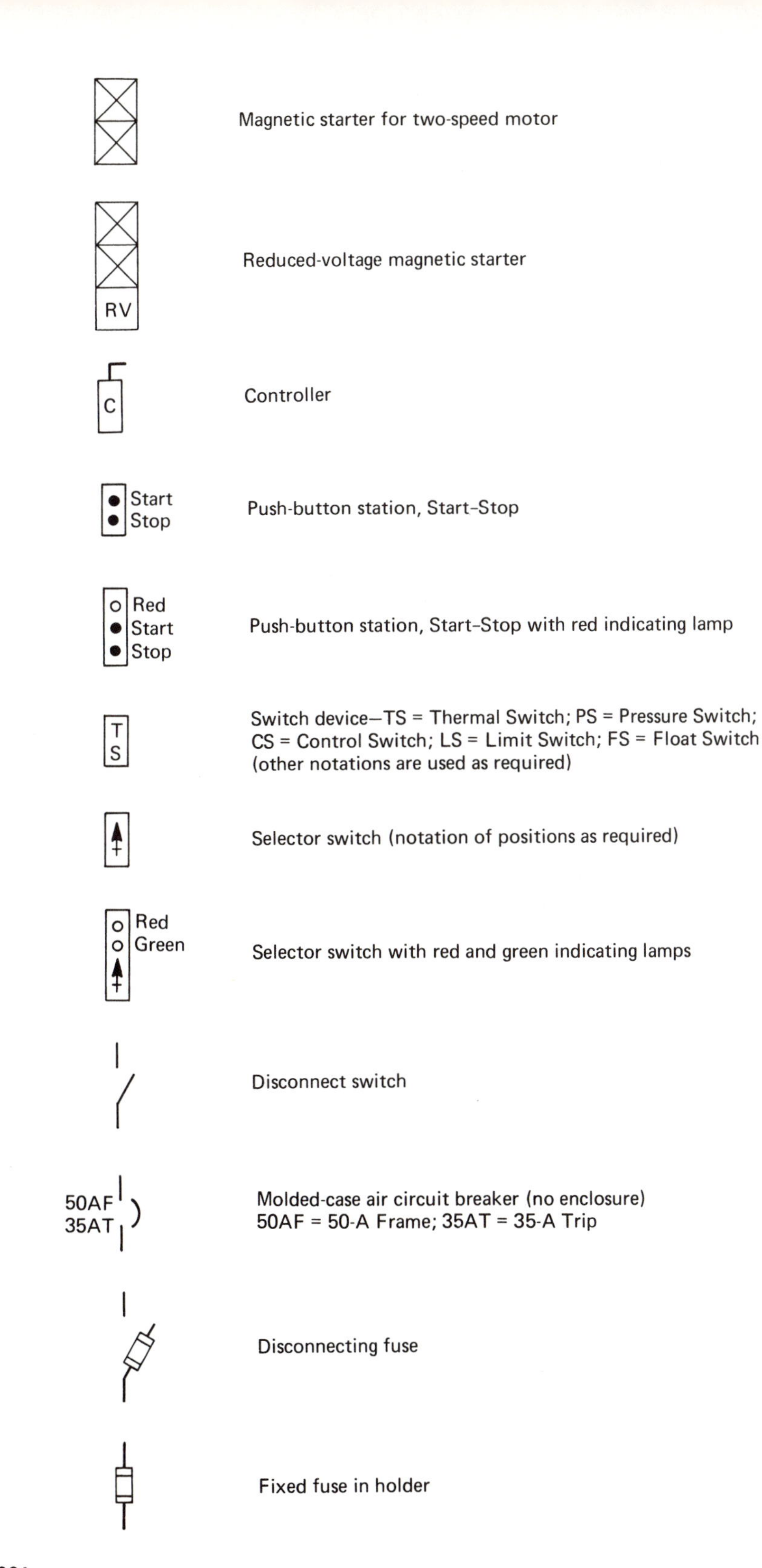
Magnetic starter for two-speed motor
RV
Reduced-voltage magnetic starter
C
Controller
Start
Stop
Push-button station, Start–Stop
Red
Start
Stop
Push-button station, Start–Stop with red indicating lamp
T
S
Switch device—TS = Thermal Switch; PS = Pressure Switch; CS = Control Switch; LS = Limit Switch; FS = Float Switch (other notations are used as required)
Selector switch (notation of positions as required)
Red
Green
Selector switch with red and green indicating lamps
Disconnect switch
50AF
35AT
Molded-case air circuit breaker (no enclosure)
50AF = 50-A Frame; 35AT = 35-A Trip
Disconnecting fuse
Fixed fuse in holder

Typical electrical symbols (cont.)

Fuse in pull-out disconnect mounting

Squirrel-cage motor—numeral denotes horsepower

Solenoid Valve (mechanical operation should be noted)

Protective relay—number is device number

Lamp—G = Green; R = Red; W = White; A = Amber, etc.

Receptacle—100-A rating

Annunciator

Bell

Buzzer

Horn

Battery

Capacitor

Air circuit breaker, manually operated with series thermal trip device

Air circuit breaker, manually operated with series magnetic trip

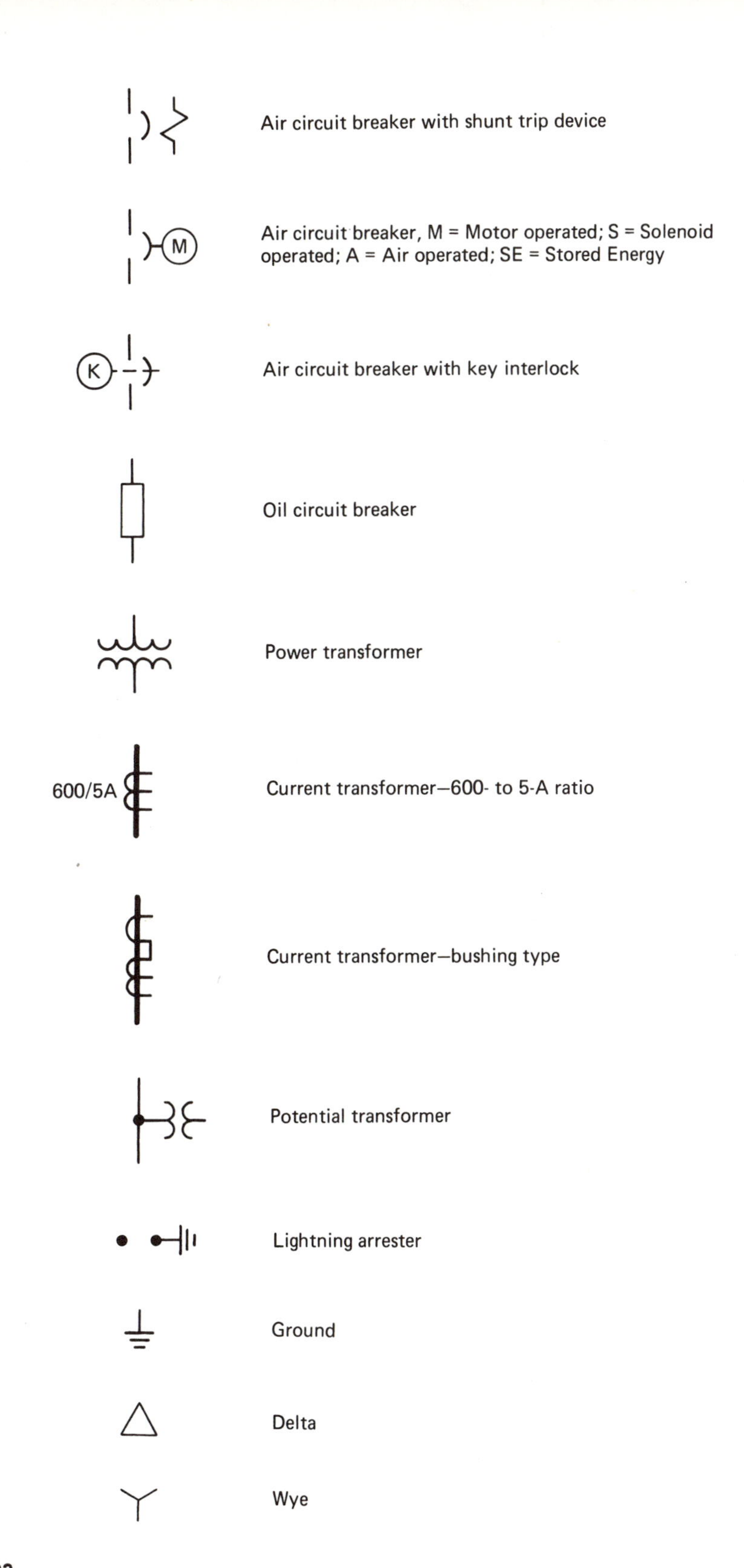
Air circuit breaker with shunt trip device
M
Air circuit breaker, M = Motor operated; S = Solenoid operated; A = Air operated; SE = Stored Energy
K
Air circuit breaker with key interlock
Oil circuit breaker
Power transformer
600/5A
Current transformer—600- to 5-A ratio
Current transformer—bushing type
Potential transformer
Lightning arrester
Ground
Delta
Wye

ILLUMINATION LEVEL GUIDE

Industrial

	Footcandles
Aircraft manufacturing	
Stock parts	
Production	100
Inspection	200*
Parts manufacturing	
Drilling, riveting, and screw fastening	70
Spray booths	100
Aircraft hangars	
Repair service only	100
Automobile manufacturing	
Frame assembly	50
Chassis assembly line	100
Final assembly and inspection line	200*
Bakeries	
Mixing room	50
Wrapping room	30
Breweries	
Brew house	30
Filling (bottles, cans, kegs)	50
Candy making	
Chocolate department	50
Hand decorating	100
Canning and preserving	
Initial grading raw material samples	50
Canning	100
Inspection	200*
Central station indoor locations	
Boiler platforms	10
Burner platform	20
Coal conveyor, crusher, feeder, scale areas, pulverizer, fan area, transfer tower	10
Control rooms	
Vertical face of switchboards	
Large	50
Ordinary	30
Bench boards (horizontal level)	50
Turbine room	30
Central station outdoor locations	
Catwalks	2
Entrances, generating or service building	
(a) Main	10
(b) Secondary	2
Substation	
(a) General horizontal	2
(b) Specific vertical (on disconnects)	2
Chemical works	
Tanks for cooking, extractors, percolators, nitrators, electrolytic cells	30
Clay products and cements	
Grinding, filter presses, kiln rooms	30
Enameling, color, and glazing	100
Cloth products	
Cloth inspection	2000*
Cutting	300*
Dairy products	
Filling: inspection	100
Pasteurizers and storage refrigerator	30
Scales	70
Electrical equipment manufacturing	
Impregnating	50
Insulating: coil winding and testing	100
Exterior areas	
Entrances	
Active (pedestrian and/or conveyance)	5
Inactive (normally locked, infrequently used)	1
Storage areas—active	20
Storage areas—inactive	1
Loading and unloading platforms	20
Flour mills	
Rolling, sifting, and purifying	50
Packing	30
Forge shops	50
Foundries	
Annealing (furnaces)	30
Core making (fine)	100
Core making (medium)	50
Inspection (fine)	500*
Inspection (medium)	100
Pouring	50
Garages—automobile and truck	
Repairs	100
Storage	5

*Obtained with combination of general plus local lighting.

ILLUMINATION LEVEL GUIDE (continued)

	Footcandles
Glove manufacturing	
Knitting	100
Sewing and inspection	500*
Iron and steel manufacturing	
Open hearth	
Stock yard	10
Charging floor	20
Hot top	30
Stripping yard	20
Skull cracker	10
Rolling mills	
Blooming, slabbing, hot strip, hot sheet	30
Cold strip, plate	30
Pipe, rod, tube, wire drawing	50
Tin plate mills	
Tinning and galvanizing	50
Inspection	
Blackplate, bloom, billet chipping	100
Laundries	
Washing	30
Flatwork ironing, weighing, listing, and marking	50
Machine and press finishing, sorting	70
Leather manufacturing	
Cleaning, tanning and stretching, vats	30
Finishing and scarfing	100
Locker rooms	20
Machine shops	
Rough bench and machine work	50
Medium bench and machine work	100
Fine bench and machine work	500*
Materials handling	
Wrapping, packing, labeling	50
Loading, trucking	20
Picking, stock, classifying	30
Meat packing	
Slaughtering	30
Cleaning, cutting, cooking, grinding, canning, packing	100
Paint shops	
Dipping, simple spraying, firing	50
Rubbing, ordinary hand painting and finishing art	50
Fine hand painting and finishing	100

	Footcandles
Paper manufacturing	
Beaters, grinding, calendering	30
Finishing, cutting, trimming, paper-making machines	50
Hand counting, wet end of paper machine	70
Paper machine reel, paper inspection, and laboratories	100
Rewinder	150
Printing industries	
Type foundries	
Matrix making, dressing type	100
Printing plants	
Color inspection and appraisal	200*
Composing room	100
Presses	70
Proofreading	150
Electrotyping	100
Photo engraving	
Etching, staging, and blocking	50
Routing, finishing, proofing, tint laying, masking	100
Rubber tire and tube manufacturing	
Banbury, plasticating, milling	30
Calendering	50
Tire building	
Solid tire	30
Pneumatic tire	50
Curing department	70
Final inspection	
Tube casing	200*
Sheet metal works	
Presses, shears, stamps, spinning, medium bench work	50
Scribing	200*
Ship yards	
General	5
Ways	10
Fabrication areas	30
Shoe manufacturing	
Cutting and stitching	300*
Stairways, washrooms, and other service areas	20
Storage rooms or warehouses	
Rough bulky	10
Medium	20
Fine	50

ILLUMINATION LEVEL GUIDE (continued)

Tobacco products	*Footcandles*
Drying, stripping, general	30
Grading and sorting	200*
Welding	
General illumination	50
Precision manual arc welding	1000*
Woodworking	
Rough sawing and bench work	30
Sizing, planing, rough sanding, medium machine and bench work, glueing, veneering, cooperage	50
Fine bench and machine work, fine sanding and finishing	100

Institutional

Art galleries	*Footcandles*
General	30
On statuary and other displays	100
Churches	
Altar, ark, reredos	100
Choir and chancel	30
Classrooms	30
Main worship area	15
Hospitals	
Autopsy room	100
Morgue, general	20
Corridor	
General	20
Operating and delivery suites and laboratories	30
Examination and treatment room	
General	50
Examining table	100
Exits, at floor	5
Fracture room	
General	50
Fracture table	200
Kitchen	70
Dishwashing	30
Laboratories	
Work tables	50
Close work	100
Lobby	30
Nurses' station	
General, day	70
General, night	30
Desk and charts	70
Parking lot	1
Private rooms and wards	*Footcandles*
General	10
Reading	30
Stairways	20
Surgery	
Instrument and sterile supply room	30
Clean-up room (instruments)	100
Scrub-up room	30
Operating room, general	100
Operating table	2500
Waiting room	
General	20
Reading	30
Hotels	
Auditoriums	
Assembly only	15
Exhibitions	30
Dancing	5
Entrance foyer	30
Lobby	
General lighting	10
Reading and working areas	30
Schools	
Reading printed material	30
Reading pencil writing	70
Drafting, benchwork	100
Auditoriums (assembly only)	15
Swimming Pools, Indoor	30
Swimming Pools, Outdoor	10
Libraries	70
Laboratories	100

Commercial

Banks	*Footcandles*
Lobby	
General	50
Writing areas	70
Tellers' stations, posting and keypunch	150
Offices	
Designing, detailed drafting	200
Accounting, auditing, bookkeeping, business machine operation	150
Regular office work	100
Stores	
Circulation areas	30
Merchandising areas	
Service	100
Self-service	200
Stockrooms	30

Table 1 (TABLE A)* CAVITY RATIOS

Room Dimensions		*Cavity Depth*																			
Width	*Length*	*1.0*	*1.5*	*2.0*	*2.5*	*3.0*	*3.5*	*4.0*	*5.0*	*6.0*	*7.0*	*8*	*9*	*10*	*11*	*12*	*14*	*16*	*20*	*25*	*30*
8	8	1.2	1.9	2.5	3.1	3.7	4.4	5.0	6.2	7.5	8.8	10.0	11.2	12.5	—	—	—	—	—	—	—
	10	1.1	1.7	2.2	2.8	3.4	3.9	4.5	5.6	6.7	7.9	9.0	10.1	11.3	12.4	—	—	—	—	—	—
	14	1.0	1.5	2.0	2.5	3.0	3.4	3.9	4.9	5.9	6.9	7.8	8.8	9.7	10.7	11.7	—	—	—	—	—
	20	0.9	1.3	1.7	2.2	2.6	3.1	3.5	4.4	5.2	6.1	7.0	7.9	8.8	9.6	10.5	12.2	—	—	—	—
	30	0.8	1.2	1.6	2.0	2.4	2.8	3.2	4.0	4.7	5.5	6.3	7.1	7.9	8.7	9.5	11.0	—	—	—	—
	40	0.7	1.1	1.5	1.9	2.3	2.6	3.0	3.7	4.5	5.3	5.9	6.5	7.4	8.1	8.8	10.3	11.8	—	—	—
10	10	1.0	1.5	2.0	2.5	3.0	3.5	4.0	5.0	6.0	7.0	8.0	9.0	10.0	11.0	12.0	—	—	—	—	—
	14	0.9	1.3	1.7	2.1	2.6	3.0	3.4	4.3	5.1	6.0	6.9	7.8	8.6	9.5	10.4	12.0	—	—	—	—
	20	0.7	1.1	1.5	1.9	2.3	2.6	3.0	3.7	4.5	5.3	6.0	6.8	7.5	8.3	9.0	10.5	12.0	—	—	—
	30	0.7	1.0	1.3	1.7	2.0	2.3	2.7	3.3	4.0	4.7	5.3	6.0	6.6	7.3	8.0	9.4	10.6	—	—	—
	40	0.6	0.9	1.2	1.6	1.9	2.2	2.5	3.1	3.7	4.4	5.0	5.6	6.2	6.9	7.5	8.7	10.0	12.5	—	—
	60	0.6	0.9	1.2	1.5	1.7	2.0	2.3	2.9	3.5	4.1	4.7	5.3	5.9	6.5	7.1	8.2	9.4	11.7	—	—
12	12	0.8	1.2	1.7	2.1	2.5	2.9	3.3	4.2	5.0	5.8	6.7	7.5	8.4	9.2	10.0	11.7	—	—	—	—
	16	0.7	1.1	1.5	1.8	2.2	2.5	2.9	3.6	4.4	5.1	5.8	6.5	7.2	8.0	8.7	10.2	11.6	—	—	—
	24	0.6	0.9	1.2	1.6	1.9	2.2	2.5	3.1	3.7	4.4	5.0	5.6	6.2	6.9	7.5	8.7	10.0	12.5	—	—
	36	0.6	0.8	1.1	1.4	1.7	1.9	2.2	2.8	3.3	3.9	4.4	5.0	5.5	6.0	6.6	7.8	8.8	11.0	—	—
	50	0.5	0.8	1.0	1.3	1.5	1.8	2.1	2.6	3.1	3.6	4.1	4.6	5.1	5.6	6.2	7.2	8.2	10.2	—	—
	70	0.5	0.7	1.0	1.2	1.5	1.7	2.0	2.4	2.9	3.4	3.9	4.4	4.9	5.4	5.8	6.8	7.8	9.7	12.2	—
14	14	0.7	1.1	1.4	1.8	2.1	2.5	2.9	3.6	4.3	5.0	5.7	6.4	7.1	7.8	8.5	10.0	11.4	—	—	—
	20	0.6	0.9	1.2	1.5	1.8	2.1	2.4	3.0	3.6	4.2	4.9	5.5	6.1	6.7	7.3	8.6	9.8	12.3	—	—
	30	0.5	0.8	1.0	1.3	1.6	1.8	2.1	2.6	3.1	3.7	4.2	4.7	5.2	5.8	6.3	7.3	8.4	10.5	—	—
	42	0.5	0.7	1.0	1.2	1.4	1.7	1.9	2.4	2.9	3.3	3.8	4.3	4.7	5.2	5.7	6.7	7.6	9.5	11.9	—
	60	0.4	0.7	0.9	1.1	1.3	1.5	1.8	2.2	2.6	3.1	3.5	3.9	4.4	4.8	5.2	6.1	7.0	8.8	10.9	—
	90	0.4	0.6	0.8	1.0	1.2	1.4	1.6	2.0	2.5	2.9	3.3	3.7	4.1	4.5	5.0	5.8	6.6	8.3	10.3	12.4
17	17	0.6	0.9	1.2	1.5	1.8	2.1	2.3	2.9	3.5	4.1	4.7	5.3	5.9	6.5	7.0	8.2	9.4	11.7	—	—
	25	0.5	0.7	1.0	1.2	1.5	1.7	2.0	2.5	3.0	3.5	4.0	4.5	5.0	5.5	6.0	7.0	8.0	10.0	12.5	—
	35	0.4	0.7	0.9	1.1	1.3	1.5	1.7	2.2	2.6	3.1	3.5	3.9	4.4	4.8	5.2	6.1	7.0	8.7	10.9	—
	50	0.4	0.6	0.8	1.0	1.2	1.4	1.6	2.0	2.4	2.8	3.1	3.5	3.9	4.3	4.5	5.4	6.2	7.7	9.7	11.6
	80	0.4	0.5	0.7	0.9	1.1	1.2	1.4	1.8	2.1	2.5	2.9	3.3	3.6	4.0	4.3	5.1	5.8	7.2	9.0	10.9
	120	0.3	0.5	0.7	0.8	1.0	1.2	1.3	1.7	2.0	2.3	2.7	3.0	3.4	3.7	4.0	4.7	5.4	6.7	8.4	10.1

*Table identification used in manufacturers' catalogs.

Table 1 (continued)

Room Dimensions		*Cavity Depth*																			
Width	*Length*	*1.0*	*1.5*	*2.0*	*2.5*	*3.0*	*3.5*	*4.0*	*5.0*	*6.0*	*7.0*	*8*	*9*	*10*	*11*	*12*	*14*	*16*	*20*	*25*	*30*
20	20	0.5	0.7	1.0	1.2	1.5	1.7	2.0	2.5	3.0	3.5	4.0	4.5	5.0	5.5	6.0	7.0	8.0	10.0	12.5	—
	30	0.4	0.6	0.8	1.0	1.2	1.5	1.7	2.1	2.5	2.9	3.3	3.7	4.1	4.5	4.9	5.8	6.6	8.2	10.3	12.4
	45	0.4	0.5	0.7	0.9	1.1	1.3	1.4	1.8	2.2	2.5	2.9	3.3	3.6	4.0	4.3	5.1	5.8	7.2	9.1	10.9
	60	0.3	0.5	0.7	0.8	1.0	1.2	1.3	1.7	2.0	2.3	2.7	3.0	3.4	3.7	4.0	4.7	5.4	6.7	8.4	10.1
	90	0.3	0.5	0.6	0.8	0.9	1.1	1.2	1.5	1.8	2.1	2.4	2.7	3.0	3.3	3.6	4.2	4.8	6.0	7.5	9.0
	150	0.3	0.4	0.6	0.7	0.8	1.0	1.1	1.4	1.7	2.0	2.3	2.6	2.9	3.2	3.4	4.0	4.6	5.7	7.2	8.6
24	24	0.4	0.6	0.8	1.0	1.2	1.5	1.7	2.1	2.5	2.9	3.3	3.7	4.1	4.5	5.0	5.8	6.7	8.2	10.3	12.4
	32	0.4	0.5	0.7	0.9	1.1	1.3	1.5	1.8	2.2	2.6	2.9	3.3	3.6	4.0	4.3	5.1	5.8	7.2	9.0	11.0
	50	0.3	0.5	0.6	0.8	0.9	1.1	1.2	1.5	1.8	2.2	2.5	2.8	3.1	3.4	3.7	4.4	5.0	6.2	7.8	9.4
	70	0.3	0.4	0.6	0.7	0.8	1.0	1.1	1.4	1.7	2.0	2.2	2.5	2.8	3.0	3.3	3.8	4.4	5.5	6.9	8.2
	100	0.3	0.4	0.5	0.6	0.8	0.9	1.0	1.3	1.6	1.8	2.1	2.4	2.6	2.9	3.1	3.7	4.2	5.2	6.5	7.9
	160	0.2	0.4	0.5	0.6	0.7	0.8	1.0	1.2	1.4	1.7	1.9	2.1	2.4	2.6	2.8	3.3	3.8	4.7	5.9	7.1
30	30	0.3	0.5	0.7	0.8	1.0	1.2	1.3	1.7	2.0	2.3	2.7	3.0	3.3	3.7	4.0	4.7	5.4	6.7	8.4	10.0
	45	0.3	0.4	0.6	0.7	0.8	1.0	1.1	1.4	1.7	1.9	2.2	2.5	2.7	3.0	3.3	3.8	4.4	5.5	6.9	8.2
	60	0.3	0.4	0.5	0.6	0.7	0.9	1.0	1.2	1.5	1.7	2.0	2.2	2.5	2.7	3.0	3.5	4.0	5.0	6.2	7.4
	90	0.2	0.3	0.4	0.6	0.7	0.8	0.9	1.1	1.3	1.6	1.8	2.0	2.2	2.5	2.7	3.1	3.6	4.5	5.6	6.7
	150	0.2	0.3	0.4	0.5	0.6	0.7	0.8	1.0	1.2	1.4	1.6	1.8	2.0	2.2	2.4	2.8	3.2	4.0	5.0	5.9
	200	0.2	0.3	0.4	0.5	0.6	0.7	0.8	1.0	1.1	1.3	1.5	1.7	1.9	2.0	2.2	2.6	3.0	3.7	4.7	5.6
36	36	0.3	0.4	0.6	0.7	0.8	1.0	1.1	1.4	1.7	1.9	2.2	2.5	2.8	3.0	3.3	3.9	4.4	5.5	6.9	8.3
	50	0.2	0.4	0.5	0.6	0.7	0.8	1.0	1.2	1.4	1.7	1.9	2.1	2.5	2.6	2.9	3.3	3.8	4.8	5.9	7.2
	75	0.2	0.3	0.4	0.5	0.6	0.7	0.8	1.0	1.2	1.4	1.6	1.8	2.0	2.3	2.5	2.9	3.3	4.1	5.1	6.1
	100	0.2	0.3	0.4	0.5	0.6	0.7	0.8	0.9	1.1	1.3	1.5	1.7	1.9	2.1	2.3	2.6	3.0	3.8	4.7	5.7
	150	0.2	0.3	0.3	0.4	0.5	0.6	0.7	0.9	1.0	1.2	1.4	1.6	1.7	1.9	2.1	2.4	2.8	3.5	4.3	5.2
	200	0.2	0.2	0.3	0.4	0.5	0.6	0.7	0.8	1.0	1.1	1.3	1.5	1.6	1.8	2.0	2.3	2.6	3.3	4.1	4.9
42	42	0.2	0.4	0.5	0.6	0.7	0.8	1.0	1.2	1.4	1.6	1.9	2.1	2.4	2.6	2.8	3.3	3.8	4.7	5.9	7.1
	60	0.2	0.3	0.4	0.5	0.6	0.7	0.8	1.0	1.2	1.4	1.6	1.8	2.0	2.2	2.4	2.8	3.2	4.0	5.0	6.0
	90	0.2	0.3	0.3	0.4	0.5	0.6	0.7	0.9	1.0	1.2	1.4	1.6	1.7	1.9	2.1	2.4	2.8	3.5	4.4	5.2
	140	0.2	0.2	0.3	0.4	0.5	0.5	0.6	0.8	0.9	1.1	1.2	1.4	1.5	1.7	1.9	2.2	2.5	3.1	3.9	4.6
	200	0.1	0.2	0.3	0.4	0.4	0.5	0.6	0.7	0.9	1.0	1.1	1.3	1.4	1.6	1.7	2.0	2.3	2.9	3.6	4.3
	300	0.1	0.2	0.3	0.3	0.4	0.5	0.5	0.7	0.8	0.9	1.1	1.3	1.4	1.5	1.7	1.9	2.2	2.8	3.5	4.2

50	50	0.2	0.3	0.4	0.5	0.6	0.7	0.8	1.0	1.2	1.4	1.6	1.8	2.0	2.2	2.4	2.8	3.2	4.0	5.0	6.0
	70	0.2	0.3	0.3	0.4	0.5	0.6	0.7	0.9	1.0	1.2	1.4	1.5	1.7	1.9	2.0	2.4	2.7	3.4	4.3	5.1
	100	0.1	0.2	0.3	0.4	0.4	0.5	0.6	0.7	0.9	1.0	1.2	1.3	1.5	1.6	1.8	2.1	2.4	3.0	3.7	4.5
	150	0.1	0.2	0.3	0.3	0.4	0.5	0.5	0.7	0.8	0.9	1.1	1.2	1.3	1.5	1.6	1.9	2.1	2.7	3.3	4.0
	300	0.1	0.2	0.2	0.3	0.3	0.4	0.5	0.6	0.7	0.8	0.9	1.0	1.1	1.3	1.4	1.6	1.9	2.3	2.9	3.5
60	60	0.2	0.2	0.3	0.4	0.5	0.6	0.7	0.8	1.0	1.2	1.3	1.5	1.7	1.8	2.0	2.3	2.7	3.3	4.2	5.0
	100	0.1	0.2	0.3	0.3	0.4	0.5	0.5	0.7	0.8	0.9	1.1	1.2	1.3	1.5	1.6	1.9	2.1	2.7	3.3	4.0
	150	0.1	0.2	0.2	0.3	0.3	0.4	0.5	0.6	0.7	0.8	0.9	1.0	1.2	1.3	1.4	1.6	1.9	2.3	2.9	3.5
	300	0.1	0.1	0.2	0.2	0.3	0.3	0.4	0.5	0.6	0.7	0.8	0.9	1.0	1.1	1.2	1.4	1.6	2.0	2.5	3.0
75	75	0.1	0.2	0.3	0.3	0.4	0.5	0.5	0.7	0.8	0.9	1.1	1.2	1.3	1.5	1.6	1.9	2.1	2.7	3.3	4.0
	120	0.1	0.2	0.2	0.3	0.3	0.4	0.4	0.5	0.6	0.8	0.9	1.0	1.1	1.2	1.3	1.5	1.7	2.2	2.7	3.3
	200	0.1	0.1	0.2	0.2	0.3	0.3	0.4	0.5	0.5	0.6	0.7	0.8	0.9	1.0	1.1	1.3	1.5	1.8	2.3	2.7
	300	0.1	0.1	0.2	0.2	0.2	0.3	0.3	0.4	0.5	0.6	0.7	0.7	0.8	0.9	1.0	1.2	1.3	1.7	2.1	2.5
100	100	0.1	0.1	0.2	0.2	0.3	0.3	0.4	0.5	0.6	0.7	0.8	0.9	1.0	1.1	1.2	1.4	1.6	2.0	2.5	3.0
	200	0.1	0.1	0.1	0.2	0.2	0.3	0.3	0.4	0.4	0.5	0.6	0.7	0.7	0.8	0.9	1.0	1.2	1.5	1.9	2.2
	300	0.1	0.1	0.1	0.2	0.2	0.2	0.3	0.3	0.4	0.5	0.5	0.6	0.7	0.7	0.8	0.9	1.1	1.3	1.7	2.0
150	150	0.1	0.1	0.1	0.2	0.2	0.2	0.3	0.3	0.4	0.5	0.5	0.6	0.7	0.7	0.8	0.9	1.1	1.3	1.7	2.0
	300	—	0.1	0.1	0.1	0.1	0.2	0.2	0.2	0.3	0.3	0.4	0.5	0.5	0.6	0.6	0.7	0.8	1.0	1.2	1.5
200	200	—	0.1	0.1	0.1	0.1	0.2	0.2	0.2	0.3	0.3	0.4	0.5	0.5	0.6	0.6	0.7	0.8	1.0	1.2	1.5
	300	—	0.1	0.1	0.1	0.1	0.1	0.2	0.2	0.2	0.3	0.3	0.4	0.4	0.5	0.5	0.6	0.7	0.8	1.0	1.2
300	300	—	—	0.1	0.1	0.1	0.1	0.1	0.2	0.2	0.2	0.3	0.3	0.3	0.4	0.4	0.5	0.5	0.6	0.7	0.8
500	500	—	—	—	—	0.1	0.1	0.1	0.1	0.1	0.1	0.2	0.2	0.2	0.2	0.2	0.3	0.3	0.4	0.5	0.6

Table 2 (Table B)* Per cent effective ceiling or floor cavity reflectance for various reflectance combinations

% Ceiling or Floor Reflectance	*90*				*80*				*70*			*50*			*30*				*10*		
% wall reflectance	*90*	*70*	*50*	*30*	*80*	*70*	*50*	*30*	*70*	*50*	*30*	*70*	*50*	*30*	*65*	*50*	*30*	*10*	*50*	*30*	*10*
0	90	90	90	90	80	80	80	80	70	70	70	50	50	50	30	30	30	30	10	10	10
0.1	90	89	88	87	79	79	78	78	69	69	68	59	49	48	30	30	29	29	10	10	10
0.2	89	88	86	85	79	78	77	76	68	67	66	49	48	47	30	29	29	28	10	10	9
0.3	89	87	85	83	78	77	75	74	68	66	64	49	47	46	30	29	28	27	10	10	9
0.4	88	86	83	81	78	76	74	72	67	65	63	48	46	45	30	29	27	26	11	10	9
0.5	88	85	81	78	77	75	73	70	66	64	61	48	46	44	29	28	27	25	11	10	9
0.6	88	84	80	76	77	75	71	68	65	62	59	47	45	43	29	28	26	25	11	10	9
0.7	88	83	78	74	76	74	70	66	65	61	58	47	44	42	29	28	26	24	11	10	8
0.8	87	82	77	73	75	73	69	65	64	60	56	47	43	41	29	27	25	23	11	10	8
0.9	87	81	76	71	75	72	68	63	63	59	55	46	43	40	29	27	25	22	11	9	8
1.0	86	80	74	69	74	71	66	61	63	58	53	46	42	39	29	27	24	22	11	9	8
1.1	86	79	73	67	74	71	65	60	62	57	52	46	41	38	29	26	24	21	11	9	8
1.2	86	78	72	65	73	70	64	58	61	56	50	45	41	37	29	26	23	20	12	9	7
1.3	85	78	70	64	73	69	63	57	61	55	49	45	40	36	29	26	23	20	12	9	7
1.4	85	77	69	62	72	68	62	55	60	54	48	45	40	35	28	26	22	19	12	9	7
1.5	85	76	68	61	72	68	61	54	59	53	47	44	39	34	28	25	22	18	12	9	7
1.6	85	75	66	59	71	67	60	53	59	52	45	44	39	33	28	25	21	18	12	9	7
1.7	84	74	65	58	71	66	59	52	58	51	44	44	38	32	28	25	21	17	12	9	7
1.8	84	73	64	56	70	65	58	50	57	50	43	43	37	32	28	25	21	17	12	9	6
1.9	84	73	63	55	70	65	57	49	57	49	42	43	37	31	28	25	20	16	12	9	6
2.0	83	72	62	53	69	64	56	48	56	48	41	43	37	30	28	24	20	16	12	9	6
2.1	83	71	61	52	69	63	55	47	56	47	40	43	36	29	28	24	20	16	13	9	6
2.2	83	70	60	51	68	63	54	45	55	46	39	42	36	29	28	24	19	15	13	9	6
2.3	83	69	59	50	68	62	53	44	54	46	38	42	35	28	28	24	19	15	13	9	6
2.4	82	68	58	48	67	61	52	43	54	45	37	42	35	27	28	24	19	14	13	9	6
2.5	82	68	57	47	67	61	51	42	53	44	36	41	34	27	27	23	18	14	13	9	6

cavity ratio

Ceiling or floor																						
2.6	82	67	56	46	66	60	50	41	53	43	35	41	34	26	27	23	18	13	13	9	5
2.7	82	66	55	45	66	60	49	40	52	43	34	41	33	26	27	23	18	13	13	9	5
2.8	81	66	54	44	66	59	48	39	52	42	33	41	33	25	27	23	18	13	13	9	5
2.9	81	65	53	43	65	58	48	38	51	41	33	40	33	25	27	23	17	12	13	9	5
3.0	81	64	52	42	65	58	47	38	51	40	32	40	32	24	27	22	17	12	13	8	5
3.1	80	64	51	41	64	57	46	37	50	40	31	40	32	24	27	22	17	12	13	8	5
3.2	80	63	50	40	64	57	45	36	50	39	30	40	31	23	27	22	16	11	13	8	5
3.3	80	62	49	39	64	56	44	35	49	39	30	39	31	23	27	22	16	11	13	8	5
3.4	80	62	48	38	63	56	44	34	49	38	29	39	31	22	27	22	16	11	13	8	5
3.5	79	61	48	37	63	55	43	33	48	38	29	39	30	22	26	22	16	11	13	8	5
3.6	79	60	47	36	62	54	42	33	48	37	28	39	30	21	26	21	15	10	13	8	5
3.7	79	60	46	35	62	54	42	32	48	37	27	38	30	21	26	21	15	10	13	8	4
3.8	79	59	45	35	62	53	41	31	47	36	27	38	29	21	26	21	15	10	13	8	4
3.9	78	59	45	34	61	53	40	30	47	36	26	38	29	20	26	21	15	10	13	8	4
4.0	78	58	44	33	61	52	40	30	46	35	26	38	29	20	26	21	15	9	13	8	4
4.1	78	57	43	32	60	52	39	29	46	35	25	37	28	20	26	21	14	9	13	8	4
4.2	78	57	43	32	60	51	39	29	46	34	25	37	28	19	26	20	14	9	13	8	4
4.3	78	56	42	31	60	51	38	28	45	34	25	37	28	19	26	20	14	9	13	8	4
4.4	77	56	41	30	59	51	38	28	45	34	24	37	27	19	26	20	14	8	13	8	4
4.5	77	55	41	30	59	50	37	27	45	33	24	37	27	19	25	20	14	8	14	8	4
4.6	77	55	40	29	59	50	37	26	44	33	24	36	27	18	25	20	14	8	14	8	4
4.7	77	54	40	29	58	49	36	26	44	33	23	36	26	18	25	20	13	8	14	8	4
4.8	76	54	39	28	58	49	36	25	44	32	23	36	26	18	25	19	13	8	14	8	4
4.9	76	53	38	28	58	49	35	25	44	32	23	36	26	18	25	19	13	7	14	8	4
5.0	76	53	38	27	57	48	35	25	43	32	22	36	26	17	25	19	13	7	14	8	4

*Table identification used in manufacturers' catalogs.

Table 3 (TABLE C)* MULTIPLYING FACTORS FOR 10% EFFECTIVE FLOOR CAVITY REFLECTANCE (20% = 1.00)

% Effective Ceiling Cavity Reflectance, ρ_{cc}	*80*				*70*				*50*			*30*			*10*		
% wall reflectance, ρ_w	*70*	*50*	*30*	*10*	*70*	*50*	*30*	*10*	*50*	*30*	*10*	*50*	*30*	*10*	*50*	*30*	*10*
Room cavity ratio																	
1	0.923	0.929	0.935	0.940	0.933	0.939	0.943	0.948	0.956	0.960	0.963	0.973	0.976	0.979	0.989	0.991	0.993
2	0.931	0.942	0.950	0.958	0.940	0.949	0.957	0.963	0.962	0.968	0.974	0.976	0.980	0.985	0.988	0.991	0.995
3	0.939	0.951	0.961	0.969	0.945	0.957	0.966	0.973	0.967	0.975	0.981	0.978	0.983	0.988	0.988	0.992	0.996
4	0.944	0.958	0.969	0.978	0.950	0.963	0.973	0.980	0.972	0.980	0.986	0.980	0.986	0.991	0.987	0.992	0.996
5	0.949	0.964	0.976	0.983	0.954	0.968	0.978	0.985	0.975	0.983	0.989	0.981	0.988	0.993	0.987	0.992	0.997
6	0.953	0.969	0.980	0.986	0.958	0.972	0.982	0.989	0.977	0.985	0.992	0.982	0.989	0.995	0.987	0.993	0.997
7	0.957	0.973	0.983	0.991	0.961	0.975	0.985	0.991	0.979	0.987	0.994	0.983	0.990	0.996	0.987	0.993	0.998
8	0.960	0.976	0.986	0.993	0.963	0.977	0.987	0.993	0.981	0.988	0.995	0.984	0.991	0.997	0.987	0.994	0.998
9	0.963	0.978	0.987	0.994	0.965	0.979	0.989	0.994	0.983	0.990	0.996	0.985	0.992	0.998	0.988	0.994	0.999
10	0.965	0.980	0.989	0.995	0.967	0.981	0.990	0.995	0.984	0.991	0.997	0.986	0.993	0.998	0.988	0.994	0.999

*Table identification used in manufacturers' catalogs.

Table 4 (Table D)* Multiplying factors for 30% effective floor cavity reflectance (20% = 1.00)

% Effective Ceiling Cavity Reflectance, ρ_{cc}	*80*				*70*				*50*			*30*			*10*		
% wall reflectance, ρ_w	*70*	*50*	*30*	*10*	*70*	*50*	*30*	*10*	*50*	*30*	*10*	*50*	*30*	*10*	*50*	*30*	*10*
Room cavity ratio																	
1	1.092	1.082	1.075	1.068	1.077	1.070	1.064	1.059	1.049	1.044	1.040	1.028	1.026	1.023	1.012	1.010	1.008
2	1.079	1.066	1.055	1.047	1.068	1.057	1.048	1.039	1.041	1.033	1.027	1.026	1.021	1.017	1.013	1.010	1.006
3	1.070	1.054	1.042	1.033	1.061	1.048	1.037	1.028	1.034	1.027	1.020	1.024	1.017	1.012	1.014	1.009	1.005
4	1.062	1.045	1.033	1.024	1.055	1.040	1.029	1.021	1.030	1.022	1.015	1.022	1.015	1.010	1.014	1.009	1.004
5	1.056	1.038	1.026	1.018	1.050	1.034	1.024	1.015	1.027	1.018	1.012	1.020	1.013	1.008	1.014	1.009	1.004
6	1.052	1.033	1.021	1.014	1.047	1.030	1.020	1.012	1.024	1.015	1.009	1.019	1.012	1.006	1.014	1.008	1.003
7	1.047	1.029	1.018	1.011	1.043	1.026	1.017	1.009	1.022	1.013	1.007	1.018	1.010	1.005	1.014	1.008	1.003
8	1.044	1.026	1.015	1.009	1.040	1.024	1.015	1.007	1.020	1.012	1.006	1.017	1.009	1.004	1.013	1.007	1.003
9	1.040	1.024	1.014	1.007	1.037	1.022	1.014	1.006	1.019	1.011	1.005	1.016	1.009	1.004	1.013	1.007	1.002
10	1.037	1.022	1.012	1.006	1.034	1.020	1.012	1.005	1.017	1.010	1.004	1.015	1.009	1.003	1.013	1.007	1.002

*Table identification used in manufacturers' catalogs.

Table 5 GENERAL SERVICE INCADESCENT LAMPS (All lamps rated for 120 V)

Watts	*Bulb*	*Base*	*Rated average life (hr)*	*Approximate initial lumens*	*Rated initial lumens per watt*
10	S-14‡	Medium	1,500	79	7.9
15	A-15‡	Medium	2,500	126	8.4
25	A-19‡	Medium	2,500	232	9.3
40*	A-19	Medium	1,500	450	11.2
50	A-19§	Medium	1,000	680	13.6
60*	A-19§	Medium	1,000	855	14.3
75*	A-19§	Medium	750	1,170	15.6
100*	A-19‖	Medium	750	1,750	17.5
150†	A-21‖	Medium	750	2,830	18.9
150	PS-25	Medium	750	2,640	17.6
200	A-23‖	Medium	750	3,940	19.7
200	PS-30	Medium	750	3,680	18.4
300	PS-25‖	Medium	750	6,240	20.8
300	PS-35	Mogul	1,000	5,750	19.2
500	PS-35‖	Mogul	1,000	10,500	21.0
750	PS-52	Mogul	1,000	15,600	20.8
750	PS-52‖	Mogul	1,000	16,700	22.2
1,000	PS-52	Mogul	1,000	21,600	21.6
1,000	PS-52‖	Mogul	1,000	23,300	23.3
1,500	PS-52	Mogul	1,000	33,000	22.0

*Available in T-19 bulb.
†Available in T-21 bulb.
‡Vacuum Lamp.
§Coiled Coil filament.
‖Vertical Coiled Coil filament.

Table 6 LUMEN VALUES FOR TYPICAL FLUORESCENT LAMPS

Lamp Watts	Bulb Designation	Base	Length (in.)	Approximate Initial Lumen Output: Cool White	Approximate Initial Lumen Output: Warm White and White	Approximate Initial Lumen Output: Deluxe Cool White	Approximate Initial Lumen Output: Deluxe Warm White	Life Hours 3 Hours/Start
			Preheat					
15	T-12	Medium bipin	18	970	770	520	505	7,500
20	T-12	Medium bipin	24	1,220	1,250	850	820	7,500
30	T-12	Medium bipin	36	2,100	2,180	1,580	1,520	7,500
90	T-17	Mogul bipin	60	6,000	6,300	4,300	—	9,000
			Instant start					
40	T-12	Medium bipin	48	3,050	3,100	2,400	2,350	9,000
40	T-17	Mogul bipin	60	2,900	2,950	—	—	9,000
			Circline					
22	T-9	Four pin	8.25*	950	980	755	745	7,500
32	T-10	Four pin	12	1,750	1,800	1,250	1,240	7,500
40	T-10	Four pin	16	2,300	2,350	1,780	1,760	7,500
			Slimline					
40	T-12	Single pin	48	2,900	3,000	2,020	1,970	12,000
55	T-12	Single pin	72	4,400	4,450	3,080	3,000	12,000
75	T-12	Single pin	96	6,200	6,250	4,300	4,100	12,000
			Rapid start (light loading, 430 MA)					
30	T-12	Medium bipin	36	2,380	2,420	1,530[†]	1,480[†]	12,000
40	T-12	Medium bipin	48	3,120	3,170	2,130	2,080	15,000
			Rapid start (medium loading, 800 MA)					
60	T-12	Recessed double contact	48	4,000	4,000	2,750	2,700	12,000
85	T-12	Recessed double contact	72	6,450	6,300	4,300	4,250	12,000
110	T-12	Recessed double contact	96	9,000	8,900	6,160	5,950	12,000
			Rapid start (high loading, 1,500 MA)					
110	T-12	Recessed double contact	48	6,900	6,900	5,050	—	9,000
165	T-12	Recessed double contact	72	12,000	12,000	7,175	—	9,000
215	T-12	Recessed double contact	96	16,000	15,500	11,000	—	9,000

These are representative 1975 values. Values differ from manufacturer to manufacturer and from year to year. For up-to-date values, current lamp catalogs should be consulted.

*Nominal outside diameter.

†9,000 hr.

Table 7 TRANSFORMER FULL-LOAD AND INTERRUPTING CAPACITY CURRENT VALUES

		600 V				480 V				240 V				208 V			
Transformer rating 3-ph kVA and impedance (%)	Max. short-circuit kVA available from primary system	Normal load continuous current (A)	Interrupting capacity current total rms amperes: Transformer alone	100% motor load	Combined	Normal load continuous current (A)	Interrupting capacity current total rms amperes: Transformer alone	100% motor load	Combined	Normal load continuous current (A)	Interrupting capacity current total rms amperes: Transformer alone	100% motor load	Combined	Normal load continuous current (A)	Interrupting capacity current total rms amperes: Transformer alone	50% motor load	Combined
	15,000										12,900		16,500		14,900		17,000
	25,000		5,850		7,300		7,300		9,100		14,600		18,200		16,700		18,800
	50,000		6,450		7,900		8,100		9,900		16,100		19,700		18,600		20,700
300 5%	100,000	289	6,850	1,450	8,300	361	8,500	1,800	10,300	722	17,000	3,600	20,600	834	19,600	2,100	21,700
	150,000		6,950		8,400		8,700		10,500		17,400		21,000		20,000		22,100
	250,000		7,050		8,500		8,800		10,600		17,600		21,200		20,300		22,400
	500,000		7,150		8,600		8,900		10,700		17,900		21,500		20,500		22,600
	Unlimited		7,250		8,700		9,000		10,800		18,100		21,700		20,800		22,900
	25,000		7,900		10,100		9,900		12,600		19,900		25,300		22,900		26,000
	50,000		9,000		11,200		11,500		14,200		22,900		28,300		26,500		29,600
	100,000		9,900		12,100		12,400		15,100		24,800		30,200		28,600		31,700
450 5%	150,000	433	10,200	2,200	12,400	542	12,700	2,700	15,400	1,083	25,400	5,400	30,800	1,250	29,400	3,100	32,500
	250,000		10,400		12,600		13,000		15,700		26,000		31,400		30,100		33,200
	500,000		10,600		12,800		13,300		16,000		26,500		31,900		30,600		33,700
	Unlimited		10,800		13,000		13,500		16,200		27,000		32,400		31,400		34,500
	25,000		8,600		11,000		10,800		13,800		21,500		27,500		24,800		28,300
	50,000		10,000		12,400		12,500		15,500		25,100		31,100		28,900		32,400
	100,000		11,000		13,400		13,700		16,700		27,300		33,300		31,500		35,000
500 5%	150,000	481	11,300	2,400	13,700	600	14,100	3,000	17,100	1,200	28,200	6,000	34,200	1,388	32,500	3,500	36,000
	250,000		11,600		14,000		14,500		17,500		28,900		34,900		33,300		36,800
	500,000		11,800		14,200		14,800		17,800		29,500		35,500		34,000		37,500
	Unlimited		12,000		14,400		15,100		18,100		30,100		36,100		34,600		38,100

Transformer	Primary																
600 5%	25,000		9,700		12,600		12,200		15,800		24,400		31,600		28,200		32,400
	50,000		11,600		14,500		14,600		18,200		29,000		36,200		33,500		37,700
	100,000		12,900		15,800		16,100		19,700		32,100		39,300		37,100		41,300
	150,000	578	13,300	2,900	16,200	722	16,700	3,600	20,300	1,443	33,300	7,200	40,500	1,668	38,500	4,200	42,700
	250,000		13,800		16,700		17,200		20,800		34,400		41,600		39,700		43,900
	500,000		14,100		17,000		17,600		21,200		35,100		42,300		40,600		44,800
	Unlimited		14,500		17,400		18,100		21,700		36,000		43,200		41,700		45,900
750 5½%	25,000		10,600		14,200		13,300		17,800		26,600		35,600		30,600		35,800
	50,000		12,900		16,500		16,100		20,600		32,300		41,300		37,100		42,300
	100,000		14,500		18,100		18,000		22,500		36,100		45,100		41,600		46,800
	150,000	722	15,100	3,600	18,700	900	18,800	4,500	23,300	1,800	37,600	9,000	46,600	2,080	43,300	5,200	48,500
	250,000		15,600		19,200		19,500		24,000		39,000		48,000		44,800		50,000
	500,000		16,000		19,600		20,000		24,500		40,000		49,000		46,100		51,300
	Unlimited		16,400		20,000		20,500		25,000		41,100		50,100		47,300		52,500
1000 5½%	25,000		12,700		17,500		15,800		21,800		31,700		43,700		36,500		43,500
	50,000		16,100		20,900		20,100		26,100		40,200		52,200		46,300		53,300
	100,000		18,500		23,300		23,200		29,200		46,300		58,300		53,400		60,400
	150,000	962	19,500	4,800	24,300	1,200	24,400	6,000	30,400	2,400	48,800	12,000	60,800	2,780	56,300	7,000	63,300
	250,000		20,500		25,300		25,500		31,500		51,000		63,000		58,900		65,900
	500,000		21,100		25,900		26,400		32,400		52,800		64,800		60,900		67,900
	Unlimited		21,900		26,700		27,400		33,400		54,700		66,700		63,200		70,200
1500 5½%	25,000		15,700		22,900		19,600		28,600		39,300		57,300				
	50,000		21,300		28,500		26,600		35,600		53,200		71,200				
	100,000		25,900		33,100		32,300		41,300		64,500		82,500				
	150,000	1,444	27,900	7,200	35,100	1,800	34,800	9,000	43,800	3,600	69,500	18,000	87,500		—		—
	250,000		29,700		36,900		37,000		46,000		74,000		92,000				
	500,000		31,200		38,400		38,900		47,900		77,900		95,900				
	Unlimited		32,900		40,100		41,100		50,100		82,000		100,000				
2000 5½%	25,000		17,700		27,300		22,100		34,100								
	50,000		25,300		34,900		31,700		43,700								
	100,000		31,600		41,200		39,400		51,400								
	150,000	1,924	35,300	9,600	44,900	2,400	44,100	12,000	56,100		—		—		—		—
	250,000		37,500		47,100		46,800		58,800								
	500,000		40,500		50,100		50,700		62,700								
	Unlimited		43,800		53,400		54,700		66,700								

*All computations are based on voltages, transformer impedances, and motor loads, as indicated, including a factor of 1.25 for the dc component. *For conditions differing from those given in these tables* the short-circuit currents should be calculated.

The motor short-circuit contributions are computed on the basis of motor characteristics that will give five times normal current. For the 208-V table, 50% motor load is assumed. For the 600-, 480-, and 240-V tables, 100% motor loads are assumed. For other percentages of motor load, the motor contribution to the short-circuit current will be in direct proportion.

Where the circuit voltage is less than 600, 480, or 240 V, the current values given should be multiplied by the ratio (600 or 480 or 240)/circuit voltage.

Table 8 FAULT CURRENT AVAILABLE (SYMMETRICAL RMS AMPERES)*
208 V

kVA Rating of Transformer	*Conductor Size Per Phase*	*Distance from Transformer to Point of Fault (ft)*								
		0	*5*	*10*	*20*	*50*	*100*	*200*	*500*	*1,000*
150	#4	11,500	10,700	10,000	8,500	5,400	3,200	1,750	720	350
	#0	11,500	11,120	10,750	10,050	8,070	5,850	3,600	1,620	860
	250 MCM	11,500	11,300	11,050	10,550	9,250	7,600	5,550	3,000	1,600
	2–250 MCM	11,500	11,400	11,250	11,050	10,300	9,240	7,600	4,820	3,000
225	#4	17,220	15,700	13,950	12,000	6,100	3,400	1,800	750	400
	#0	17,220	16,450	15,600	14,100	10,400	6,750	3,600	1,700	900
	250 MCM	17,220	16,700	16,200	15,200	12,600	9,750	6,500	3,200	1,700
	2–250 MCM	17,220	17,000	16,700	16,200	14,700	12,700	9,600	5,600	3,250
	2–500 MCM	17,220	17,100	16,900	16,500	15,300	13,700	11,300	7,200	4,500
300	#4	23,000	20,400	17,100	12,600	6,500	3,500	1,800	750	400
	#0	23,000	21,600	20,200	17,500	13,950	7,500	4,000	1,750	900
	250 MCM	23,000	22,100	21,200	19,500	15,300	12,200	7,300	3,350	1,750
	2–250 MCM	23,000	22,500	22,000	21,200	18,500	15,300	11,300	6,000	3,300
	2–500 MCM	23,000	22,750	22,450	21,700	19,550	16,800	13,300	7,900	4,550
500	#4	38,200	30,800	24,000	15,400	6,900	3,500	1,800	800	400
	#0	38,200	34,400	30,400	24,000	14,200	8,000	4,000	1,800	1,000
	250 MCM	38,200	36,000	33,800	29,400	20,100	13,600	8,000	3,400	1,800
	2–250 MCM	38,200	36,900	35,700	33,300	27,000	20,100	13,200	6,400	3,500
	2–500 MCM	38,200	37,400	36,500	34,600	29,400	23,800	17,000	9,000	5,000

	#4	47,200	35,800	26,000	16,000	6,900	3,400	1,700	800	400
	#0	47,200	41,900	36,300	27,300	14,800	8,000	4,100	1,800	950
750	250 MCM	47,200	43,600	40,000	34,300	23,000	14,000	8,000	3,200	1,700
	2–250 MCM	47,200	45,100	43,300	40,000	31,700	22,800	14,400	6,900	3,500
	2–500 MCM	47,200	45,900	44,300	41,700	34,600	27,000	18,300	9,200	5,000
	#4	62,700	43,000	29,100	17,000	7,800	3,700	1,800	700	400
	#0	62,700	53,500	44,300	31,200	16,000	8,500	4,400	1,800	950
1,000	250 MCM	62,700	56,600	51,000	42,000	26,000	15,900	8,800	3,400	1,870
	2–250 MCM	62,700	59,900	56,300	50,400	37,800	25,900	15,500	6,900	3,500
	2–500 MCM	62,700	61,800	58,200	54,700	42,400	31,500	21,000	10,000	5,300
	#4	92,400	53,000	33,000	18,100	7,800	3,900	2,000	800	600
	#0	92,400	73,500	57,000	36,500	17,800	9,200	4,600	2,000	1,000
1,500	250 MCM	92,400	80,000	69,500	52,000	30,000	17,400	9,200	3,800	2,000
	2–250 MCM	92,400	85,700	79,500	68,500	46,000	30,000	17,600	7,000	3,800
	2–500 MCM	92,400	88,000	83,000	74,000	57,000	38,000	23,800	11,000	6,000
	#4	121,800	58,000	33,800	18,200	7,200	3,800	1,800	600	—
	#0	121,800	88,000	63,700	38,000	17,000	8,800	4,200	1,800	800
2,000	250 MCM	121,800	100,200	83,800	60,000	31,000	17,000	8,500	3,200	1,800
	2–250 MCM	121,800	110,800	100,500	83,000	50,000	30,000	17,000	6,800	3,500
	2–500 MCM	121,800	114,200	106,000	91,000	62,000	40,000	23,900	10,000	5,000

*The fault currents listed are maximum available symmetrical rms values based on liquid filled transformers, with nominal impedances of $4\frac{1}{2}\%$ for ratings up to and including 500 kVA and $5\frac{1}{2}\%$ for ratings above 500 kVA, and include motor contribution based on 100% motor load.

Table 9 Fault current available (symmetrical rms amperes)*
480 V

kVA Rating of Transformer	*Conductor Size Per Phase*	*Distance from Transformer to Point of Fault (ft)*								
		0	*5*	*10*	*20*	*50*	*100*	*200*	*500*	*1,000*
150	#4	4,990	4,930	4,880	4,770	4,420	3,800	2,800	1,480	790
	#0	4,990	4,940	4,920	4,880	4,700	4,400	3,850	2,650	1,680
	250 MCM	4,990	4,960	4,930	4,910	4,800	4,600	4,250	3,350	2,500
	2–250 MCM	4,990	4,970	4,940	4,920	4,900	4,800	4,600	4,050	3,350
225	#4	7,470	7,380	7,240	7,000	6,140	4,880	3,300	1,600	840
	#0	7,470	7,400	7,320	7,200	6,800	6,200	5,100	3,180	1,860
	250 MCM	7,470	7,420	7,360	7,300	7,040	6,640	5,900	4,400	3,000
	2–250 MCM	7,470	7,440	7,400	7,350	7,220	7,000	6,600	5,580	4,300
	2–500 MCM	7,470	7,460	7,450	7,400	7,300	7,100	6,800	6,000	5,000
300	#4	9,985	9,800	9,600	9,100	7,600	5,600	3,560	1,620	840
	#0	9,985	9,840	9,750	9,520	8,800	7,650	5,900	3,400	1,920
	250 MCM	9,985	9,880	9,800	9,660	9,240	8,500	7,300	5,000	3,240
	2–250 MCM	9,985	9,920	9,825	9,790	9,580	9,200	8,450	6,800	5,020
	2–500 MCM	9,985	9,950	9,850	9,800	9,660	9,400	8,820	7,500	5,880
500	#4	16,550	16,000	15,400	14,000	10,250	6,800	3,800	1,600	800
	#0	16,550	16,200	15,950	15,250	13,250	10,500	7,400	3,500	1,900
	250 MCM	16,550	16,300	16,050	15,700	14,500	12,700	10,000	5,900	3,500
	2–250 MCM	16,550	16,350	16,250	16,100	15,450	14,400	12,500	9,000	6,000
	2–500 MCM	16,550	16,400	16,350	16,300	15,700	14,800	13,400	10,500	7,500

	#4	20,450	19,700	18,700	16,800	11,700	7,500	4,000	1,600	800
	#0	20,450	20,000	19,500	18,700	16,000	12,400	8,100	3,800	2,000
750	250 MCM	20,450	20,200	19,800	19,250	17,500	15,000	11,500	6,600	3,800
	2–250 MCM	20,450	20,250	20,200	19,700	19,000	17,500	15,000	10,500	6,600
	2–500 MCM	20,450	20,400	20,250	19,900	19,300	18,200	16,300	12,000	8,400
	#4	27,200	26,000	24,200	21,000	13,400	7,900	4,400	1,800	800
	#0	27,200	26,700	25,900	24,300	20,000	14,400	9,000	4,100	2,200
1,000	250 MCM	27,200	26,900	26,400	25,300	22,400	18,600	13,600	7,200	4,000
	2–250 MCM	27,200	27,000	26,700	26,200	24,500	22,200	18,500	12,100	7,200
	2–500 MCM	27,200	27,100	26,800	26,500	25,300	23,300	20,300	14,500	9,500
	#4	40,050	37,000	33,100	26,000	14,400	8,200	4,000	1,400	600
	#0	40,050	38,800	36,800	33,200	24,500	16,000	9,200	4,000	2,000
1,500	250 MCM	40,050	39,100	37,800	35,600	29,900	23,000	15,200	7,500	4,000
	2–250 MCM	40,050	39,600	39,000	37,900	34,100	29,000	22,500	13,000	7,400
	2–500 MCM	40,050	39,700	39,200	38,200	35,500	31,600	25,900	16,400	10,100
	#4	52,800	47,400	40,700	30,000	15,100	8,200	4,200	1,900	1,000
	#0	52,800	50,200	47,000	41,200	28,000	17,000	9,700	4,200	2,400
2,000	250 MCM	52,800	51,000	49,000	45,400	36,200	26,500	16,500	8,000	4,200
	2–250 MCM	52,800	51,800	50,900	48,900	43,100	36,000	26,700	14,000	8,000
	2–500 MCM	52,800	52,100	51,300	49,900	45,100	39,200	30,800	18,500	11,000

*The fault currents listed are maximum available symmetrical rms values based on liquid filled transformers, with nominal impedances of $4\frac{1}{2}\%$ for ratings up to and including 500 kVA and $5\frac{1}{2}\%$ for ratings above 500 kVA, and include motor contribution based on 100% motor load.

Table 10 TABLE 430-147.* FULL-LOAD CURRENT IN AMPERES, DIRECT-CURRENT MOTORS.

The following values of full-load currents† are for motors running at base speed.

HP	*Armature Voltage Rating†*					
	90 V	*120 V*	*180 V*	*240 V*	*500 V*	*550 V*
1/4	4.0	3.1	2.0	1.6		
1/3	5.2	4.1	2.6	2.0		
1/2	6.8	5.4	3.4	2.7		
3/4	9.6	7.6	4.8	3.8		
1	12.2	9.5	6.1	4.7		
1 1/2		13.2	8.3	6.6		
2		17	10.8	8.5		
3		25	16	12.2		
5		40	27	20		
7 1/2		58		29	13.6	12.2
10		76		38	18	16
15				55	27	24
20				72	34	31
25				89	43	38
30				106	51	46
40				140	67	61
50				173	83	75
60				206	99	90
75				255	123	111
100				341	164	148
125				425	205	185
150				506	246	222
200				675	330	294

*Table numbers marked with an asterisk are reproduced from the National Electrical Code, 1975 Edition, copyright National Fire Protection Association, 470 Atlantic Avenue, Boston, Mass.

†These are average direct-current quantities.

Table 11 TABLE 430-148.* FULL-LOAD CURRENTS IN AMPERES SINGLE-PHASE ALTERNATING-CURRENT MOTORS

The following values of full-load currents are for motors running at usual speeds and motors with normal torque characteristics. Motors built for especially low speeds or high torques may have higher full-load currents, and multispeed motors will have full-load current varying with speed, in which case the nameplate current ratings shall be used.

To obtain full-load currents of 208- and 200-V motors, increase corresponding 230-V motor full-load currents by 10 and 15%, respectively.

The voltages listed are rated motor voltages. The currents listed shall be permitted for system voltage ranges of 110 to 120 and 220 to 240.

HP	*115 V*	*230 V*	*HP*	*115 V*	*230 V*	*HP*	*115 V*	*230 V*
1/6	4.4	2.2	1	16	8	5	56	28
1/4	5.8	2.9	1 1/2	20	10	7 1/2	80	40
1/3	7.2	3.6	2	24	12	10	100	50
1/2	9.8	4.9	3	34	17			
3/4	13.8	6.9						

*Table numbers marked with an asterisk are reproduced from the National Electrical Code, 1975 Edition, copyright National Fire Protection Association, 470 Atlantic Avenue, Boston, Mass.

Table 12 TABLE 430-150.* FULL-LOAD CURRENT† THREE-PHASE ALTERNATING-CURRENT MOTORS

	Induction-Type Squirrel-Cage and Wound Rotor (amperes)					*Synchronous-Type Unity Power Factor‡ (amperes)*			
HP	*115 V*	*230 V*	*460 V*	*575 V*	*2,300 V*	*220 V*	*440 V*	*550 V*	*2,300 V*
½	4	2	1	.8					
¾	5.6	2.8	1.4	1.1					
1	7.2	3.6	1.8	1.4					
1½	10.4	5.2	2.6	2.1					
2	13.6	6.8	3.4	2.7					
3		9.6	4.8	3.9					
5		15.2	7.6	6.1					
7½		22	11	9					
10		28	14	11					
15		42	21	17					
20		54	27	22					
25		68	34	27		54	27	22	
30		80	40	32		65	33	26	
40		104	52	41		86	43	35	
50		130	65	52		108	54	44	
60		154	77	62	16	128	64	51	12
75		192	96	77	20	161	81	65	15
100		248	124	99	26	211	106	85	20
125		312	156	125	31	264	132	106	25
150		360	180	144	37		158	127	30
200		480	240	192	49		210	168	40

For full-load currents of 208- and 200-V motors, increase the corresponding 230-V motor full-load current by 10 and 15%, respectively.

*Table numbers marked with an asterisk are reproduced from the National Electrical Code, 1975 Edition, copyright National Fire Protection Association, 470 Atlantic Avenue, Boston, Mass.

†These values of full-load current are for motors running at speeds usual for belted motors and motors with normal torque characteristics. Motors built for especially low speeds or high torques may require more running current, and multispeed motors will have full-load current varying with speed, in which case the nameplate current rating shall be used.

‡For 90 and 80% power factor the figures above shall be multipled by 1.1 and 1.25, respectively.

The voltages listed are rated motor voltages. The currents listed shall be permitted for system voltage ranges of 110 to 120, 220 to 240, 440 to 480, and 550 to 600 V.

Table 13 Branch-circuit protection of direct-current motors

120 V				*240 V*				*550 V*			
Size of Motor		***Branch-Circuit Fuse Size***		***Size of Motor***		***Branch-Circuit Fuse Size***		***Size of Motor***		***Branch-Circuit Fuse Size***	
HP	*F.L. ampere rating*	*Non-time delay Fuse*	*Time delay fuse*	*HP*	*F.L. ampere rating*	*Non-time delay fuse*	*Time delay fuse*	*HP*	*F.L. ampere rating*	*Non-time delay fuse*	*Time delay fuse*
$\frac{1}{4}$	2.9	15	5	$\frac{1}{4}$	1.5	15	3	$\frac{3}{4}$	1.6	15	$2\frac{1}{2}$
$\frac{1}{3}$	3.6	15	5	$\frac{1}{3}$	1.8	15	3	1	2	15	$3\frac{2}{10}$
$\frac{1}{2}$	5.2	15	8	$\frac{1}{2}$	2.6	15	4	$1\frac{1}{2}$	2.7	15	$4\frac{1}{2}$
$\frac{3}{4}$	7.4	15	12	$\frac{3}{4}$	3.7	15	7	2	3.6	15	$6\frac{1}{4}$
1	9.4	15	15	1	4.7	15	8	3	5.2	15	9
$1\frac{1}{2}$	13.2	25	20	$1\frac{1}{2}$	6.6	15	10	5	8.3	15	15
2	17	30	25	2	8.5	15	15	$7\frac{1}{2}$	12	20	20
3	25	40	40	3	12.2	20	20	10	16	25	25
5	40	60	60	5	20	30	30	15	23	35	35
$7\frac{1}{2}$	58	90	90	$7\frac{1}{2}$	29	45	45	20	31	45	45
10	76	125	125	10	38	60	60	25	38	60	60
				15	55	90	80	30	46	70	70
				20	72	110	100	40	61	90	90
				25	89	150	125	50	75	125	110
				30	106	175	150	60	90	150	150
				40	140	225	200	75	111	175	175
				50	173	300	250	100	148	225	225
				60	206	350	300	125	184	300	300
				75	255	400	400	150	220	350	350
				100	341	600	500	200	295	450	450
				125	425		600				

Branch circuit protection of single-phase alternating-current motors

115 V				*230 V*			
$\frac{1}{6}$	4.4	15	8	$\frac{1}{6}$	2.2	15	4
$\frac{1}{4}$	5.8	20	10	$\frac{1}{4}$	2.9	15	5
$\frac{1}{3}$	7.2	25	12	$\frac{1}{3}$	3.6	15	6
$\frac{1}{2}$	9.8	30	15	$\frac{1}{2}$	4.9	15	8
$\frac{3}{4}$	13.8	45	20	$\frac{3}{4}$	6.9	25	12
1	16	50	25	1	8	25	15
$1\frac{1}{2}$	20	60	30	$1\frac{1}{2}$	10	30	15
2	24.0	80	35	2	12	40	20
				3	17	60	25
				5	28	90	40
				$7\frac{1}{2}$	40	125	60
				10	50	150	80

Note: All fuse ratings are maximum permitted.

Table 14 BRANCH-CIRCUIT PROTECTION OF THREE-PHASE MOTORS (SEE NOTE 2)

208 V				
Size and Class of Motor			*Branch-Circuit Fuse Size*	
HP	*F.L. ampere rating*	*Class*	*Non-time delay*	*Time delay*
$\frac{1}{2}$	2.2	Any	15	4
$\frac{3}{4}$	3.1	"	15	5
1	4	"	15	8
$1\frac{1}{2}$	5.7	"	15	10
2	7.5	1	25	10
		2	20	10
		3–4	15	10
3	10.6	1	35	15
		2	30	15
		3	25	15
		4	20	15
5	16.7	1	50	25
		2	45	25
		3	35	25
		4	25	25
$7\frac{1}{2}$	24.2	1	80	35
		2	70	35
		3	50	35
		4	40	35
10	30.8	1	100	45
		2	80	45
		3	70	45
		4	50	45
15	46.2	1	150	70
		2	125	70
		3	100	70
		4	70	70
20	59.4	1	200	90
		2	150	90
		3	125	90
		4	90	90
25	74.8	1	225	100
		2	200	100
		3	150	100
		4	125	100
30	88	1	300	125
		2	225	125
		3	175	125
		4	150	125
40	114	1	350	175
		2	300	175
		3	250	175
		4	175	175
50	143	1	450	200
		2	400	200
		3	300	200
		4	225	200
60	169	1	600	250
		2	450	250
		3	350	250
		4	300	250
75	211	1		300
		2	600	300
		3	450	300
		4	350	300
100	273	1		400
		2		400
		3	600	400
		4	450	400
125	343	1–2–3		500
		4	600	500
150	396	1–2–3		600
		4	600	600

Classification of Motors Based on Type or Code Marking	
Class	*Motors with code letter marking (430-152)*
1	F to V full voltage start or resistor or reactor start.
2	F to V with auto transformer start. B to E full voltage start or resistor or reactor start.
3	B to E with auto transformer start.
4	A—

Class	*Motors without code letter marking, such as ordinary squirrel cage, synchronous, high-reactance squirrel cage or wound rotor motors (430-152)*
1	Ordinary squirrel cage or synchronous with full voltage start or resistor or reactor start.
2	Ordinary squirrel cage or synchronous with auto-transformer start 30 A or less. High-reactance squirrel cage 30 A or less.
3	Ordinary squirrel cage or synchronous with auto-transformer start more than 30 A. High-reactance squirrel cage more than 30 A. Sealed (hermetic-type) refrigeration compressor—400-kVA locked-rotor or less.
4	Wound rotor or slip ring motor.

Notes:
1. All fuse ratings are maximum permitted.
2. For tables covering 230-, 460-, and 575-V motor, see table 15.

Table 15 Branch-circuit protection of three-phase motors (see notes)

230 V

Size and Class of Motor			Branch-Circuit Fuse Size	
HP	*F.L. ampere rating*	*Class*	*Nontime delay*	*Time delay*
$\frac{1}{2}$	2	Any	15	4
$\frac{3}{4}$	2.8	″	15	4
1	3.6	″	15	6
$1\frac{1}{2}$	5.2	″	15	8
2	6.8	1	25	10
		2	20	10
		3–4	15	10
3	9.6	1	30	15
		2	25	15
		3	20	15
		4	15	15
5	15.2	1	50	25
		2	40	25
		3	35	25
		4	25	25
$7\frac{1}{2}$	22	1	70	35
		2	60	35
		3	45	35
		4	35	35
10	28	1	90	40
		2	70	40
		3	60	40
		4	45	40
15	42	1	125	60
		2	110	60
		3	90	60
		4	70	60
20	54	1	175	80
		2	150	80

460 V

Size and Class of Motor			Branch-Circuit Fuse Size	
HP	*F.L. ampere rating*	*Class*	*Nontime delay*	*Time delay*
$\frac{1}{2}$	1	Any	15	2
$\frac{3}{4}$	1.4	″	15	3
1	1.8	″	15	3
$1\frac{1}{2}$	2.6	″	15	4
2	3.4	″	15	5
3	4.8	Any	15	8
5	7.6	1	25	15
		2	20	15
		3–4	15	15
$7\frac{1}{2}$	11	1	35	20
		2	30	20
		3	25	20
		4	20	20
10	14	1	45	20
		2	35	20
		3	30	20
		4	25	20
15	21	1	70	30
		2	60	30
		3	45	30
		4	35	30
20	27	1	90	40
		2	70	40
		3	60	40
		4	45	40
25	34	1	110	50
		2	90	50
		3	70	50
		4	60	50

575 V

Size and Class of Motor			Branch-Circuit Fuse Size	
HP	*F.L. ampere rating*	*Class*	*Nontime delay*	*Time delay*
$\frac{1}{2}$	.8	Any	15	$1\frac{4}{10}$
$\frac{3}{4}$	1.1	″	15	2
1	1.4	″	15	3
$1\frac{1}{2}$	2.1	″	15	4
2	2.7	″	15	4
3	3.9	Any	15	7
5	6.1	1	20	10
		2–3–4	15	10
$7\frac{1}{2}$	9	1	30	15
		2	25	15
		3	20	15
		4	15	15
10	11	1	35	20
		2	30	20
		3	25	20
		4	20	20
15	17	1	60	25
		2	45	25
		3	35	25
		4	30	25
20	22	1	70	30
		2	60	30
		3	45	30
		4	35	30
25	27	1	90	40
		2	70	40
		3	60	40
		4	45	40
30	32	1	100	50

		3	110	80
		4	90	80
25	68	1	225	100
		2	175	100
		3	150	100
		4	110	100
30	80	1	250	125
		2	200	125
		3	175	125
		4	125	125
40	104	1	350	150
		2	300	150
		3	225	150
		4	175	150
50	130	1	400	200
		2	350	200
		3	300	200
		4	200	200
60	154	1	500	250
		2	400	250
		3	350	250
		4	250	250
75	192	1	600	300
		2	500	300
		3	400	300
		4	300	300
100	248	1		400
		2		400
		3	500	400
		4	400	400
125	312	1		450
		2		450
		3		450
		4	500	450
150	360	1–2–3		500
		4	600	500
200	480	Any		

30	40	1	125	60
		2	100	60
		3	80	60
		4	60	60
40	52	1	175	80
		2	150	80
		3	110	80
		4	80	80
50	65	1	200	100
		2	175	100
		3	150	100
		4	100	100
60	77	1	250	125
		2	200	125
		3	175	125
		4	125	125
75	96	1	300	150
		2	250	150
		3	200	150
		4	150	150
100	124	1	400	200
		2	350	200
		3	250	200
		4	200	200
125	156	1	500	250
		2	400	250
		3	350	250
		4	250	250
150	180	1	600	300
		2	450	300
		3	400	300
		4	300	300
200	240	1		400
		2	600	400
		3	500	400
		4	400	400

		2	80	50
		3	70	50
		4	50	50
40	41	1	125	60
		2	100	60
		3	80	60
		4	60	60
50	52	1	175	80
		2	150	80
		3	110	80
		4	80	80
60	62	1	200	100
		2	175	100
		3	125	100
		4	100	100
75	77	1	250	125
		2	200	125
		3	175	125
		4	125	125
100	99	1	300	150
		2	250	150
		3	200	150
		4	150	150
125	125	1	400	200
		2	350	200
		3	250	200
		4	200	200
150	144	1	450	225
		2	400	225
		3	300	225
		4	225	225
200	192	1	600	300
		2	500	300
		3	400	300
		4	300	300

Note: For class of motor, see Table 19; for 208-V motors see Table 19.

Table 16 TABLE 310-16.* ALLOWABLE AMPACITIES OF INSULATED COPPER CONDUCTORS

Not more than three conductors in raceway or cable or direct burial (based on ambient temperature of 30°C, 86°F)

Size	*Temperature Rating of Conductor (See Table 310-13)*							
AWG MCM	*60°C (140°F)*	*75°C (167°F)*	*85°C (185°F)*	*90°C (194°F)*	*110°C (230°F)*	*125°C (257°F)*	*200°C (392°F)*	*250°C (482°F)*
	Types RUW (14-2), T, TW, UF	*Types RH, RHW, RUH (14-2), THW, THWN, XHHW, USE*	*Types V, MI*	*Types TA, TBS, SA, AVB, SIS, FEP, FEPB, RHH, THHN, XHHW‡*	*Types AVA, AVL*	*Types AI (14-8), AIA*	*Types A (14-8), AA, FEP† FEPB†*	*Type TFE (Nickel or nickel-coated copper only)*
18	—	—	—	21	—	—	—	—
16	—	—	22	22	—	—	—	—
14	15	15	25	25§	30	30	30	40
12	20	20	30	30§	35	40	40	55
10	30	30	40	40§	45	50	55	75
8	40	45	50	50	60	65	70	95
6	55	65	70	70	80	85	95	120
4	70	85	90	90	105	115	120	145
3	80	100	105	105	120	130	145	170
2	95	115	120	120	135	145	165	195
1	110	130	140	140	160	170	190	220
1/0	125	150	155	155	190	200	225	250
2/0	145	175	185	185	215	230	250	280
3/0	165	200	210	210	245	265	285	315
4/0	195	230	235	235	275	310	340	370
250	215	255	270	270	315	335	—	—
300	240	285	300	300	345	380	—	—
350	260	310	325	325	390	420	—	—
400	280	335	360	360	420	450	—	—
500	320	380	405	405	470	500	—	—
600	355	420	455	455	525	545	—	—
700	385	460	490	490	560	600	—	—
750	400	475	500	500	580	620	—	—
800	410	490	515	515	600	640	—	—
900	435	520	555	555	—	—	—	—
1000	455	545	585	585	680	730	—	—
1250	495	590	645	645	—	—	—	—
1500	520	625	700	700	785	—	—	—
1750	545	650	735	735	—	—	—	—
2000	560	665	775	775	840	—	—	—

†Special use only. See Table 310-13.

‡For dry locations only. See Table 310-13.

These ampacities relate only to conductors described in Table 310-13.

§The ampacities for Types FEP, FEPB, RHH, THHN, and XHHW conductors for sizes 14, 12, and 10 shall be the same as designated for 75°C conductors in this table.

For ambient temperatures over 30°C, see Correction Factors, Note 13.

Table 17 TABLE 310-18.* ALLOWABLE AMPACITIES OF INSULATED ALUMINUM AND COPPER-CLAD ALUMINUM CONDUCTORS

Not more than three conductors in raceway or cable or direct burial (based on ambient temperature of 30°C, 86°F)

Size	*Temperature Rating of Conductor (See Table 310-13)*							
AWG MCM	*60°C (140°F)*	*75°C (167°F)*	*85°C (185°F)*	*90°C (194°F)*	*110°C (230°F)*	*125°C (257°F)*	*200°C (392°F)*
	Types RUW (12-2), T, TW, UF	*Types RH, RHW, RUH (12-2), THW THWN XHHW, USE*	*Types V, MI*	*Types TA, TBS, SA, AVB, SIS, RHH THHN XHHW†*	*Types AVA, AVL*	*Types AI (12-8), AIA*	*Types A (12-8), AA*
12	15	15	25	25‡	25	30	30
10	25	25	30	30‡	35	40	45
8	30	40	40	40‡	45	50	55
6	40	50	55	55	60	65	75
4	55	65	70	70	80	90	95
3	65	75	80	80	95	100	115
2	75	90	95	95	105	115	130
1	85	100	110	110	125	135	150
1/0	100	120	125	125	150	160	180
2/0	115	135	145	145	170	180	200
3/0	130	155	165	165	195	210	225
4/0	155	180	185	185	215	245	270
250	170	205	215	215	250	270	—
300	190	230	240	240	275	305	—
350	210	250	260	260	310	335	—
400	225	270	290	290	335	360	—
500	26C	310	330	330	380	405	—
600	285	340	370	370	425	440	—
700	310	375	395	395	455	485	—
750	320	385	405	405	470	500	—
800	330	395	415	415	485	520	—
900	355	425	455	455	—	—	—
1000	375	445	480	480	560	600	—
1250	405	485	530	530	—	—	—
1500	435	520	580	580	650	—	—
1750	455	545	615	615	—	—	—
2000	470	560	650	650	705	—	—

These ampacities relate only to conductors described in Table 310-13.

*Table numbers marked with an asterisk are reproduced from the National Electrical Code, 1975 Edition copyright National Fire Protection Association, 470 Atlantic Avenue, Boston, Mass. For references not reproduced in this text see N.E.C.

†For dry locations only. See Table 310-13.

‡The ampacities for Types RHH, THHN, and XHHW conductors for sizes 12 and 10 shall be the same as designated for 75°C conductors in this table.

For ambient temperatures over 30°C, see Correction Factors, Note 13.

Table 18 TABLE 3A.* MAXIMUM NUMBER OF CONDUCTORS IN TRADE SIZES OF CONDUIT OR TUBING

Conduit Trade Size (in.)		$\frac{1}{2}$	$\frac{3}{4}$	***1***	$1\frac{1}{4}$	$1\frac{1}{2}$	***2***	$2\frac{1}{2}$	***3***	$3\frac{1}{2}$	***4***	$4\frac{1}{2}$	***5***	***6***
Type letters	*Conductor size AWG, MCM*													
TW, T, RUH,	14	9	15	25	44	60	99	142						
RUW,	12	7	12	19	35	47	78	111	171					
XHHW (14 thru 8)	10	5	9	15	26	36	60	85	131	176				
	8	2	4	7	12	17	28	40	62	84	108			
RHW and RHH	14	6	10	16	29	40	65	93	143	192				
(without outer	12	4	8	13	24	32	53	76	117	157				
covering),	10	4	6	11	19	26	43	61	95	127	163			
THW	8	1	3	5	10	13	22	32	49	66	85	106	133	
TW,	6	1	2	4	7	10	16	23	36	48	62	78	97	141
T,	4	1	1	3	5	7	12	17	27	36	47	58	73	106
THW,	3	1	1	2	4	6	10	15	23	31	40	50	63	91
RUH (6 thru 2),	2	1	1	2	4	5	9	13	20	27	34	43	54	78
RUW (6 thru 2),	1		1	1	3	4	6	9	14	19	25	31	39	57
FEPB (6 thru 2), RHW and	0		1	1	2	3	5	8	12	16	21	27	33	49
RHH (without	00		1	1	1	3	5	7	10	14	18	23	29	41
outer covering)	000		1	1	1	2	4	6	9	12	15	19	24	35
	0000			1	1	1	3	5	7	10	13	16	20	29
	250			1	1	1	2	4	6	8	10	13	16	23
	300			1	1	1	2	3	5	7	9	11	14	20
	350				1	1	1	3	4	6	8	10	12	18
	400				1	1	1	2	4	5	7	9	11	16
	500				1	1	1	1	3	4	6	7	9	14
	600					1	1	1	3	4	5	6	7	11
	700					1	1	1	2	3	4	5	7	10
	750					1	1	1	2	3	4	5	6	9

*Notes and tables indicated with an asterisk are reproduced from the National Electrical Code, 1975 Edition, copyright National Fire Protection Association, 470 Atlantic Avenue, Boston, Mass. For references not reproduced in this text see N.E.C.

Table 19 Table 3B.* Maximum number of conductors in trade sizes of conduit or tubing

Type letters	Conductor size AWG, MCM	½	¾	1	1¼	1½	2	2½	3	3½	4	4½	5	6
	14	13	24	39	69	94	154							
THWN	12	10	18	29	51	70	114	164						
	10	6	11	18	32	44	73	104	160					
	8	3	5	9	16	22	36	51	79	106	136			
THHN	6	1	4	6	11	15	26	37	57	76	98	125	154	
	4	1	2	4	7	9	16	22	35	47	60	75	94	137
FEP (14 thru 2),	3	1	1	3	6	8	13	19	29	39	51	64	80	116
FEPB (14 thru 8)	2	1	1	3	5	7	11	16	25	33	43	54	67	97
	1		1	1	3	5	8	12	18	25	32	40	50	72
XHHW (4 thru	0		1	1	3	4	7	10	15	21	27	33	42	61
500MCM)	00		1	1	2	3	6	8	13	17	22	28	35	51
	000		1	1	1	3	5	7	11	14	18	23	29	42
	0000		1	1	1	2	4	6	9	12	15	19	24	35
	250			1	1	1	3	4	7	10	12	16	20	28
	300			1	1	1	3	4	6	8	11	13	17	24
	350			1	1	1	2	3	5	7	9	12	15	21
	400				1	1	1	3	5	6	8	10	13	19
	500				1	1	1	2	4	5	7	9	11	16
	600				1	1	1	1	3	4	5	7	9	13
	700					1	1	1	3	4	5	6	8	11
	750					1	1	1	2	3	4	6	7	11
XHHW	6	1	3	5	9	13	21	30	47	63	81	102	128	185
	600				1	1	1	1	3	4	5	7	9	13
	700					1	1	1	3	4	5	6	7	11
	750					1	1	1	2	3	4	6	7	10

*Tables marked with an asterisk are reproduced from the National Electrical Code, 1975 Edition, copyright National Fire Protection Association, 470 Atlantic Avenue, Boston, Mass. For references not reproduced in this text see N.E.C.

Table 20 TABLE 3C.* MAXIMUM NUMBER OF CONDUCTORS IN TRADE SIZES OF CONDUIT OR TUBING

Conduit Trade Size (in.)		½	¾	*1*	*1¼*	*1½*	*2*	*2½*	*3*	*3½*	*4*	*4½*	*5*	*6*
Type letters	*Conductor size AWG, MCM*													
	14	3	6	10	18	25	41	58	90	121	155			
RHW	12	3	5	9	15	21	35	50	77	103	132			
	10	2	4	7	13	18	29	41	64	86	110	138		
	8	1	2	4	7	9	16	22	35	47	60	75	94	137
RHH	6	1	1	2	5	6	11	15	24	32	41	51	64	93
	4	1	1	1	3	5	8	12	18	24	31	39	50	72
(with outer covering)	3	1	1	1	3	4	7	10	16	22	28	35	44	63
	2		1	1	3	4	6	9	14	19	24	31	38	56
	1		1	1	1	3	5	7	11	14	18	23	29	42
	0		1	1	1	2	4	6	9	12	16	20	25	37
	00			1	1	1	3	5	8	11	14	18	22	32
	000			1	1	1	3	4	7	9	12	15	19	28
	0000			1	1	1	2	4	6	8	10	13	16	24
	250				1	1	1	3	5	6	8	11	13	19
	300				1	1	1	3	4	5	7	9	11	17
	350				1	1	1	2	4	5	6	8	10	15
	400				1	1	1	1	3	4	6	7	9	14
	500				1	1	1	1	3	4	5	6	8	11
	600					1	1	1	2	3	4	5	6	9
	700					1	1	1	1	3	3	4	6	8
	750						1	1	1	3	3	4	5	8

*Tables marked with an asterisk are reproduced from the National Electrical Code, 1975 Edition, copyright National Fire Protection Association, 470 Atlantic Avenue, Boston, Mass. For references not reproduced in this text see N.E.C.

Table 21 TABLE 4.* DIMENSIONS AND PERCENT AREA OF CONDUIT AND OF TUBING

		Area (sq. in.)								
			Not Lead-Covered			*Lead-Covered*				
Trade size	*Internal diameter (in.)*	*Total 100%*	*Two cond. 31%*	*Over two cond. 40%*	*One cond. 53%*	*One cond. 55%*	*Two cond. 30%*	*Three cond. 40%*	*Four cond. 38%*	*Over four cond. 35%*
$\frac{1}{2}$	0.622	0.30	0.09	0.12	0.16	0.17	0.09	0.12	0.11	0.11
$\frac{3}{4}$	0.824	0.53	0.16	0.21	0.28	0.29	0.16	0.21	0.20	0.19
1	1.049	0.86	0.27	0.34	0.46	0.47	0.26	0.34	0.33	0.30
$1\frac{1}{4}$	1.380	1.50	0.47	0.60	0.80	0.83	0.45	0.60	0.57	0.53
$1\frac{1}{2}$	1.610	2.04	0.63	0.82	1.08	1.12	0.61	0.82	0.78	0.71
2	2.067	3.36	1.04	1.34	1.78	1.85	1.01	1.34	1.28	1.18
$2\frac{1}{2}$	2.469	4.79	1.48	1.92	2.54	2.63	1.44	1.92	1.82	1.68
3	3.068	7.38	2.29	2.95	3.91	4.06	2.21	2.95	2.80	2.58
$3\frac{1}{2}$	3.548	9.90	3.07	3.96	5.25	5.44	2.97	3.96	3.76	3.47
4	4.026	12.72	3.94	5.09	6.74	7.00	3.82	5.09	4.83	4.45
$4\frac{1}{2}$	4,506	15.94	4.94	6.38	8.45	8.77	4.78	6.38	6.06	5.56
5	5.047	20.00	6.20	8.00	10.60	11.00	6.00	8.00	7.60	7.00
6	6.065	28.89	8.96	11.56	15.31	15.89	8.67	11.56	10.98	10.11

*Table numbers marked with an asterisk are reproduced from the National Electrical Code, 1975 Edition, copyright National Fire Protection Association, 470 Atlantic Avenue, Boston, Mass. For references not reproduced in this text see N.E.C.

Table 22 TABLE 5.[*] DIMENSIONS OF RUBBER-COVERED AND THERMOPLASTIC-COVERED CONDUCTORS

	Types RFH-2, RH, RHH,[‖] RHW,[‖] SF-2		*Types TF, T, THW,[§] TW, RUH,[‡] RUW[‡]*		*Types TFN, THHN, THWN*		*Types[¶] FEP, FEPB, TFE, PF, PGF, PTF*		*Type XHHW*	
Size AWG MCM	*Approx. diam. (in.)*	*Approx. area (sq. in.)*	*Approx. diam. (in.)*	*Approx. area (sq. in.)*	*Approx. diam. (in.)*	*Approx. area (sq. in.)*	*Approx. area (in.)*	*Approx. area (sq. in.)*	*Approx. diam. (in.)*	*Approx. area (sq. in.)*
Col. 1	Col. 2	Col. 3	Col. 4	Col. 5	Col. 6	Col. 7	Col. 8	Col. 9	Col. 10	Col. 11
18	0.146	0.0167	0.106	0.0088	0.089	0.0064	0.081	0.0052	—	—
16	0.158	0.0196	0.118	0.0109	0.100	0.0079	0.092	0.0066	—	—
14	30 mils 0.171	0.0230	0.131	0.0135	0.105	0.0087	0.105 0.105	0.0087 0.0087		
14	45 mils 0.204[†]	0.0327[†]	—	—	—	—	— —	— —		
14	— —	—	0.162[§]	0.0206[§]	—	—	— —	— —	0.129	0.0131
12	30 mils 0.188	0.0278	0.148	0.0172	0.122	0.0117	0.121 0.121	0.0115 0.0115		
12	45 mils 0.221[†]	0.0384[†]	—	—	—	—	— —	— —		
12	— —	—	0.179[‡]	0.0251[‡]	—	—	— —	— —	0.146	0.0167
10	— 0.242	0.0460	0.168	0.0224	0.153	0.0184	0.142 0.142	0.0159 0.0159		
10	— —	—	0.199[‡]	0.0311[‡]	—	—	— —	— —	0.166	0.0216
8	— 0.328	0.0854	0.245	0.0471	0.218	0.0373	0.206 0.186	0.0333 0.0272		
8	— —	—	0.276[‡]	0.0598[‡]	—	—	— —	— —	0.241	0.0456
6	0.397	0.1238	0.323	0.0819	0.257	0.0519	0.244 0.302	0.0467 0.0716	0.282	0.0625
4	0.452	0.1605	0.372	0.1087	0.328	0.0845	0.292 0.350	0.0669 0.0962	0.328	0.0845
3	0.481	0.1817	0.401	0.1263	0.356	0.0995	0.320 0.378	0.0803 0.1122	0.356	0.0995
2	0.513	0.2067	0.433	0.1473	0.388	0.1182	0.352 0.410	0.0973 0.1316	0.388	0.1182
1	0.588	0.2715	0.508	0.2027	0.450	0.1590	0.420 —	0.1385 —	0.450	0.1590

0	0.629	0.3107	0.549	0.2367	0.491	0.1893	0.462	—	0.1676	—	0.491	0.1893
00	0.675	0.3578	0.595	0.2781	0.537	0.2265	0.498	—	0.1974	—	0.537	0.2265
000	0.727	0.4151	0.647	0.3288	0.588	0.2715	0.560	—	0.2463	—	0.588	0.2715
0000	0.785	0.4840	0.705	0.3904	0.646	0.3278	0.618	—	0.2999	—	0.646	0.3278
250	0.868	0.5917	0.788	0.4877	0.716	0.4026	—		—		0.716	0.4026
300	0.933	0.6837	0.843	0.5581	0.771	0.4669	—		—		0.771	0.4669
350	0.985	0.7620	0.895	0.6291	0.822	0.5307	—		—		0.822	0.5307
400	1.032	0.8365	0.942	0.6969	0.869	0.5931	—		—		0.869	0.5931
500	1.119	0.9834	1.029	0.8316	0.955	0.7163	—		—		0.955	0.7163
600	1.233	1.1940	1.143	1.0261	1.058	0.8792	—		—		1.073	0.9043
700	1.304	1.3355	1.214	1.1575	1.129	1.0011	—		—		1.145	1.0297
750	1.339	1.4082	1.249	1.2252	1.163	1.0623	—		—		1.180	1.0936
800	1.372	1.4784	1.282	1.2908	1.196	1.1234	—		—		1.210	1.1499
900	1.435	1.6173	1.345	1.4208	1.259	1.2449	—		—		1.270	1.2668
1000	1.494	1.7531	1.404	1.5482	1.317	1.3623	—		—		1.330	1.3893
1250	1.676	2.2062	1.577	1.9532	—	—	—		—		1.500	1.7672
1500	1.801	2.5475	1.702	2.2748	—	—	—		—		1.620	2.0612
1750	1.916	2.8895	1.817	2.5930	—	—	—		—		1.740	2.3779
2000	2.021	3.2079	1.922	2.9013	—	—	—		—		1.840	2.6590

*Table numbers marked with an asterisk are reproduced from the National Electrical Code, 1975 Edition, copyright National Fire Protection Association, 470 Atlantic Avenue, Boston, Mass. For references not reproduced in this text see N.E.C.

†The dimensions of Types RHH and RHW.

‡No. 14 to No. 2.

§Dimensions of THW in sizes 14 to 8. No. 6 THW and larger is the same dimension as T.

‖Dimensions of RHH and RHW without outer covering are the same as THW.

No. 18 to No. 10, solid; No. 8 and larger, stranded.

¶In Col. 8 and 9 the values shown for sizes No. 1 through 0000 are for TFE only. The right-hand values in Col. 8 and 9 are for FEPB only.

INDEX

INDEX

D

E

F

G

S

W